Advancing Safety, Service, and Performance in Organizations With Impactful Leadership

Darcia Ann Marie Roache
Capilano University, Canada

IGI Global
Scientific Publishing
Publishing Tomorrow's Research Today

Vice President of Editorial	Melissa Wagner
Director of Acquisitions	Mikaela Felty
Director of Book Development	Jocelynn Hessler
Production Manager	Mike Brehm
Cover Design	Jose Rosado

Published in the United States of America by
IGI Global Scientific Publishing
701 East Chocolate Avenue
Hershey, PA, 17033, USA
Tel: 717-533-8845
Fax: 717-533-7115
Website: https://www.igi-global.com E-mail: cust@igi-global.com

Library of Congress Cataloging-in-Publication Data

Names: Roache, Darcia Ann-Marie, 1970- editor
Title: Advancing safety, service, and performance in organizations with impactful leadership / edited by Darcia Roache.
Description: Hershey, PA : IGI Global Scientific Publishing, [2026] | Includes bibliographical references and index. | Summary: "This book explores the lasting impact of meaningful leadership practices on organizational development. It examines various practices in business safety, service, and performance, and outlines the challenges of impactful leadership"-- Provided by publisher.
Identifiers: LCCN 2025028491 (print) | LCCN 2025028492 (ebook) | ISBN 9798337356860 hardcover | ISBN 9798337356877 paperback | ISBN 9798337356884 ebook
Subjects: LCSH: Industrial management | Leadership | Transformational leadership
Classification: LCC HD31.2 .A385 2026 (print) | LCC HD31.2 (ebook)
LC record available at https://lccn.loc.gov/2025028491
LC ebook record available at https://lccn.loc.gov/2025028492

British Cataloguing in Publication Data
A Cataloguing in Publication record for this book is available from the British Library.

I, Dr. Darcia Roache, dedicate the book entitled *Advancing Safety, Service, and Performance in Organizations with Impactful Leadership*, to all those who contribute to making the world a safer and more effective place—leaders, professionals, and changemakers. You champion safety, elevate service, and pursue excellence in performance daily.

Your unwavering commitment to lead with integrity, empathy, and purpose continues to inspire meaningful transformation and strengthen the organizations and communities you serve.

Dr. Darcia Roache, PhD. DBA. MBA. BSc

Table of Contents

Detailed Table of Contents

This chapter focuses on the importance of proof and vigilance in organizations and their relation to the concept of transformational leadership practices and legal governance. Exploring leadership theories and analysing organisations' examples from the healthcare, business, and education context, it also demonstrates how leadership fosters safety culture, accountability, and ethical decision-making. The chapter focuses on the communication processes, ethical theories and measures of effectiveness, pertaining to compliance with the legal requirements. Finally, it gives the leaders the recommendations on how to strengthen the organizations to make them responsible for their actions. This combination of leadership and law provides a course that leads to organizational success in the creation of safety and services.

The rapid evolution of technology, pedagogy, and societal demands places educational leadership at a critical crossroads. This chapter explores transformative trends—AI integration, personalized learning, and competency-based education—through a review of studies (2019–2025). It identifies systemic barriers such as the digital divide (40% of low-income students lack internet), resource shortages, and inadequate teacher training (only 25% feel prepared). Ethical challenges, including algorithmic bias (35% of schools lack AI bias policies) and data privacy concerns, complicate leadership. Case studies from Finland and Summit Public Schools highlight strategies like collaborative governance, professional development, and data-driven decision-making. Recommendations include investing in infrastructure, creating ethical AI frameworks, and fostering inclusive learning environments. By linking theory (Kotter, Fullan) to practice, this research offers actionable guidance for leaders to build equitable, adaptive systems that promote student agency, and pedagogical integrity.

This study aims to propose a research model of organizational performance focused on the self-managed work team environment (SWTE), work team management and self-management, based on the assumption that self-management of work teams facilitates and accelerates growth in organizational performance. The method used is based on meta-analysis and meta-cognitive-descriptive based on the conceptual, theoretical and empirical review of the literature. The results of the analysis are conclusive in that the management of the self-managed work team environment contributes to facilitating the growth and development of organizational performance.

This study examines the crisis management and enterprise risk management (ERM) challenges faced by General Motors (GM) following a series of fatal accidents in 2024 involving one of its sedan models. The incidents, which resulted in vehicle explosions and passenger fatalities, underscored significant weaknesses in GM's risk management strategies and quality assurance processes. Using a narrative literature review, the research synthesizes past safety incidents in large organizations and explores theoretical frameworks like COSO and RIMS Risk Maturity Models. Key recommendations include adopting a mature ERM framework, refocusing on quality assurance, reimagining engineering and financial risk strategies, and fostering cultural change through Kotter's change model and other organizational theories. The study emphasizes that a robust ERM strategy and effective change management are critical for restoring stakeholder trust, mitigating financial and reputational risks, and aligning GM's operations with its long-term objectives.

This chapter delves into the essential communication strategies that foster organizational excellence and ensure long-term success in today's dynamic business environment. It highlights the significance of effective communication in cultivating a strong organizational culture, enhancing stakeholder engagement, and promoting collaboration among teams. Key concepts such as leadership, trust, feedback, and clarity are explored, demonstrating their critical roles in creating an adaptive and resilient organization. By synthesizing contemporary research and established frameworks, this chapter serves as a comprehensive guide for leaders and practitioners seeking to implement effective communication strategies that drive continuous improvement and organizational longevity. The practical implications discussed lay a foundation for organizations to thrive, emphasizing that strong communication is not just a skill but a vital cornerstone of sustainable success.

This scoping review systematically maps the literature on police victimization. Searches across EBSCO, PubMed, and Web of Science yielded 41 articles, with 11 included after selection and manual searching. The review identified working conditions, inadequate protective equipment, and exposure to violence as key contributors to aggression against police officers. It explored aspects such as prevalence, firearm attacks, organizational structure impact, and the effectiveness of Body-Worn Cameras (BWCs), which were linked to reduced assaults. Findings revealed troubling statistics on threats, assaults, and psychological impacts on officers, along with demographic patterns of perpetrators and victims. Studies emphasized the role of organizational structure in assaults and highlighted the cultural and contextual influences on police safety. Longitudinal and comparative analyses underscored effective measures and trends in reducing police victimization globally. These findings stress the need for addressing diverse factors to enhance law enforcement safety and well-being.

Employee engagement is now a key driver of organizational success, linked directly to productivity, retention, and performance. As workplaces evolve, organizations recognize the need for clear strategies to enhance engagement by aligning objectives with individual motivations, fostering transparent communication, and cultivating a culture of inclusion, recognition, and trust. This chapter provides a framework integrating research and theory to help leaders build strategies that address current employee needs and future trends. It examines the role of transformational leadership, employee autonomy, and work-life balance in maintaining engagement. Practical applications for organizations of various sizes and industries are discussed, with actionable insights for leaders and HR professionals. Recommendations emphasize the need for ongoing evaluation of strategies in light of remote work trends and increasing workforce diversity. This holistic approach contributes to knowledge on employee engagement and offers a roadmap to improve employee experience and achieve lasting business success.

The cybersecurity industry has historically lacked diversity among its professional workforce. Women and minorities have been underrepresented in STEM for over ten years. Statistics conclude that, in the United States, only 3% of cybersecurity analysts are African American, 11% are women, 5% are Asian, and 2% are Hispanic. A diverse workforce promotes critical thinking with different perspectives, allowing organizations to use unique methods for problem-solving, which positively impacts the success of an organization. This research uses an exploratory research design to define areas within the literature that frequently ignore underrepresented communities. The failure to explore this underserved community demonstrates gaps in cross-sector cyber intelligent markets. The cyber intelligent market historically and currently hinders diversity, equity, and inclusion of individuals in underserved communities. The goal of the research is to gain actionable and applicable methodologies and ideals needed to influence change in the cyber intelligence community.

It is estimated that 16% of American workers have a hidden disability. The large share of employees with a hidden disability highlights the need to provide support to these employees. Supporting employees with hidden disabilities can lead to positive reputational and economic benefits for organizations. This study outlines a framework that organizations can use to establish and administer an effective performance support program for employees with hidden disabilities. Through a review of the literature on organizational obligations to disabled employees and factors that constitute effective employee performance support programs, the research shows that an effective program is framed around: 1. Understanding the legal, moral, and ethical obligations 2. Defining the scope of the program appropriate for the organization 3. Engaging all employees through training, outreach, and change management 4. Developing a confidential disclosure process 5. Investing in appropriate accommodations and applying where required 6. Monitoring and evaluating the effectiveness of the program

This paper examines transformative developments in Artificial Intelligence (AI), focusing on advancements in technical methodologies, ethical challenges, and real-world applications. Innovations like transformer models and generative AI have revolutionized fields such as natural language processing (NLP) and image synthesis, leading to breakthroughs in language translation and content creation. These advancements benefit sectors like healthcare, where AI aids diagnostics and personalized treatment, and finance, where it enhances fraud detection and risk assessment. However, ethical concerns, particularly around bias, accountability, and transparency, have emerged. Biases from unrepresentative datasets risk discriminatory outcomes, highlighting critical ethical issues. The paper argues that embedding fairness and accountability in AI is essential for trust and responsible use. It emphasizes the need for comprehensive ethical frameworks to guide socially sound AI development and recommends transparent, fair practices and further research into regulatory strategies.

Foreword

I commend Dr. Darcia Roache and her collaborators in their authorship of this well-named book which is generous in its insights for enhancing safety, security and service in organizations. I am guessing that most educated adolescents would provide at least vague recitation of Abraham Maslow's Hierarchy of Human Needs. Because of the pervasive appreciations of those needs, people in our organizations also will understand and experience various articulations of human needs. It must also be recognized that in our efforts to rush forward in pursuit of production, we sometimes forget these basic needs. We forget to provide self-care, to think of the "least among us," and appreciate the benefits of creating environments of grace, integrity and consideration. We know that the needs for safety and security of person, community, and workplaces are foundational to our exercise of liberty, operational efficiencies, wholesome development of social relationships, and honouring of diversity. There is an Asian saying that "as the load of the ox increased, the ox bowed its head to the ground." I wonder if this means that the oxen failed to gain the benefits and economies of seeing the larger picture and fundamental facets of the constitution of the pathways to successful performance. Is it possible that organizations, as their best, are ones that habituate the awareness and practices associated with fulfilling fundamental human needs?

As organizations create safe places for work and relationships, they move forward in their efforts to achieve their mandates. Recall that "driving fear out of the workplace" was one of the cornerstone facets of Deming's early total quality management thinking. It is difficult to sustain performance, retain talented persons and attract resources in fear-laced, risky and toxic environments. Likewise, service that is grounded in principled mission and purpose will be more likely to be experienced as authentic and regenerative for organizations. Of course, the problem with communications in our organizations is that "it" (communications) is too often assumed to have take place. In challenging times, the importance of truth and trust is often forsaken. Communication and conspiring (breathing together) are essential to thriving and productive human workplaces. So, as the authors of this collection reinforce, we need to be explicit in our attention to safety, security, service, and communications to perform well. Leaders help organizations to define their realities, to safeguard the organizational community, and make space for caring and good work.

Leading well is a daunting and even tricky role. We want to show production and yet we need to pay constant attention to organizational production capacity. This is an ever-present tension of our attentions. There is a need to maximize efficiency in performance but to do so without diminishing relational capital, dignity and inclusion, community (belongingness) or freedoms. You will see these themes in the chapters of this book. Efforts to enhance communications, to provide timely and systemic support for employees, to acknowledge diversity and unconscious bias are critical leader activities.

How do we adapt the way we do things in our organizations in a fashion that makes visible our buying into the positive organizational researchers' findings that wellbeing, health, sense of safety are correlated with preferred performance outcomes? The authors commend strategic and adaptive frames of mind and heart that orient leaders' work with their constituents and industries such that they humanize, innovate, initiate and institutionalize to achieve well-doing and wellbeing. The character and integrity of an organization is marked by how key leaders "treat" followers and how the organization's leaders deal with chronic or episodic disruptions. Insights and reflexive practice with respect to compassion in the workplace-under-stress are invaluable and readers will benefit from chapters which address these circumstances.

How do organizations contend with the myriad technological challenges of the 21st Century, with its threats and uncertainties? I commend the authors in this book for providing concrete descriptions and their expert exegesis of cases that pertain to these difficult and even wicked issues. How do we serve, communicate, and operationalize the transformations we seek in the organizations we lead or that we send our students into? These are questions that require practical wisdom and well-considered understandings. You will find that the international scholars and practitioners assembled to contribute to this book have much to offer as we seek to enhance safety, security and service in ways that matter to us and our organizations.

Thank you Dr. Roache and contributing authors for this wide-ranging but intentional set of chapters.

Keith D. Walker

University of Saskatchewan, Canada

Preface

As the editor *of Advancing Safety, Service, and Performance in Organizations With Impactful Leadership,* I am elated to present this edited reference book that presents a multidisciplinary exploration of how organizations across diverse sectors can navigate complex challenges while cultivating environments of safety, equity, and operational excellence. This edited volume draws together research and applied insights from scholars and practitioners across the globe to address pressing organizational concerns from employees' safety and security, leadership, team management, change, communication and employee engagement, diversity, equity, inclusion, and artificial intelligence in organization

Amidst global shifts triggered by health crises, digital transformation, and evolving social expectations, the need for resilient, inclusive, and strategically agile organizations has never been more urgent. This book synthesizes emerging theories, real-world case studies, and innovative frameworks to support leaders, managers, and policy-makers in creating high-performing, people-centered institutions.

CHAPTER OVERVIEW

Chapter 1: Transformational Leadership and Legal Duties- Accountability and Vigilance in Organizations

Dr. SHEIKH INAM UL MANSOOR, School of Law, Dayananda Sagar University; Ms. Shahida Qadir, Ahmad ibrahim kulliyah of Law, International Islamic university of Malaysia; Dr. HINA KAUSAR, School of Law, University of Petroleum and Energy Studies; Ms. Amthul Naseeb, School of Law, Dayananda Sagar University

Transformational Leadership and Legal Duties – Accountability and Vigilance in Organizations explores the critical link between transformational leadership and legal governance, emphasizing how leadership can drive accountability, ethical decision-making, and a culture of safety. Using examples from healthcare, business, and education, the chapter highlights the role of communication, ethical frameworks, and compliance in strengthening organizational responsibility. It underscores the importance of vigilance and evidence in meeting legal obligations and offers practical strategies for leaders to build ethically grounded and legally compliant organizations. By integrating leadership practices with legal duties, the chapter outlines a pathway to sustainable organizational success.

Chapter 2: Future Trends in Educational Leadership, Tech Advances, Evolving Paradigms, Personalized Learning: Future Trends, Tech, Paradigms & Personalized Learning

Professor AIEMAN AHMAD AL-OMARI, Hashemite University.

In *Future Trends in Educational Leadership, Tech Advances, Evolving Paradigms, Personalized Learning*, the authors examine how rapid advancements in technology and shifting educational paradigms are reshaping school leadership. Drawing on studies from 2019 to 2025, the chapter explores key trends such as AI integration, personalized learning, and competency-based education. It identifies major challenges including the digital divide (affecting 40% of low-income students), under-resourced schools, and insufficient teacher training (with only 25% feeling adequately prepared). Ethical issues like algorithmic bias (with 35% of schools lacking relevant policies) and data privacy—further complicate the landscape. Case studies from Finland and Summit Public Schools illustrate successful approaches, including collaborative leadership, ongoing professional development, and data-informed strategies. The authors recommend investing in digital infrastructure, developing ethical AI policies, and fostering inclusive, student-centered environments. Grounded in leadership theories such as Kotter's and Fullan's, the chapter offers practical guidance for building adaptive, equitable education systems.

Chapter 3: Self-Managed Work Team Environment (SMT), Work Team Management and Self-Management Focusing on an Organizational Performance Model

Dr. José G Vargas-Hernandez, Financial Literacy Research Center, Universidad Cristóbal Colón; Dr. Absalón J. Salmerón-Zapata, Universidad Cristóbal Colón; Dr. Carlos A. Rodriguez-Maillard, Universidad Cristóbal Colon.

In *Self-Managed Work Team Environment (SMT), Work Team Management and Self-Management Focusing on an Organizational Performance Model*, the authors propose a research model linking self-managed work teams to improved organizational performance. Using a meta-analytic and meta-cognitive descriptive approach grounded in conceptual, theoretical, and empirical literature, the study confirms that effective management of self-managed work team environments significantly enhances organizational growth and development. The findings support the assumption that self-management within teams plays a critical role in accelerating performance outcomes.

Chapter 4: Ignition Switch to Change - The Journey to Risk Management Reform

Mrs. Megan Pekol-Evans, Marymount University

In *Ignition Switch to Change – The Journey to Risk Management Reform*, the author analyzes General Motors' response to a 2024 crisis involving fatal accidents linked to vehicle malfunctions. The study highlights critical failures in GM's risk management and quality assurance systems. Through a narrative literature review and frameworks like COSO and RIMS Risk Maturity Models, the research identifies key areas for reform. Recommendations include implementing a mature ERM framework, enhancing quality controls, revamping engineering and financial risk strategies, and driving cultural change using Kotter's model. The study underscores the importance of integrated risk and change management to rebuild trust, reduce risk, and align operations with long-term goals.

Chapter 5: Communication Strategies for Organization Excellence and Longevity

Mrs. Marcia K. Reid, Grand Canyon University, USA.

In *Communication Strategies for Organizational Excellence and Longevity*, the author examines the pivotal role of communication in building resilient, high-performing organizations. The chapter explores how effective communication enhances organizational culture, stakeholder engagement, and team collaboration. It emphasizes key elements such as leadership, trust, feedback, and clarity as drivers of adaptability and long-term success. Drawing on contemporary research and established frameworks, the chapter offers a practical guide for leaders aiming to implement communication strategies that support continuous improvement. It underscores that strong communication is not just a competency but a foundational element of sustainable organizational success.

Chapter 6: Revealing Police Officer's Victimization: A Scoping Review

Prof. Ana Isabel Sani, University Fernando Pessoa; Dr. Sara Correia Gonçalves, University Fernando Pessoa; Prof. Laura M. Nunes, University Fernando Pessoa; Prof. Maria Alzira Pimenta Dinis, Fernando Pessoa Research, Innovation and Development Institute (FP-I3ID), University Fernando Pessoa.

In their scoping review on *Police Victimization*, the authors systematically examine existing literature on violence against police officers. Drawing from 41 articles identified through databases like EBSCO, PubMed, and Web of Science. With 11 studies ultimately included, the review highlights critical factors contributing to officer victimization, including poor working conditions, lack of protective equipment, and routine exposure to violence. It investigates the prevalence of assaults, firearm-related incidents, and the psychological toll on officers, while also analyzing the impact of organizational structures and tools like Body-Worn Cameras (BWCs), which are associated with reduced assault rates. Demographic patterns of both victims and perpetrators, along with cultural and contextual influences, are also explored. Longitudinal and comparative studies offer insights into effective strategies for reducing police victimization worldwide. The authors underscore the urgent need for systemic changes to improve the safety and well-being of law enforcement personnel.

Chapter 7: Developing Strategy for Employee Engagement

Lord Dordunoo, Marymount University

In *Developing Strategy for Employee Engagement*, the author presents a comprehensive framework for enhancing employee engagement as a driver of organizational success. The chapter integrates research and theory to guide leaders in aligning organizational goals with individual motivations, promoting transparent communication, and fostering a culture of inclusion, recognition, and trust. It explores key factors such as transformational leadership, employee autonomy, and work-life balance, offering practical strategies adaptable across industries and organization sizes. With a focus on evolving workplace trends, like remote work and workforce diversity, the chapter provides actionable insights and emphasizes continuous evaluation to sustain engagement and improve overall employee experience.

Chapter 8: A Focus-Group Study on Diversity, Equity, and Inclusion in Cross-Sector Cyber Intelligent Markets in Relationship to Underrepresented Minorities

Allison Fullwood, Capitol Technology University; Dr. Eugene J Lewis, Capitol Technology University

This study explores the persistent lack of diversity, equity, and inclusion (DEI) within the cybersecurity industry, particularly affecting underrepresented minorities and women. Despite the known benefits of diverse workforces, African Americans, Hispanics, Asians, and women remain significantly underrepresented in cybersecurity roles. Using an exploratory research design and focus groups, the study highlights overlooked gaps in the literature and practice regarding DEI in cross-sector cyber intelligence markets. The aim is to develop actionable strategies to drive meaningful change and promote inclusion in the cyber intelligence field.

Chapter 9: Towards a Framework to Support Employees with Hidden Disabilities

Mr. Jim W. Tebrake, School of Business, Marymount University.

This study presents a framework to help organizations effectively support employees with hidden disabilities—conditions that affect an estimated 16% of the American workforce. Recognizing and addressing the needs of these employees not only fulfills legal and ethical responsibilities but also enhances organizational reputation and performance. Drawing from existing research on disability inclusion and employee support, the framework emphasizes six key components: understanding legal and ethical obligations, defining a program scope tailored to the organization, engaging all staff through training and change management, creating a confidential disclosure process, providing appropriate accommodations, and continuously monitoring program effectiveness. This structured approach aims to foster a more inclusive and high-performing workplace.

Chapter 10: Building Trust in Artificial Intelligence

Lord Dordunoo, Marymount University

Building Trust in Artificial Intelligence explores recent technical advances in AI—such as transformer models and generative AI—and their transformative impact across industries like healthcare and finance. While these innovations have driven breakthroughs in language processing, diagnostics, and fraud detection, the paper highlights pressing ethical challenges, including bias, accountability, and transparency. It stresses that embedding fairness and responsibility into AI systems is crucial for building trust. The chapter calls for robust ethical frameworks, transparent practices, and ongoing research into effective regulation to ensure AI's socially responsible development and deployment.

These chapters illuminate how communication, leadership, and strategic planning can converge to foster safe, inclusive, and high-performing organizations. This book is a timely contribution for academics, professionals, and policy-makers invested in the future of organizational development across sectors.

As the editor, together with the contributing authors and co-authors, we are confident that the book will make a significant impact on the academic community, business organizations, and industry professionals. Emphasizing the vital role of communication alongside leadership and performance, the book provides practical insights and research-driven strategies to foster safer, more secure, and service-oriented workplaces. We believe it will serve as a valuable resource for those striving to create resilient, high-performing organizations through effective communication and informed leadership.

Darcia Ann Marie Roache
Capilano University, Canada

Acknowledgements

It is with great gratitude and humility that I acknowledge the many individuals and institutions whose contributions made this volume, *Advancing Safety, Service, and Performance in Organizations with Impactful Leadership* possible.

First and foremost, I give all glory and thanks to my Lord and Savior, Jesus Christ. It is through His grace, strength, and divine guidance that this work was both envisioned and brought to completion. At every stage of the journey, I have been reminded of the sustaining power of faith, the value of perseverance, and the clarity of purpose. May this book be used to foster deeper understanding and inspire meaningful, lasting impact.

I extend my heartfelt appreciation to my esteemed authors. Your commitment to academic rigor, critical analysis, and thoughtful reflection is evident in each chapter. The depth and diversity of your contributions have significantly enriched this volume, offering multifaceted perspectives on one of the most pressing concerns in contemporary organizational life. Your work has not only enhanced the intellectual quality of this book but also helped advance the ongoing conversation around leadership, communication, and workplace safety.

I am deeply honored by the contribution of Professor Keith Walker, who graciously accepted the invitation to write the foreword. His ability to frame the themes of this book with such clarity, wisdom, and academic insight has set the stage for the discussions that follow. Thank you for lending your voice to this work and for your long-standing dedication to advancing scholarship in the field. As my former PhD Supervisor, you have molded my research and writing interest.

A heartfelt note of appreciation is extended to Mr. Richard Muschette, Captain and Aerospace Engineer. Our frequent and thought-provoking discussions, especially those focused on safety and security within the aviation industry—served as the conceptual groundwork for this book. Your extensive expertise, practical insights, and unwavering passion for the field have been both inspiring and invaluable. These conversations were pivotal in shaping the original vision of this work, and for that, I offer my deepest gratitude.

To the team at IGI Global, thank you for your continued support, professionalism, and guidance throughout the editorial and production process. Your commitment to publishing scholarship that bridges research and practice is commendable. From manuscript preparation to final proof, your attention to detail and collaborative approach ensured that this work reached its full potential.

I would also like to acknowledge the many research participants who contributed their time, experiences, and insights to the study at the core of this project. Their candid reflections and perspectives provided essential grounding to the theoretical frameworks and arguments explored in the book. Without their voices, the research would have lacked the depth and authenticity that make it truly meaningful.

To my family, thank you for your unwavering support, patience, and encouragement throughout this demanding yet fulfilling journey. Your understanding during late nights, tight deadlines, and moments of uncertainty sustained me and made this endeavor possible.

To all who contributed in ways both large and small, thank you. This book is as much yours as it is mine.

Dr. Darcia Roache, PhD. DBA. MBA. BSc.

Editor

Section 1
Leadership, Team Management, Change

Chapter 1
Transformational Leadership and Legal Duties:
Accountability and Vigilance in Organizations

Sheikh Inam Ul Mansoor
https://orcid.org/0000-0001-8636-4769
Dayananda Sagar University, India

Shahida Qadir
https://orcid.org/0009-0006-2053-8950
International Islamic University of Malaysia, Malaysia

Amthul Naseeb
https://orcid.org/0009-0001-2724-254X
Dayananda Sagar University, India

Hina Kausar
https://orcid.org/0000-0002-7009-153X
University of Petroleum and Energy Studies, India

Qurat Ul Ain
https://orcid.org/0009-0002-0798-1270
University of Kashmir, India

ABSTRACT

This chapter focuses on the importance of proof and vigilance in organizations and their relation to the concept of transformational leadership practices and legal

DOI: 10.4018/979-8-3373-5686-0.ch001

governance. Exploring leadership theories and analysing organisations' examples from the healthcare, business, and education context, it also demonstrates how leadership fosters safety culture, accountability, and ethical decision-making. The chapter focuses on the communication processes, ethical theories and measures of effectiveness, pertaining to compliance with the legal requirements. Finally, it gives the leaders the recommendations on how to strengthen the organizations to make them responsible for their actions. This combination of leadership and law provides a course that leads to organizational success in the creation of safety and services.

INTRODUCTION

Leadership patterns and legal liability are more relevant in modern organizations, and their relationships are growing more significant with contexts that demand safety and security measures (Deng et al., 2022). Transformational leadership, which is based on the belief that organizational change and organization development is most effectively facilitated by leadership which inspires followers to look at the bigger picture and reach their full potential, provides a rather attractive solution to the problem of creating a highly conscious and responsible organisational culture (Bass & Avolio, 1993). This leadership goes beyond simple transactional processes, or the call for governance by the establishment of clear moral visions, calling on leaders to develop cultures that would advocate for ethical behaviour, teamwork and ongoing learning. In today's ambiguous settings of global legal structures of organizations there are explicit sensibilities and legal imperatives that go beyond simple operational requirements but define the culture and credibility of organizations (Ford et al., 2021).

On this view, transformational leadership fundamentally assumes that leaders who positively influence the accomplishment of objectives at the same time positively influence the psychological and emotional states of the subordinates (Pulido-Martos et al., 2024). This approach emphasizes four key components: organizational influence, personal communication, inspirational encouragement, innovative inspiration, and personal attentiveness. Such leaders lead with these personality characteristics and set standards by being motivational figures with high levels of organizational commitment. It motivates the people to work towards the vision and enables people to play constructive role in achievement of organizational objectives (Farahnak et al., 2020). Furthermore, through the concepts of proposal and perspectival switch, transformational leaders foster challenge for innovation to create an organisational culture that values creativeness and logical thinking to make sure that change on observed/identified and new practices and processes occur (Ford et al., 2021).

Nevertheless, let us emphasize that the transformational leadership paradigm is closely connected with the legal responsibility or legal liability as a fundamental form of the organization's control (Juan & Khan, 2023). In most regions, the leaders are required to comply with legal requirements that prevent and protect their employees and clients from hazards. For example, under the Occupational Safety and Health Act OSHA in the United States the employer is legally required to ensure that an employee's workplace is reasonable free from hazards that can be recognized that can cause death or serious physical harm (Michaels & Barab, 2020). This legal duty of care further means that leaders not only develop relevant safety measures, but also make sure that their organisations follow legal requirements. Ordinarily, failure to perform the various tasks leads to important legal consequences such as penalties, legal suits, and adverse effects on the image of the organization (Ta et al., 2022).

Management and leadership are very close but different concepts that perform differently in the organization. The management deals mostly with the planning, organizing and execution of processes to keep efficiency and consistency. It is an emphasis on structure, control, task completion (Ford et al., 2021). While leadership is about inspiration, vision, guiding people in doing things together, leadership is about making them work together for a vison, be it a vision word, a vision sentence, a vision image or vision number, either as an individual or as a team. Innovation is something leaders develop, teams are motivated, and long term growth is fostered, and sometimes even challenged. Management settles down and is ordered whereas leadership brings in change and growth (Deng et al., 2022).

This research is specifically targeting leadership and more specifically the aspects of transformational leadership than the transactional leadership. Transformational leaders lift the mood of staff, enable them and help them make more than they could otherwise. Transactional leadership works things through structured rewards and punishments, with predetermined roles and rewards for maintaining a status quo and order (Heimstädt & Dobusch, 2020). Since elements of transactional and transactional leadership may overlap some management or practices, transformational leadership is in essence dedicated to change, inspiration and the general improvement of employees (Melo et al., 2020). This research focuses on how to separate the two (management versus leadership) and the two leadership skills (transactional versus transformational) so that it is clear how leadership, and particularly transformational leadership, boost employee engagement and organization success as opposed to just managing tasks and processes.

There are two basic categories of legal responsibility: compliance assurance, employee protection and stakeholder protection. Many prescriptive measures to promote security and performance have to be taken by managers in sectors like healthcare, education and corporate governance (Waldman & Balven, 2014). For example, healthcare managers need to adhere to regulations that evaluate organizations on their compliance with safety and quality standards of the patient's care (Young & Smith, 2022). Similarly, school administrators are also given responsibility of making sure that institutional policies are consistent with legal mandates made to prevent discrimination and sexual harassment in educational institutions (Cherry et al., 2021). Legal requirements covered here greatly emphasise the importance of managers to incorporate legal concerns within its strategic vision and operational activities (Sonmez Cakir & Adiguzel, 2020). Transformational leadership is important for shaping an environment for answerability, so managers and employees perform legally and morally correct (Wood et al., 2005). Considering how transformational leadership has a payoff of organizational accountability through increased transparency, effective communication, and shared responsibility between employees (Heimstädt & Dobusch, 2020), this chapter will take the exploration further.

Clear communication of what legal requirements and ethical standards are, use of necessary tools and resources for maintaining workplace safety and regular review of the adherence to safety policies (Babi et al. 2021) can reinforce compliance of managers. In addition, accountability generation involves employees participating in explaining legal and safety issues due to which, compliance is emphasized and individual responsibility is developed to maintain safety standards (Melo et al., 2020). Thus, participatory leadership grounds in traditional leadership principles in that it induces employees to participate in making decisions and reinforce that they are necessary participants in the action to achieve the organization's goals (Wang et al., 2022).The importance of this research area is rooted in the author's focus at delineating how the approach of transformational leadership can best fill the gap between leadership measures and legal responsibility. Given the current focus on looking into organizational behaviour, pointing out how leadership might possibly encourage accountability as well as vigilance is central. This research's goals are to identify the aspects of transformational leadership that support legal integrity, assess the concerns for leadership that legal responsibilities offer, and offer sensible recommendations for leaders who encourage legal and compliant, yet safe and service-oriented employee behaviours.

The purpose of this study is to advance knowledge on organisational leadership and management by exploring the relationship between transformational leadership behaviour and legal liability. So, this chapter aims to explain how leadership might address the legal requirements and protect people at the same time. Moreover, this research aims to offer the organizational leaders, policymakers, and scholars with guidelines on the application of the transformational leadership concepts with focus on the legal requirements, to increase the organizational performance and stakeholder confidence.

Since organisations are still experiencing diversity of challenges ranging from legal compliance and the dynamic complexities of the work environment particularly in safety, the need for managers to be transformational as they discharge their legal responsibilities grows sharper (Patrick & Kumar, 2012). Thus focusing on accountability and ethical behaviours, managers contribute to Legal risk management as well as improving the odds of employee and stakeholder satisfaction. The aim to achieve the organizational management's vision of excellence places emphasis on organization-wide commitment to learning, action, and conscientiousness of the operating legal frameworks (Endenich & Trapp, 2020). Thus, the role of the transformational leadership becomes indispensable as a means to promote positive change and guarantee organizational practice compliance with the relevant legislation for sustainability (Kalifa, 2024).

TRANSFORMATIONAL LEADERSHIP AND LEGAL OBLIGATIONS

The subjects of transformational leadership and legal requirements are one of the most significant complexes in current organizational theories and applications that enhance the organizational culture of safety and responsibility (Bendak et al., 2020). The legal requirement is only part of what leaders have to do in the new challenging context; leaders have to make an impact on the quality of ethical practices and safety in their organisations. The theoretical framework adopted in this context is an understanding and analysis of the conceptual structure that may be used to understand and analyse the relationships between transformational leadership and legal requirements with organizational safety and responsibility. It logically integrates different components including leadership theories, ethical and moral standards, legal obligations, etc. (Buil et al., 2019). This framework provides that transformational leaders are not only required to fulfill the legal duties but also develop the organizational culture by inspiring ethical practices, safety measures and compliance. A combination of theories of leadership such as Burns' transformational leadership theory compared to transactional leadership that works on the moral and ethical transformation of followers, is taken here. Also, the framework contains accountability, alertness and the role of leadership in producing an organizational success with the help of ethical motivation models. They combine to instruct organizations on how to fabricate a culture of safety, responsibility and ethical conduct (Walker, 1984). This theoretical framework sought to titrate identification of effective practice of transformational leadership, presentation of the duties that a leader has legally assumed in relation to safety and security, and prescription of theoretical models on accountability and alertness (Sonmez Cakir & Adiguzel, 2020). Studying these components assist organizations approaches the specifics of governance while fostering compliances and ethical standards. This school of leadership helps in changing the motivation level and morality of the leaders and the followers respectively. This style is marked by leaders who are not only involved in transactional exchanges but are also interested in changing the follower's orientation towards inspiring values. Burns' work is seminal in the study of transformational leadership and establishing the difference between this kind of leadership and transactional leadership through the application of ethical and moral theory. Heroes mobilize people into embracing new ideas which will enhance the achievement of organizational objectives and personal dreams (Reid & Dold, 2018).

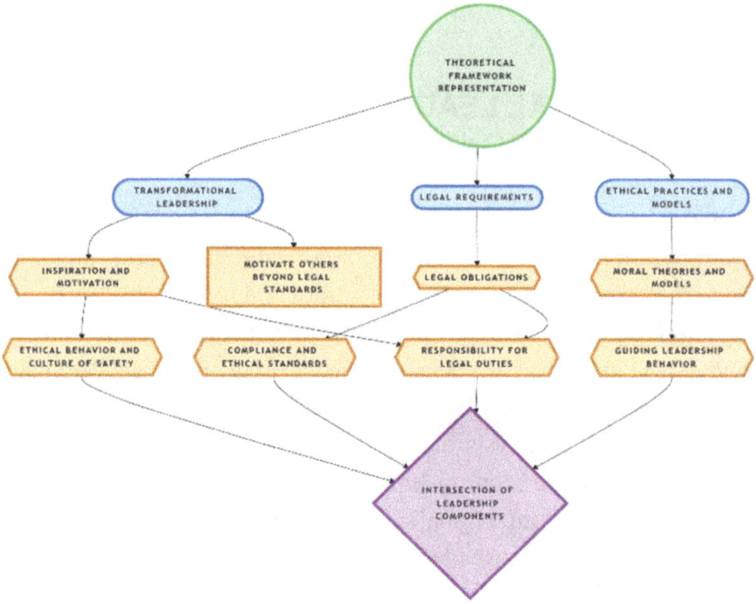

The essence of transformational leadership lies in the four components i.e. principle of identification, inspirational communication, intellectual encouragement and individual regard. Idealized influence is making the leaders to act as example to emulate in issues to do with ethical standard and organizational values (Bass, 1999). This aspect ensures that those that follow the leadership of a particular leader develop a positive attitude towards the leader and such an environment breeds free discussion and ideas sharing. Inspirational motivation implies that the leaders convey a message that captivate the followers to perform beyond their expectations. This way, transformational leaders bring energy and confidence in their abilities that compel the employees to go for the best (Hancock et al., 2023). Another characteristic of transformational leadership is the engagement of staff with ideas by stimulating the members of the organization to engage in problem solving. This characteristic is crucial in enabling lower levels of employees to feel free to express their problems and strategies being geared towards increasing organizational creativity (Buil et al., 2019). Finally, the individual consideration speaks about the tenure of the leader towards understanding strength and developmental acumen of each of the followers. This dimension brings support where there is identification of the employee with the need to support him/her to be able to execute the tasks in the organization (Tang et al., 2024).

In organizational contexts, transformational leadership is conditional on legal regulation that determines the obligations of managers in terms of safety and security. Legal requirements are basic forms of compliance requirement that are critical in governance since they set out leadership expectations regarding compliance with laws and regulations (Odom & Green, 2003). Most of such requirements are widely seen in fields like health, learning institutions, and business management since failure to meet those results in serious repercussions. For example under the Occupational Safety and Health Act in the United States, employers are legally required to provide a workplace which is safe as is reasonable, foreseeably practicable in as far as any activity that is recognized as being hazardous to health or likely to cause grievous injury (Fairchild & Ede, 1978). This legal requirement means that leaders also must positively supervise safety measures as well as enforce safe culture as prescribed by law on health Safety. Non-compliance with these standards can lead to serious consequences, legal consequences and even loss of confidence from employees. Therefore, the legal requirements play a rather important role in defining how transformational leaders manage their leadership activities (Walker, 1984).

In the India healthcare organizations leaders are to obey a number of rules and regulation which are provided by the National Accreditation Board for Hospitals & Healthcare Providers and Indian Public Health Standards (Wu et al., 2024). These organizations establish very high standards for the safety of patients and quality services most especially in emergency and casualty units. Negligence to the above standards puts patient lives at risk besides attracting penalties and legal cases besides the suspension of institutional licenses (Young & Smith, 2024). In schools institutions, other legal requirements which the leaders have to factor include the Protection of Children from Sexual Offences (POCSO) Act and Right to Education (RTE) Act where children are protected in school from all forms of harm as well as from discrimination (Fiorvanti & Brassard, 2020). Further, procedures such as the UGC guidelines on ragging and sexual harassment demand heads of institutions to safeguard policies and actions that make environment for all students, faculty and employees safe and free from disrespect and harassment. Such legal requirements call for leaders to put and maintain standards in educational institutions on how they protect, promote the rights and the best interests of every learner (Rao et al., 2018).

Leaders cannot just rely on simple legal requirements; they bear a general duty and responsibility of ensuring that the cultures that govern an organization are safe, ethical, and accountable. This has to do with articulating corporate expectations of legal spirit and efforts to ensure that employees have the knowledge and tools to enforce the requirements (Shakeel et al., 2019). It is on this regard that transformational leaders take the whole process a notch higher by being responsible for integration of legal obligations into the organizational fabric. This is done through encouraging discussion in the corporate environment concerning the legal obligations and encouraging them to independently accept legal accountability for safety measures (Odom & Green, 2003). In addition, the theoretical concepts of accountability and vigilance provide understanding of the relation between transformational leadership and legal requirements. Corporate accountability means that leaders are answerable for their decisions can, and the decisions of their subordinates to legal and ethical standards (Wood & Winston, 2005). This concept is related to the concept of organization alertness which in essence insist that organizations must remain cautious and sensitive to potentials threats and risks. Performance can be kept responsive and keen through measures set by the transformational leaders, in this case, the legal benchmarks, and safety measures. This demands the development of the ways and means of how compliance and performance can be intelligently monitored, accessed and communicated to the employees about their role in supporting and underlining organizational safety and security (Errida & Lotfi, 2021).

Different theoretical frameworks are used to explain organization's operation in accountability and vigilance. According to the follower theory, organizations are by mutual understanding so leaders should act for the best interest of followers to create an ethical leadership and responsibility to fulfil legal obligation to create safe organizational culture (Fan, 2024). According to stakeholder theory which follows transformational leadership principles, it suggests that the leaders have to look at all concerned stakeholders including the employees, customers and the community and create an atmosphere where they can participate in making decisions (Brown & Treviño, 2006). In systems theory, leadership and legal requirements are related to the organizational culture, because organizations are treated as system that interacts among each other, where leadership, legislations and the behaviour of the employees influence a safe and efficient organizational setting (Mele et al., 2010). The understanding as explained above assists leaders to respond to legal compliance while encouraging and promoting collaboration and communication in the structure of the organization; that way, they change legal risks to functionality (Adams, 2012).

Accountabilities also include action plans with regard to risks and emergencies in organizational settings where leaders are expected to formulate and advise on the risk management policies and elaboration of measures for desirable organizational safety measures. Organizations in high risk business sectors like construction, health, and education require keen observation of risk factors, risk management check-ups, and offering the required facilities and instruction to handle with the premature signals of risks (Zwikael & Ahn, 2011). For example, in the case involving a healthcare organization, the violation of safety regulations involves much more than legal consequences, which are loss of revenue as patients experience harm as a result of operations from the organization they trust. Besides, by making ethical decisions, the concept of continuous improvement fosters accountability and vigilance resulting from the transformational leadership. It's expected that leaders with transform the company or organization should be of high integrity and provide a positive change on the ethical values of the organization. Thus, the development of guidelines for ethical decision making that corresponds to legal requirements guarantee that the workers are trained on how to operate in intricate situations safely and legally (Scholl et al., 2016). When ethical values and standards are incorporated as a framework into decision-making in an organization, transformational leaders positively influence organizational adaptability to legal and ethical issues.

The connection between transformational leadership, legal responsibilities and culture is further expanded by the exertion of training and development activities. Legal compliance and ethics training also act as one of the major tools for creating awareness and improving accountability and parternetwork (Deng et al., 2022). In this way, employees receive information and learn what potential risks might occur and how they should be handled, so the organization creates a safety and ethical-oriented culture. These training initiatives can only be driven by change since transformational leaders show their readiness to foster this kind of culture in their organizations to promote adoptions of legal compliance and occupational safety measures (Jacobsen et al., 2022). So it will be critical for any organisation that wants to enhance the organisational culture to comprehend abstract theory linked to transformational leadership as well as legal requirements. The research argues that among the types of leadership it is transformational leadership that, with its focus on the followers, can be instrumental in enacting legal responsibilities throughout the organization (Xenikou, 2022). Managing legal requirements with creating the culture that focuses on safety and ethical performance improves organisational performance and reduces legal liabilities. Theoretical concepts of accountability and watchfulness add a different understanding of the process, as well as features of transforming leadership as a driver of culture change and compliance with the legislation (Radu, 2023).

As organizations are exposed to numerous challenges in the various legal environments, more emphasis is placed on transformational actions among leaders as they discharge their responsibilities responsibly while observing the required laws (Mair & Seelos, 2021). Combining the use of transformational leadership with a comprehensive appreciation of legal duties empowers leaders with the right tools to manage the challenges of current leadership while creating the right environment of safety for even the most vulnerable members of society (Lai et al., 2020). Therefore, it becomes crucial for organisations to invest in formation of transformational leaders that are able to factor Ethics besides the law in cases of enforcing accountability hence in turn bringing safer environment within organisations.

Cultivating a Culture of Accountability

The idea and practice of workplace compliance that responds to legal imperative as well as critical standards of ethics demands that organizational cultures embrace accountability. The role which transformational leadership occupy in relation with accountability is to determine how this type of leadership, which focuses on the inspiration and mobilization of followers, can contribute to the establishment of an organizational culture that embraces this activity (Roy et al., 2024). Top management emphasis on culture of accountability not only ensures that letter and spirit of the law are followed but also everyone feels that they are party to something special. Therefore, the approach involves the improvement of the effective communication patterns, encouragement of ethical behaviours, as well as the matching of the performance measures with the legal requirements that are related to the mentioned transformation (Tadesse Bogale & Debela, 2024). By this way, the transformational leaders can foster specific culture, which is based on legal and ethical behaviour and constant enhancement.

From this perspective, it is important to maintain that the approximate element that acts as the main pillar of organizational communication is the issue of legal compliance. Managers need to provide clear avenues through which they and others can pass information concerning the law and safety measures. This calls for the establishment of highly elaborate staff awareness training involving the law, company policies and the standards governing the respective industry (Hillmann, 2021). When the employees have the knowledge of their legal liabilities, the leaders assist them in making good decisions and avoiding possible negative impacts. In addition, more importantly, transformational leaders have to employ different forms of communication in echoing the compliance with laws. These practices may refer to periodical meetings, seminars, training sessions or informative sessions which are used in asserting the organizations commitment to legal compliance (Mishra et al., 2023). In this way integration of practical scenarios and case studies into the training and development programmes can help the employees to gain knowledge about the actual implementation of legal liabilities. For example, information on HIPAA training for employees is crucial for it to be aware and able to follow the healthcare rules controlling use of images and information of the patient. Due to increased use of communication strategies such as artificial intelligence, it becomes easier for leaders to ensure that legal issues are discussed in the organization. Also, leaders should encourage the employees to give their feedback concerning their knowledge of the legal requirements and the difficulties which they are likely to experience in implementing these standards (Zhang et al., 2017). Opportunity to freely share information with the management through suggestion boxes, reporting to managers or via anonymous reporting procedures compels employees to report violating incidents without being penalized. This way it does not only show compliance to the legal requirements but also engages the employees into accepting full responsibility of their roles as it relates to compliance (Yeboah, 2023). According to the program, accountability helps lead to the improvement of culture, which in turn, ensures that employees embrace legal and ethical requirements.

Another important factor that need to be addressed hence leading to accountability is encouraging ethical practice. The importance of organizational transformational leader can be summed up in the fact that such a leader provides ethical example and promptly sets the ethical standard of the company. There should be a clear codification of the expected conduct in ethical matters that gives the employee the direction they need to take as they make a decision. Leadership frameworks need to be communicated since by so doing, people in an organization can be motivated to undertake their responsibilities in ethically sound manners (Mitchell et al., 2020). Besides, the presence of ethical code of the organization and its transformations should be accompanied by the leaders' actions aimed at ensuring shared responsibility for ethical behaviors. This Seventh Code means that employees should be motivated to report errant colleagues and prevent such individuals from continuing with their unethical conduct (Ahmad et al., 2024). Several points can be made about how transformational leaders can help to support this process of ethical decision making and to reduce the incidence of unethical behaviour: For instance, by awarding the outstanding ethical performances, an organization demonstrates good accountability and ensures that others follow the same (Deng et al., 2022). In addition, there are best practises that leaders can adopt involve the establishment of mentorship and coaching to ensure that people in the organization behave ethically. By identifying the strategic relationships between the experienced managerial leaders and young talents, organizations are in a position to align themselves with appropriate models as well as ethical expertise (Sonmez Cakir & Adiguzel, 2020). Such relationships can enable the employees to provide solutions that address ethical challenges while making decisions for the firm. Further, use of ethical factors in assessing the performance of employees also promotes ethics within the organizational ideals of the firm. If performance appraisals are done based on performance indicators as well as unethical behavior, managers can eliminate one problem while addressing the other as a way of making the solution almost perfect (Curado et al., 2022).

One of the most important ways to enforce accountability throughout the culture of an organization is by linking performance measurement with legal requirements. It is the article in that it discusses how transformational leadership is important in the successful adoption of legal and safety systems through these performance assessment mechanisms. One of the most critical roles that transformational leadership can play is in the direction of organizations towards establishing clear and measurable legal and ethical standards with respect to the regulatory frameworks (Plsek, 2013). This leader style encourages employees to become an element of the value of the corporate culture, such as safety and legal compliance. A practical example of this is that in financial industry, financial institutions establish KPIs regarding how they comply with the anti money laundering laws. These are KPIs that will measure both how much students and employees go along with the norms of the law and how much students and employees understand and help promote the trends of the law (Ahmad et al., 2024). To achieve such measures, the transformational leaders have to not only assure the regulatory compliance but also motivate employees to become actively involved in preventing the financial crime. Organizations achieve this by incorporating ethical as well as legal concerns in performance evaluations that bring the element of responsibility and accountability at all levels. The transformational leadership and its emphasis on ethical decision making and the continuous improvement fit with this approach. By knowing the significance of their actions, it enables employees to understand the business outcomes as well as legal and social responsibilities that are important for the long run sustainability (Vuong & Nguyen, 2022).

In addition, leaders should constantly remind stakeholders on the improvement of performance indicators. This include periodic checking and revision of relative performance standards to meet legal requirements and organizational goals. When promoting the culture of learning and organization transformation, the leaders should entail elements that will keep the employees motivated in following legal requirements. This unstemmed drive to increase organizational efficiency not only improves performance but also decreases the possibility of legal non-adherence (Sonmez Cakir & Adiguzel, 2020). However, organization transformational leaders should also ensure adequate performance standards that can check compliance with the legal requirement when it comes to regulation enforcement. This certain may encompass crucial ties such as administration and assessments in order to determine compliance with legal compliance (Lai et al., 2020). For example, the healthcare organizations may schedule periodic compliance check to consider EMTALA law compliance. These audits can be an excellent source of finding out where organizations and managers stand in terms of compliance and can help leaders take steps to rectify compliance deficiencies. In addition, leaders should ensure; there is the disclosure of performance and compliance reports. Communicating compliance information to employees fosters organization commitment and makes the employees take personal responsibility in the compliance process. Through frequency in reporting and reviewing the compliance performance as well as acknowledging the teams who are exemplary in the compliance performance it becomes easier to remind the organization that compliance is everybody's business (Sonmez Cakir & Adiguzel, 2020).

When transformational leadership and legal compliance are at play, the only viable strategy to foster responsibility throughout an organization is to be precautious. It is crucial to foster good communication practices in order so that the employees will comprehend their legal responsibilities and will be willing to act according to these rules (Ayoko, 2022). Simply put, organization transformation fosters the notion of 'we are all in it together' and ethically motivated leaders increase people's desire to act accountably. Furthermore, the use of performance metrics in conformance with legal requirements constitutes a critical means of tracking performance and effecting change. This requirement of aligning the performance evaluations to the legal obligation means that if organisations design an environment that aligns to their goals and objectives, then those legal requirements constitute the outer legal perimeters in containing organisational risks. Thus, such activities help transformational leaders build the organization's positive climate based on accountability and ethical, as well as, legal and safe environment for organization's relative improvement (Bourne et al., 2018). In order to meet these goals, these organizations need to constantly provide information to staff and other stakeholders regarding legal requirements and ethical considerations. This commitment to professional development supports the work of clarifying accountability expectations and provides employees the tools they need to successfully dissect and work within frequently convoluted legal situations. Director-driven paradigms should support such activities, and underscore the importance of personal integrity and lawfulness for organizational success.

Case Studies: Transformational Leadership in Practice

Transformational leadership has never been reported to be used and failed to bring about positive change in healthcare settings (with regards to patient safety/ legal requirements and /or organizational productivity). The Virginia Mason Production System (VMPS) was first made operational at Virginia Mason Medical Center (VMMC) in Seattle Washington. Transformational leadership calls for these principles and this approach aims for automating processes, and continuously increasing performance through multi team strategy. The organization wanted to strengthen its culture toward one that is safety based, whilst maintaining and continually improving the standard and quality of patient care, as well as organizational outcomes in general (Nelson-Peterson & Leppa, 2007). When Dr. Gary Kaplan took charge of the VMMC, he embarked on a transformational process that aimed at accomplishing transformation through achieving patient safety which required everyone in the VMMC organization to be involved. It comprises large scale education courses that provide staff information on legal bearings of patients' care and safety measures (Kaplan, 2006). This chapter demonstrates how VMMC continuously encouraged workers to report possible risks leading to their development of an organizational safety culture that decreased adverse occurrences and increased patients' well-being. This phenomenon was like the light bulb moment at VMMC in a sense that is demonstrated how the relational aspect of communication can be effective in delivering the messages that relate to legal tactical measures for compliance and ethical standards. Communications used by the leaders to pass information on safety measures and legal requirements included briefings, newsletter and e-learning (Brooks et al., 2019). These endeavours also helped in making sure that all employees were fully acquainted with the roles they have in the promotion of the safety policy and the legal consequences they would face in case of failing to observe the laid down safety policies. Therefore, VMMC saw an enhanced success of patient safety with a decrease of hospital acquired infections by 50% and overall patient harm incidents (Brooks et al., 2019). In this context the presence of the connection with transformational leadership is evidenced in continuous improvement, employee engagement and creating a safety focused organizational culture. VMMC transformational leadership facilitated team collaboration across teams, and provided reasons for employees to be active in supporting the safety of patients. Staff were led by leaders who created a sense of responsibility and accountability and made sure that everyone knew what was expected of them to uphold safe policies (Ayoko, 2022). The result of this approach was a decrease in hospital acquired infections and patient harm incidents as evidence of the role transformational leadership can play in making both organizational change and compliance with legal and safety.

This demonstrates how Virginia Mason Medical Center has adopted exceptional organizational leadership strategies in realizing a successful organization. The performance measurement system that was put in place in this organization entailed safety and legal performance that had to be met all the time (Plsek, 2013). It showed that needs for development and training, as well as areas for resources in audits, really needed careful consideration in addition to it. In principle three a relationship between the identified performance indicators and legal requirements to recommend that ethical benchmarking enhanced accountability and deliver ethical as well as legal VMMC (Truitt, 2011). However, not all organizational stories are as positive as the above examples, in that they always reflect the organization's best image. The company that is worth noting especially in the Antecedent part of the film is the Enron Corporation, which is a classic example of the corporate world's shocking mecca of dreadful debacle borne out of folly leadership and irresponsibility. 2001 Enron scandal was therefore caused by fraud and the reason was anchored on corporate greed and aggression spearheaded by Enron's then Chief Executive Officer Skilling (Nguyen, 2011). Fostered by Houston culture, organizational culture at Enron promoted unnecessary risks an inadequate disclosure and fraudulent accounting. People mostly motivated through the self-checking financial benefits disregarded any ethical consideration and engaged in fraud and unlawful practices. Erratic communication of legal obligations and ethical standards were also missed which worsened these problems by hiding unethical behaviour (Di Miceli da Silveira, 2013).

The ethically-corruptive practices of the Enron organization shows exactly why role of transformational leadership is so significant. For that reason, the omission of other leadership practices like ethical communication, shared responsibility, and legal compliance were a green light to unethical behaviour (Ahmad et al., 2024). As a result, different legal changes that means and ends of improving corporate governance and accountability such as the Sarbanes-Oxley Act were made. These reforms focused on the culture for ethical and legal behaviour within organizations, and its enhancement by strong, visionary organizational leadership (Wasdani et al., 2021).

CONCLUSION

In today's environment the relationship between the concept of transformational leadership and possibility of legal actions against the manager is one of the essential vectors along which effective practices coalesce. Investigation of this interface leads to understanding of some basic, but significant, truths about leadership and practice of management at the company and corporate level where organizational culture may influence legal requirements which are vital for compliance with the applicable law. From the case studies under review in this chapter, it is clear that a stated and well-actualized commitment to the promotion of accountability and ethical standards, as well as adherence to the law, is crucial for corporate entities aspiring to raise the security and efficiency of their service delivery. From the study, it is clear that the transformational leadership creates awareness regarding the culture that is supposed to be fostered not only to meet the legal requirements of the company but also to strive for ethical conduct in all organizational levels. Literatures that focus on the application of transformational leadership principles, where health care particularly the Virginia Masons Medical Centre has built its foundation, show that encouraging the employees in the process of the improvement journey is critical.

Enabling staff at health care organizations take more responsibilities to work on safety and components of the legal systems contribute to the development of a strong safety program that will improve the safety of the patients and the functionality of the organization. The involvement of the employees in this process helps to ensure commitment from the employees that accountability is not a top management edict.

In addition, business failures, especially the Enron case, are equally viewed as a pointer to the possible calamities tangible whenever there are no standard measures to ensure accountability and ethics practice. The focus upon Enron's leadership therefore demonstrates the need for the formalisation of safeguards which encourage corporate responsibility and accountability. The facts stated in this case can be utilized to educate modern organizations on the necessity for developing and maintaining ethical standards that are obligatory for all organizational members. This can be by providing required specialized training that makes employees understand legal obligations and ethical standard expected of them, having clear means through which unethical behaviours are reported. Schools in particular have also provided a good example of how leadership can bring about changes and promote an organisational culture that is accountable and legal. The University of Virginia's efforts in accreditation of legal education and ethical training show how leaders of education can equip the next generation to confront legal requirement and ethical paradoxes. In passing the relevant knowledge and skills necessary to meet all the legal requirements, educational organizations fulfil an essential function of forming ethical culture in future enterprises.

The insights communicated in this study should inspire organisations to value transformational leadership as a key to travelling the middle distance between law and high standards. Leadership that is transformational places organisations in a better place than those which do not because of what is contained in modern regulations to do with safety of stakeholders. The promise to build an environment that focuses on maintaining the sense of accountability, ethical and legal compliance is not only an organizational principle but a social responsibility in the contemporary world. While leaders endeavour to achieve increased safety and service quality by attaining leadership and legal goals, there is a need to remind people that a better future depends on organisational commitment to safety and service. There is transformational leadership that can act as a framework when it comes to such a process and make leaders foster and encourage their subordinates to think and act ethically and in accordance with the law. When the day-to-day working objectives of an organization are made to reflect the legal and ethical concerns all-round the target, then it sets forth the guidelines to kick off organizational achievements as well as societal welfare.

Several recommendations for future research as well as for practice can be identified for developing the theory and improving practical application of transformational leadership considering legal requirements. First, there is a clear dearth of theoretical and empirical evidence about the endurances and proximal effects of transformational leadership on organisational culture and lawfulness across industries. Specifically, longitudinal qualitative and quantitative research on leadership interventions aimed to determine the maintenance of such practices and overall organisational performance outcomes would contribute significantly to the body of knowledge. In the same way, comparative research across different cultures would allow understanding of how ideas of transformational leadership have to be adjusted to the different legal requirements and permissive enactments of various cultures. Second, organisations should find ways of supporting leadership development interventions that mainstream legal and ethical perspectives into leadership training. They are a way of strengthening organization's ability to drive accountability by providing leaders with the requisite knowledge and skills regarding multiple legal regimes. Thematising realistic training approaches, the leaders train for real-life scenarios and ethical and legal controversies that entrench ethics as an essential part of organisational operations.

Also, it is critical for organizations to have procedures for on-going feedback and updates in the accountability procedures. Organizational culture audit, legal standards appraisal, and ethical analysis should all be conducted periodically to determine whether one needs to make any changes to the organizational culture or not, which will in turn help the leadership. With regards to information gathering and communication, the following may be used: Self completed employee satisfaction and perception questionnaires, focus group discussions, and /or identified and recommended anonymous reporting systems. Of heightened importance is the way an organisation can follow up the implementation of the accountability and ethical standard by adding on further the commitment to create an employee-friendly environment and allow the employees to define the company standards when it comes to reporting the irregularities in the organisation so as to gain a liking to the outcome. Furthermore, the legal environment also seems to change with the organisational practices, to continuously promote the principles of the transformational leadership and responsibility. Measures to enhance transparency and ethic must therefore form part of regulations governing the companies and other legal entities. These regulations can be used to motivate proper organizations into practicing transformational leader ship practices in relation to legal requirement of transformational leadership and therefore the relation between leadership and law. For example, increasing the standards governing corporate governance and ethics will force organizations into leadership practices that emphasize ethical decision making thus increasing accountability.

Just as important is the function of managing technology in terms of bringing the concept of accountability into organizations. The expertise of combining data analytics and performance metrics helps in developing better ideas about the functioning of an organization. In this manner, an appropriate use of the information technology is the continuous monitoring of the levels of adherence to legal requirements and ethical norms and the provision of information on the possible weaknesses in the system and their subsequent correction. It is freely accepting the principles of transformational leadership as the approach bespeaks of their acceptance of data and facts as instruments to ensure a successful change in the organizations.

As we look to the future, it is important not to lose sight of the big picture with regard to increasing safety and service through leadership and law. An organization is interactive, and its measures towards ensuring that it is accountable and ethical will go a long way in determining its image within the society. In other words, by promoting these values, organisations can help creating the culture of an ethical society in which the numerous stakeholders are appreciated and can feel protected. This, in turn, may lead to increased reputation and hence increase chances of organizational sustainability. Moreover, considerable emphasis has been placed on the stakeholders' management. To ensure that everyone at the workplace, customers, members of the community, and the law recognize and embrace safety, service, and accountability, stakeholders should be drawn into the processes. There is required for organisations to ensure that, all people around the world be respected and offer their opinions in any organisation. Employing this multiplicative model can extend the strength and relevancy of safety procedures and legal guidelines by integrating specific features of all user groups.

REFERENCES

Adams, K. M. G. (2012). Systems theory: A formal construct for understanding systems. *International Journal of System of Systems Engineering*, *3*(3/4), 209. DOI: 10.1504/IJSSE.2012.052684

Ayoko, O. B. (2022). Leadership, ethics and corporate social responsibility. *Journal of Management & Organization*, *28*(1), 1–8. DOI: 10.1017/jmo.2022.13

Babri, M., Davidson, B., & Helin, S. (2021). An updated inquiry into the study of corporate codes of ethics: 2005–2016. *Journal of Business Ethics*, *168*(1), 71–108. DOI: 10.1007/s10551-019-04192-x

Bass, B. M. (1999). Two decades of research and development in transformational leadership. *European Journal of Work and Organizational Psychology*, *8*(1), 9–32. DOI: 10.1080/135943299398410

Bass, B. M., & Avolio, B. J. (1993). Transformational leadership and organizational culture. *Public Administration Quarterly*, *17*(1), 112–121. https://www.jstor.org/stable/40862298

Bendak, S., Shikhli, A. M., Abdel-Razek, R. H., & Ardito, L. (2020). How changing organizational culture can enhance innovation: Development of the innovative culture enhancement framework. *Cogent Business & Management*, *7*(1), 1712125. Advance online publication. DOI: 10.1080/23311975.2020.1712125

Bourne, M., Melnyk, S., & Bititci, U. S. (2018). Performance measurement and management: Theory and practice. *International Journal of Operations & Production Management*, *38*(11), 2010–2021. DOI: 10.1108/IJOPM-11-2018-784

Brooks, L. A., Manias, E., & Bloomer, M. J. (2019). Culturally sensitive communication in healthcare: A concept analysis. *Collegian (Royal College of Nursing, Australia)*, *26*(3), 383–391. DOI: 10.1016/j.colegn.2018.09.007

Brown, M. E., & Treviño, L. K. (2006). Ethical leadership: A review and future directions. *The Leadership Quarterly*, *17*(6), 595–616. DOI: 10.1016/j.leaqua.2006.10.004

Buil, I., Martínez, E., & Matute, J. (2019). Transformational leadership and employee performance: The role of identification, engagement, and proactive personality. *International Journal of Hospitality Management*, *77*, 64–75. DOI: 10.1016/j.ijhm.2018.06.014

Cherry, J., Aquino, F., Fabelico, F., & Afalla, B. (2021). Managing educational institutions: School heads' leadership practices and teachers' performance. [IJERE]. *International Journal of Evaluation and Research in Education, 10*(4), 1325–1333. DOI: 10.11591/ijere.v10i4.21518

Curado, C., Henriques, P. L., Jerónimo, H. M., & Azevedo, J. (2022). The contribution of communication to employee satisfaction in service firms: A causal configurational analysis. *Vision (Basel), 0*(0), 09722629221101157. Advance online publication. DOI: 10.1177/09722629221101157

Deng, C., Gulseren, D., Isola, C., Grocutt, K., & Turner, N. (2022). Transformational leadership effectiveness: An evidence-based primer. *Human Resource Development International, 26*(5), 627–641. DOI: 10.1080/13678868.2022.2135938

Di Miceli da Silveira, A. (2013). The Enron scandal a decade later: Lessons learned? *SSRN*. https://doi.org/DOI: 10.2139/ssrn.2310114

Endenich, C., & Trapp, R. (2020). Ethical implications of management accounting and control: A systematic review of the contributions from the Journal of Business Ethics. *Journal of Business Ethics, 163*(2), 309–328. DOI: 10.1007/s10551-018-4034-8

Errida, A., & Lotfi, B. (2021). The determinants of organizational change management success: Literature review and case study. *International Journal of Engineering Business Management, 13*, 18479790211016273. Advance online publication. DOI: 10.1177/18479790211016273

Fairchild, E. J., & Ede, L. (1978). Occupational safety and health: United States of America. *Sozial- und Praventivmedizin, 23*(1), 20–26. DOI: 10.1007/BF02073950 PMID: 695968

Fan, Y. (2024). Accountability in public organization: A systematic literature review and future research agenda. *Public Organization Review*. Advance online publication. DOI: 10.1007/s11115-024-00792-y

Farahnak, L. R., Ehrhart, M. G., Torres, E. M., & Aarons, G. A. (2020). The influence of transformational leadership and leader attitudes on subordinate attitudes and implementation success. *Journal of Leadership & Organizational Studies, 27*(1), 98–111. DOI: 10.1177/1548051818824529

Fiorvanti, C. M., & Brassard, M. R. (2020). Child protection: A child rights approach for schools. In Nastasi, B. K., Hart, S. N., & Naser, S. C. (Eds.), *International handbook on child rights and school psychology*. Springer., DOI: 10.1007/978-3-030-37119-7_15

Ford, J., Ford, L., & Polin, B. (2021). Leadership in the implementation of change: Functions, sources, and requisite variety. *Journal of Change Management, 21*(1), 87–119. DOI: 10.1080/14697017.2021.1861697

Hancock, A. J., Gellatly, I. R., Walsh, M. M., Arnold, K. A., & Connelly, C. E. (2023). Good, bad, and ugly leadership patterns: Implications for followers' work-related and context-free outcomes. *Journal of Management*, *49*(2), 640–676. DOI: 10.1177/01492063211050391 PMID: 36484084

Heimstädt, M., & Dobusch, L. (2020). Transparency and accountability: Causal, critical and constructive perspectives. *Organization Theory*, *1*(4), 2631787720964216. Advance online publication. DOI: 10.1177/2631787720964216

Hillmann, J. (2021). Disciplines of organizational resilience: Contributions, critiques, and future research avenues. *Review of Managerial Science*, *15*(4), 879–936. DOI: 10.1007/s11846-020-00384-2

Jacobsen, C. B., Andersen, L. B., Bøllingtoft, A., & Eriksen, T. L. M. (2022). Can leadership training improve organizational effectiveness? Evidence from a randomized field experiment on transformational and transactional leadership. *Public Administration Review*, *82*(1), 117–131. DOI: 10.1111/puar.13356

Juan, D., & Khan, D. (2023). Ensuring corporate accountability: Environmental responsibility and legal liability.

Kalifa, G. (2024). The role of transformational leadership in enhancing organizational innovation. *International Journal of Financial, Administrative, and Economic Sciences.*

Kaplan, G. (2006). Interview with Gary Kaplan. *Quality & Safety in Health Care*, *15*(4), 228–231. DOI: 10.1136/qshc.2005.017771 PMID: 16751462

Lai, F.-Y., Tang, H.-C., Lu, S.-C., Lee, Y.-C., & Lin, C.-C. (2020). Transformational leadership and job performance: The mediating role of work engagement. *SAGE Open*, *10*(1), 2158244019899085. Advance online publication. DOI: 10.1177/2158244019899085

Mair, J., & Seelos, C. (2021). Organizations, social problems, and system change: Invigorating the third mandate of organizational research. *Organization Theory*, *2*(4), 26317877211054858. Advance online publication. DOI: 10.1177/26317877211054858

Mele, C., Pels, J., & Polese, F. (2010). A brief review of systems theories and their managerial applications. *Service Science*, *2*(1-2), 126–135. DOI: 10.1287/serv.2.1_2.126

Melo, P., Martins, A., & Pereira, M. (2020). The relationship between leadership and accountability: A review and synthesis of the research. *Leadership and Organization Development Journal, 23*, 1–10.

Michaels, D., & Barab, J. (2020). The Occupational Safety and Health Administration at 50: Protecting workers in a changing economy. *American Journal of Public Health, 110*(5), 631–635. DOI: 10.2105/AJPH.2020.305597 PMID: 32191515

Mishra, D., Dashora, J., & Dubey, D. (2023). The impact of transformational leadership on employee engagement and performance: A comprehensive review. *3*, 2581–9429.

Mitchell, M. S., Reynolds, S. J., & Treviño, L. K. (2020). The study of behavioral ethics within organizations: A special issue introduction. *Personnel Psychology, 73*(1), 5–17. DOI: 10.1111/peps.12381

Nelson-Peterson, D. L., & Leppa, C. J. (2007). Creating an environment for caring using lean principles of the Virginia Mason production system. *The Journal of Nursing Administration, 37*(6), 287–294. DOI: 10.1097/01.NNA.0000277717.34134.a9 PMID: 17563521

Nguyen, C. H. (2011). Factors causing Enron's collapse: An investigation into corporate governance and company culture. *Corporate Ownership & Control Journal, 8*(3), 22-36. Available at *SSRN*: https://ssrn.com/abstract=1785199

Odom, L., & Green, M. T. (2003). Law and the ethics of transformational leadership. *Leadership and Organization Development Journal, 24*(2), 62–69. DOI: 10.1108/01437730310463251

Patrick, H. A., & Kumar, V. R. (2012). Managing workplace diversity: Issues and challenges. *SAGE Open, 2*(2), 2158244012444615. Advance online publication. DOI: 10.1177/2158244012444615

Plsek, P. E. (2013). *Accelerating health care transformation with lean and innovation: The Virginia Mason experience* (1st ed.). Productivity Press., DOI: 10.1201/b15817

Pulido-Martos, M., Gartzia, L., Augusto-Landa, J., & Lopez-Zafra, E. (2024). Transformational leadership and emotional intelligence: Allies in the development of organizational affective commitment from a multilevel perspective and time-lagged data. *Review of Managerial Science, 18*(8), 2229–2253. DOI: 10.1007/s11846-023-00684-3

Radu, C. (2023). *Fostering a positive workplace culture: Impacts on performance and agility*. IntechOpen., DOI: 10.5772/intechopen.1003259

Rao, M., Sonpar, S., Sen, A., Seshadri, S. P., Agarwal, H., & Padalia, D. (2018). A study on the prevalence and nature of ragging practices in selected educational institutions in India. *Contemporary Education Dialogue*, *15*(2), 187–202. DOI: 10.1177/0973184918781213

Reid, W., & Dold, C. (2018). Burns, Senge, and the study of leadership. *Open Journal of Leadership*, *7*(1), 89–116. DOI: 10.4236/ojl.2018.71006

Roy, A., Newman, A., Round, H., & Bhattacharya, S. (2024). Ethical culture in organizations: A review and agenda for future research. *Business Ethics Quarterly*, *34*(1), 97–138. DOI: 10.1017/beq.2022.44

Scholl, J. A., Mederer, H. J., & Scholl, R. W. (2016). Leadership, ethics, and decision-making. In Farazmand, A. (Ed.), *Global Encyclopedia of Public Administration, Public Policy, and Governance*. Springer., DOI: 10.1007/978-3-319-31816-5_2407-1

Shakeel, F., Kruyen, P. M., & Van Thiel, S. (2019). Ethical leadership as process: A conceptual proposition. *Public Integrity*, *21*(6), 613–624. DOI: 10.1080/10999922.2019.1606544

Sonmez Cakir, F., & Adiguzel, Z. (2020). Analysis of leader effectiveness in organization and knowledge sharing behavior on employees and organization. *SAGE Open*, *10*(1), 2158244020914634. Advance online publication. DOI: 10.1177/2158244020914634

Ta, M. T. D., Kim, T. E., & Gausdal, A. H. (2022). Leadership styles and safety performance in high-risk industries: A systematic review. *Safety and Reliability*, *41*(1), 10–44. DOI: 10.1080/09617353.2022.2035627

Tadesse Bogale, A., & Debela, K. L. (2024). Organizational culture: A systematic review. *Cogent Business & Management*, *11*(1), 2340129. Advance online publication. DOI: 10.1080/23311975.2024.2340129

Tang, W.-G., Stordeur, S., Vandenberghe, C., & D'hoore, W. (2024). Leaders' individualized consideration, team commitment and patient loyalty: The role of social and task-related contexts. *Journal of Occupational and Organizational Psychology*, *97*(3), 1185–1211. DOI: 10.1111/joop.12511

Truitt, D. L. (2011). The effect of training and development on employee attitude as it relates to training and work proficiency. *SAGE Open*, *1*(3), 2158244011433338. Advance online publication. DOI: 10.1177/2158244011433338

Vuong, T. D. N., & Nguyen, L. T. (2022). The key strategies for measuring employee performance in companies: A systematic review. *Sustainability (Basel)*, *14*(21), 14017. DOI: 10.3390/su142114017

Waldman, D. A., & Balven, R. M. (2014). Responsible leadership: Theoretical issues and research directions. *The Academy of Management Perspectives, 28*(3), 224–234. https://www.jstor.org/stable/43822065. DOI: 10.5465/amp.2014.0016

Walker, B., Jr. (1984). The Occupational Safety and Health Act: The national policy for the prevention of work-related illness. *Journal of Environmental Health, 47*(1), 20–23. National Environmental Health Association. https://www.jstor.org/stable/44538684

Wang, Q., Hou, H., & Li, Z. (2022). Participative leadership: A literature review and prospects for future research. *Frontiers in Psychology, 13*, 924357. Advance online publication. DOI: 10.3389/fpsyg.2022.924357 PMID: 35719563

Wasdani, K. P., Vijaygopal, A., Manimala, M. J., & Verghese, A. K. (2021). Impact of corporate governance on organisational performance of Indian firms. *Indian Journal of Corporate Governance, 14*(2), 180–208. DOI: 10.1177/09746862211047396

Wood, J. A.Jr, & Winston, B. E. (2005). Toward a new understanding of leader accountability: Defining a critical construct. *Journal of Leadership & Organizational Studies, 11*(3), 84–94. DOI: 10.1177/107179190501100307

Wu, Y., Awang, S. R., Ahmad, T., & You, C. (2024). A systematic review of leadership styles in healthcare sector: Insights and future directions. *Geriatric Nursing, 59*, 48–59. DOI: 10.1016/j.gerinurse.2024.06.033 PMID: 38986429

Xenikou, A. (2022). Leadership and organizational culture. In *Handbook of research methods for organisational culture*. Edward Elgar Publishing., DOI: 10.4337/9781788976268.00009

Yeboah, A. (2023). Knowledge sharing in organization: A systematic review. *Cogent Business & Management, 10*(1), 2195027. Advance online publication. DOI: 10.1080/23311975.2023.2195027

Young, M., & Smith, M. A. (2022). Standards and evaluation of healthcare quality, safety, and person-centered care. In *StatPearls* [Internet]. StatPearls Publishing., https://www.ncbi.nlm.nih.gov/books/NBK576432/

Zhang, X., Qian, J., Wang, B., Jin, Z., Wang, J., & Wang, Y. (2017). Leaders' behaviors matter: The role of delegation in promoting employees' feedback-seeking behavior. *Frontiers in Psychology, 8*, 920. DOI: 10.3389/fpsyg.2017.00920 PMID: 28638357

Zwikael, O., & Ahn, M. (2011). The effectiveness of risk management: An analysis of project risk planning across industries and countries. *Risk Analysis, 31*(1), 25–37. DOI: 10.1111/j.1539-6924.2010.01470.x PMID: 20723146

Chapter 2
Future Trends in Educational Leadership: Tech Advances, Evolving Paradigms, and Personalized Learning

Aieman Ahmad Al-Omari
https://orcid.org/0000-0001-6577-7281
Hashemite University, Jordan

ABSTRACT

The rapid evolution of technology, pedagogy, and societal demands places educational leadership at a critical crossroads. This chapter explores transformative trends—AI integration, personalized learning, and competency-based education—through a review of studies (2019–2025). It identifies systemic barriers such as the digital divide (40% of low-income students lack internet), resource shortages, and inadequate teacher training (only 25% feel prepared). Ethical challenges, including algorithmic bias (35% of schools lack AI bias policies) and data privacy concerns, complicate leadership. Case studies from Finland and Summit Public Schools highlight strategies like collaborative governance, professional development, and data-driven decision-making. Recommendations include investing in infrastructure, creating ethical AI frameworks, and fostering inclusive learning environments. By linking theory (Kotter, Fullan) to practice, this research offers actionable guidance for leaders to build equitable, adaptive systems that promote student agency, and pedagogical integrity.

DOI: 10.4018/979-8-3373-5686-0.ch002

INTRODUCTION

The landscape of education is undergoing a seismic transformation, driven by the convergence of rapid technological innovation, evolving societal needs, and the urgent demand for personalized, future-ready learning experiences. Educational leadership, once largely concerned with operational management and academic oversight, must now reimagine its role within a digitally saturated, unpredictable global environment. While these shifts present unprecedented opportunities, they also introduce complex challenges requiring leaders to act with agility, foresight, and an unwavering commitment to equity and inclusion (Fullan, 2015; Harris & Jones, 2020).

The necessity for transformational leadership in education has never been greater. Scholars argue that today's leaders must navigate not only the traditional challenges of educational management but also the complexities of integrating emerging technologies such as artificial intelligence (AI), big data analytics, and adaptive learning platforms into teaching and learning practices (Zhao, 2009; Selwyn, 2019). Yet, despite the growing discourse around technological integration, there remains a notable paucity of empirical studies exploring how educational leaders are practically addressing these shifts on the ground, particularly regarding the ethical, systemic, and equity-related implications of digital innovation.

The urgency of these transformations is underscored by stark projections about the future of work. The World Economic Forum (2020) estimates that over 65% of children entering primary school today will ultimately work in professions that do not yet exist. This statistic illuminates a profound uncertainty in workforce demands and highlights the critical need for educational systems to cultivate adaptable, innovative learners. The COVID-19 pandemic further exposed systemic vulnerabilities and accelerated the adoption of digital tools, with UNESCO (2021) reporting that 95% of countries transitioned to some form of remote learning during school closures. While this shift demonstrated the potential of technology to sustain educational access, it also revealed significant gaps in infrastructure, teacher preparedness, and equity, particularly for marginalized communities (Reimers & Schleicher, 2020).

However, much of the existing literature continues to frame technological innovation in education primarily through the lens of tool adoption or pedagogical trends, often neglecting the systemic leadership practices necessary for sustainable and equitable transformation (Williamson & Hogan, 2020; Trust & Whalen, 2020). Thus, a crucial knowledge gap persists: there is limited understanding of how educational leaders can effectively integrate emerging technologies while simultaneously promoting ethical practices, ensuring equity, and fostering systemic resilience in their institutions. This study seeks to address this gap by examining leadership strategies that balance technological advancement with critical considerations of inclusivity, adaptability, and ethics.

Moreover, while broad calls for "adaptive leadership" have become common (Heifetz et al., 2009; Fullan, 2015), few recent studies provide concrete frameworks or evidence-based practices tailored to the contemporary realities of AI-driven education. Educational leaders are often advised to be flexible, visionary, and responsive, but there is a lack of consensus on what specific competencies, strategies, and systemic supports are necessary to navigate an increasingly complex educational ecosystem (Busa et al., 2024). Addressing this void is vital for guiding future leadership development programs and informing policy initiatives that support meaningful, equitable educational change.

This chapter delves into three pivotal dimensions shaping the future of educational leadership: the integration of emerging technologies, the paradigm shift from traditional to progressive educational models, and the implementation of personalized learning strategies. Each area not only offers opportunities for innovation but also presents significant ethical, practical, and systemic challenges requiring informed and intentional leadership.

The integration of technology into education is perhaps the most visible aspect of current transformations. Emerging technologies such as AI, machine learning, and data analytics promise to revolutionize instructional design, student assessment, and administrative operations (Luckin et al., 2016; Holmes et al., 2019). AI-driven tools, for instance, offer personalized learning pathways, predictive analytics for student performance, and automated administrative efficiencies. However, these innovations also introduce serious concerns related to data privacy, algorithmic bias, and digital inequity. Holmes et al. (2019) found that 60% of educators harbor significant concerns about the data privacy implications of AI applications in classrooms, a figure that signals deep unease about the ethical stewardship of educational technologies.

Furthermore, disparities in digital access remain a pressing issue. According to the Pew Research Center (2021), 40% of students from low-income households lack reliable internet access, exacerbating educational inequities. These digital divides threaten to widen achievement gaps unless leaders adopt a proactive stance that prioritizes equitable access to technological resources and supports (Krutka et al., 2025). Yet, a 2023 study by Smith and Lee revealed that 70% of educational institutions lack clear policies addressing algorithmic bias in AI systems, suggesting that many schools are ill-equipped to manage the ethical complexities of emerging technologies (Smith & Lee, 2023).

In addition to technological integration, educational leadership must contend with a broader philosophical shift away from traditional, teacher-centered models toward progressive, student-centered paradigms. Traditional models emphasizing rote memorization, standardization, and teacher authority are increasingly viewed as ill-suited for preparing students for complex, dynamic global contexts (Colby, 2017; Darling-Hammond et al., 2019). Progressive paradigms emphasize competencies such as critical thinking, creativity, collaboration, and lifelong learning—skills that research shows are critical for success in the 21st-century workforce (Trilling & Fadel, 2009).

Despite strong theoretical support, transitioning to competency-based education (CBE) remains challenging. Patrick et al. (2013) found that schools implementing CBE models reported a 20% increase in student engagement and a 15% improvement in academic outcomes. Yet, systemic barriers—including limited teacher training, rigid accountability structures, and cultural resistance—continue to impede widespread adoption. The National Education Association (2023) reported that only 25% of teachers feel adequately prepared to implement competency-based education, with many citing a lack of institutional support and professional development opportunities as primary obstacles. These findings underscore the critical leadership task of fostering environments that support pedagogical innovation while managing the risks and resistances inherent in systemic change.

Finally, the movement toward personalized learning—tailoring instruction to the individual needs, interests, and pace of students—represents both a promising frontier and a formidable leadership challenge. Pane et al. (2015) demonstrated that personalized learning approaches can significantly improve student achievement, particularly among historically marginalized groups. However, successful implementation demands more than pedagogical adjustments; it requires substantial infrastructure investments, sophisticated data management capabilities, and coherent stakeholder engagement strategies.

Despite its potential, personalized learning remains under-implemented. Bishop et al. (2019) reported that only 30% of schools possess the necessary resources and training to deploy personalized learning models effectively. Leadership support plays a crucial role: a 2023 study by Johnson et al. found that schools with strong leadership backing for personalized learning initiatives experienced a 25% higher student retention rate compared to those without such support (Johnson et al., 2023). Thus, the role of leadership in orchestrating personalized learning environments is central to the success of these initiatives.

This chapter synthesizes contemporary research, empirical findings, and theoretical frameworks to provide a comprehensive understanding of how educational leaders can navigate these transformational currents. By examining the complex interplay between technological innovation, pedagogical evolution, and personalized learning, it aims to offer actionable insights for leaders seeking to drive equitable, ethical, and effective change in education.

Ultimately, the significance of this study lies in its response to an urgent scholarly and practical need: the need for an evidence-based, leadership-centered framework that addresses the ethical, systemic, and equity dimensions of educational innovation in the digital age. Without intentional, research-informed leadership, the promise of new technologies and pedagogical models may remain unrealized—or worse, exacerbate existing inequalities. Educational leaders must act as stewards of both innovation and justice, navigating the tensions between opportunity and risk, efficiency and equity, novelty and tradition.

By clarifying the leadership practices necessary for sustainable digital transformation, this study contributes to closing the existing knowledge gap and advancing the discourse on educational leadership for the 21st century. Through this exploration, it positions itself as a vital intervention in the ongoing effort to reimagine education in ways that are not only technologically sophisticated but also ethically grounded, inclusive, and future-ready.

Problem Statement

Educational leadership stands at a critical crossroads, challenged by the accelerating integration of technology, the shift toward personalized learning, and persistent systemic inequities. While digital tools and AI-driven learning models hold the promise of transforming education, existing leadership structures are often ill-equipped to manage these complex changes effectively. Research indicates that 40% of students from low-income households lack consistent internet access, perpetuating digital inequities that limit access to personalized learning innovations (Pew Research Center, 2021). Furthermore, a 2024 National Center for Education Statistics (NCES) report revealed that 65% of underserved schools lack sufficient funding to implement technology-facilitated personalized learning programs, deepening disparities in educational opportunity (NCES, 2024).

At the same time, ethical challenges have emerged, with a 2025 UNESCO report finding that 35% of global schools lack policies to prevent algorithmic bias in AI-facilitated learning platforms, risking the amplification of systemic inequalities (UNESCO, 2025). Compounding these challenges, only 30% of schools report having adequate resources and training to implement personalized learning effectively (Bishop et al., 2019; Johnson et al., 2023), and merely 25% of teachers feel prepared to transition to competency-based educational models, citing a lack of institutional support (NEA, 2023),and that 60% of teachers are concerned about the ethics of using student data within AI-based systems, particularly issues regarding privacy and fairness that undermine trust within these systems (Fernandes & Kumar, 2025).

These intersecting issues illuminate a profound leadership gap: current educational leadership models are insufficiently prepared to navigate the technological, pedagogical, and ethical demands of the modern learning environment. Without intentional leadership strategies that prioritize equity, ethical technology use, and systemic adaptability, the promise of personalized, technology-enhanced education will remain out of reach for many learners.

This study addresses this urgent need by investigating how educational leaders can effectively manage the integration of emerging technologies and personalized learning systems while ensuring equity and ethical responsibility. Grounded in recent empirical research from 2020–2025, this inquiry seeks to illuminate best practices and strategic frameworks that can empower leaders to transform educational institutions into inclusive, future-ready environments.

The Nature of the Study

This study employs a **narrative literature review** methodology to explore emerging trends in educational leadership, with a focus on the integration of technology, personalized learning, and competency-based education. By synthesizing existing research, theoretical frameworks, and case studies, the study seeks to provide a comprehensive analysis of how educational leaders can navigate the evolving educational landscape while promoting equity, innovation, and ethical practice.

The narrative literature review approach was chosen to allow a flexible yet systematic exploration of diverse sources, offering both breadth and depth in examining current challenges and opportunities. Unlike a systematic review that strictly quantifies results, a narrative review enables a thematic synthesis, connecting patterns, gaps, and best practices across the selected body of research.

Research Question

The central research question guiding this study is:

- **How can educational leaders effectively integrate technology and personalized learning systems while addressing issues of equity, infrastructure, ethical responsibility, and institutional readiness?**

Sub-questions include:

- What systemic barriers hinder equitable access to technological innovations in education?
- How do emerging ethical concerns, such as algorithmic bias and data privacy, affect leadership practices?

- What best practices and leadership strategies support the successful implementation of personalized and competency-based education models?

Data Collection Methods

A structured search strategy was employed to identify relevant literature, adhering to the following steps:

- **Databases Searched**: ERIC, JSTOR, Google Scholar, ProQuest Education Database.
- **Keywords Used**:
  - "educational leadership AND technology integration"
  - "personalized learning AND equity"
  - "competency-based education AND leadership"
  - "AI in education AND ethical challenges"
  - "digital equity AND school leadership"
- **Boolean Operators**: "AND" was used to combine related concepts; "OR" was used to broaden the search when necessary (e.g., "personalized learning OR adaptive learning").

Inclusion and Exclusion Criteria

- **Inclusion Criteria**:
  - Peer-reviewed journal articles, research reports, government publications, and theoretical frameworks.
  - Studies published between **2019 and 2025** to ensure scholarly currency and relevance.
  - Literature explicitly addressing educational leadership, technology integration, personalized learning models, ethical challenges in education, or systemic equity.
- **Exclusion Criteria**:
  - Non-peer-reviewed blogs, editorials, and opinion pieces without empirical evidence.
  - Studies focused solely on technology design without application to leadership or systemic change.

Data Analysis Techniques

Thematic analysis was utilized to identify and categorize patterns across the literature. Articles were coded into key themes such as digital equity, ethical AI use, leadership strategies, barriers to implementation, and personalized learning outcomes. Case studies were comparatively analyzed to extract best practices and lessons learned, particularly from institutions like Finland's education system and Summit Public Schools.

Ensuring Rigor

To ensure the credibility and trustworthiness of the review:

- Multiple sources from diverse contexts (K-12, higher education, international examples) were synthesized.
- Preference was given to empirical studies, meta-analyses, and longitudinal evaluations when available.
- A conscious effort was made to include studies addressing underserved communities and marginalized populations, ensuring that equity remained a central focus.

Significance of the Study

The future of education hinges on visionary leadership capable of navigating unprecedented technological, ethical, and equity challenges. This study is significant because it addresses a critical and urgent gap: current educational leadership models are not adequately equipped to manage the complex integration of emerging technologies, personalized learning systems, and competency-based pedagogies in ways that promote equity and ethical integrity.

Recent data highlight the pressing need for reform. Over 40% of students from low-income households still lack reliable internet access (Pew Research Center, 2021), and 65% of schools in underserved communities report insufficient funding for technology integration (NCES, 2024). Moreover, 70% of educational institutions globally lack clear policies to prevent algorithmic bias in AI-driven learning platforms, risking the amplification of existing inequities (Smith & Lee, 2023; UNESCO, 2025). Simultaneously, only 25% of teachers feel adequately prepared to implement competency-based and personalized learning models (NEA, 2023). These figures reflect systemic barriers that must be urgently addressed to ensure all students have access to innovative, future-ready education.

This study contributes to practice by offering evidence-based strategies that educational leaders, district policymakers, and teacher preparation programs can adopt to foster inclusive, technology-enhanced learning environments. By synthesizing best practices from institutions like Summit Public Schools and the Finnish education system, and grounding its recommendations in the most recent empirical literature (2019–2025), the study provides a roadmap for building equitable, ethical, and adaptive educational systems.

For policymakers and educational funders, this study offers a clear blueprint for investment: supporting leadership development, digital equity initiatives, ethical technology integration, and progressive pedagogical reforms. These strategic actions are not only critical for closing achievement gaps but also essential for preparing students to thrive in a dynamic global economy increasingly shaped by artificial intelligence, data science, and complex problem-solving demands.

Ultimately, this study empowers educational leaders to become architects of transformative change—leaders who can design learning environments that are innovative, equitable, and ethically sound for the 21st century.

LITERATURE REVIEW AND THEORETICAL GROUNDING

Educational leadership is undergoing a profound transformation as rapid technological advancements, evolving pedagogical models, and demands for greater equity reshape the global educational landscape. This literature review explores the foundational theories and emerging research that inform leadership practices in this changing environment, drawing on recent studies (2020–2025) to ensure scholarly currency and relevance.

Foundational Theories of Educational Leadership and Change

Several theoretical frameworks provide critical lenses for understanding leadership in times of systemic educational change:

- **John Dewey's Experiential Learning Theory (1916)** emphasizes learning through experience and reflection. Dewey's ideas continue to influence contemporary approaches to personalized and competency-based learning, where active, student-centered education fosters critical thinking and adaptability.
- **Kotter's Change Management Model (1996)** outlines an eight-stage process for implementing organizational change, emphasizing the importance of building urgency, creating coalitions, and embedding new practices into the culture. In the context of education, Kotter's model provides a roadmap for

leaders navigating the complex transitions associated with technology adoption and pedagogical reform (Kotter, 1996).

- **Darling-Hammond's Framework on Equity in Education (2010):** This framework is used to evaluate the impact of technology integration on underserved communities. It highlights the systemic barriers to equity in education and provides a foundation for understanding how technology can either exacerbate or mitigate these disparities.
- **Fullan's Change Leadership Theory (2015)** stresses the moral purpose of leadership in fostering deep learning and systemic improvement. Fullan et al. (2016) argue that successful educational transformation requires coherence across vision, relationships, and processes—a principle highly relevant for integrating personalized learning and technological tools today.

Together, these theories highlight that effective leadership during disruptive change must prioritize both the "technical" aspects of reform (e.g., new systems, structures) and the "human" aspects (e.g., relationships, trust, and shared values).

What systemic barriers hinder equitable access to technological innovations in education?

The promise of technological innovation in education is vast, yet its benefits remain unevenly distributed due to deeply embedded systemic barriers. These barriers are multifaceted, rooted in socioeconomic disparities, institutional neglect, and policy shortcomings that together perpetuate inequities in access and opportunity. Understanding these obstacles is critical for educational leaders seeking to foster inclusive and transformative digital learning environments.

Digital Inequity: One of the most prominent systemic barriers is digital inequity, often referred to as the "digital divide." According to the Pew Research Center (2021), approximately 40% of students from low-income households lack reliable internet access at home, severely limiting their ability to participate in online learning initiatives. Similarly, the National Center for Education Statistics (NCES, 2024) reports that 65% of underserved schools are unable to afford the infrastructure needed to implement technology-driven personalized learning programs. These disparities are particularly pronounced in rural and historically marginalized urban communities, where broadband infrastructure is outdated, insufficient, or altogether absent (Darling-Hammond, 2010).

For instance, many schools in impoverished regions struggle to maintain even basic digital connectivity, making it impossible to leverage cloud-based platforms like Google Classroom or deploy adaptive AI tools that require stable, high-speed internet (Johnson et al., 2023). Consequently, students in these communities are excluded from accessing the kinds of personalized, flexible, and engaging learning experiences that technology can offer, further exacerbating achievement gaps along socioeconomic and racial lines.

Resource Limitations: Beyond internet access, systemic resource limitations also pose significant challenges. Underserved schools often lack sufficient quantities of modern devices, technical support staff, and up-to-date software applications. The NCES (2024) highlights that many low-income districts are forced to repurpose obsolete hardware, which is frequently incapable of running contemporary educational programs or participating in emerging learning ecosystems like virtual reality (VR) labs or AI-driven tutoring systems. Darling-Hammond (2010) argues that such disparities in resource availability reflect broader systemic inequities in school funding, whereby affluent districts continually outpace their underfunded counterparts in technology adoption.

This technological lag is not merely a question of hardware but of access to an entire ecosystem of innovation, opportunity, and advanced pedagogical practices. Schools with insufficient resources are thus trapped in a cycle of digital deprivation, where outdated technologies reinforce educational inequities rather than alleviate them.

Inadequate Teacher Preparation: Another critical systemic barrier is the inadequate preparation of educators to effectively integrate technology into their teaching practices. A report by the National Education Association (NEA, 2023) found that only 25% of teachers feel adequately prepared to implement competency-based or personalized learning models. Much of the professional development currently offered is fragmented, focusing on tool-specific workshops without embedding these tools within broader pedagogical frameworks (Darling-Hammond et al., 2017).

Teachers in underserved schools, in particular, often receive minimal or inconsistent support. While they may attend occasional training sessions introducing new digital tools, they typically lack ongoing coaching and collaborative opportunities to adapt these tools for diverse classroom contexts (Johnson et al., 2023). Without sustained, meaningful professional development, technology risks becoming an add-on rather than a transformative force in teaching and learning, further entrenching existing disparities.

Policy and Ethical Oversight Gaps: Compounding the challenges of digital inequity and resource scarcity are significant gaps in policy and ethical oversight. According to UNESCO's Global Education Monitoring Report (2025), 35% of schools globally lack formal policies to mitigate algorithmic bias in AI-driven educational platforms. This regulatory vacuum leaves students vulnerable to discriminatory outcomes when predictive analytics or automated grading systems misidentify marginalized students as "at-risk," thereby limiting their access to advanced coursework or enrichment opportunities (Smith & Lee, 2023).

Similarly, weak data privacy protections expose students' sensitive information to potential exploitation by third-party vendors, undermining trust in educational institutions (Fernandes & Kumar, 2025). In the absence of strong ethical frameworks and regular audits, technology risks replicating—and even amplifying—the biases and inequalities that educational innovation purports to solve.

Recognizing these systemic barriers highlights the essential role of educational leaders as agents of equity and innovation. Leaders must not only advocate for improved funding, infrastructure, and teacher preparation but also implement proactive ethical frameworks that ensure technology serves all students fairly. Addressing digital inequity, resource gaps, inadequate training, and policy vacuums requires a multifaceted leadership approach—one rooted in strategic vision, inclusive stakeholder engagement, and a deep commitment to social justice. As the following sections explore, transformational leadership practices are vital for dismantling these systemic obstacles and cultivating technologically enriched learning environments that are accessible, ethical, and resilient in an era of rapid change.

How do emerging ethical concerns, such as algorithmic bias and data privacy, affect leadership practices?

Emerging ethical concerns—particularly algorithmic bias and data privacy—are profoundly reshaping leadership practices in educational settings. As the adoption of AI-driven tools and data analytics accelerates, leaders must move beyond mere technological enthusiasm to critically engage with the ethical, legal, and social implications of these innovations. These challenges demand that educational leaders adopt proactive, equity-centered strategies to balance the imperatives of innovation with the moral responsibility of safeguarding student welfare and promoting fairness.

- **Policy Development**: Leadership today requires establishing clear, transparent, and enforceable ethical guidelines to govern the use of AI, machine learning algorithms, and data analytics within educational environments. Without strong policy frameworks, technology adoption risks exacerbating existing inequities. For example, Summit Public Schools has become a model for ethical technology use by implementing stringent privacy protections within its Summit Learning platform, utilizing encryption, strict data access controls, and full compliance with the Family Educational Rights and Privacy Act (FERPA) and the Children's Online Privacy Protection Act (COPPA) (Summit Public Schools, 2023). Aoun (2017) underscores that effective policies must also address algorithmic bias explicitly by mandating regular audits of AI grading and recommendation systems to prevent disproportionate disciplinary referrals or misidentification of students of color and other marginalized groups.

Furthermore, policies should require transparency in how algorithms operate and ensure that decision-making processes remain accountable to human oversight. Heavily relying on opaque, "black-box" algorithms without explainability mechanisms can obscure biases and create structural barriers to student success (Williamson & Hogan, 2020). Therefore, ethical policy design becomes a central leadership responsibility, not an ancillary consideration.

- **Stakeholder Engagement:** Inclusive stakeholder engagement is critical for building trust and ensuring that ethical considerations reflect the values and concerns of the communities served. According to UNESCO (2025), schools that fail to involve families, students, and community members in conversations around AI ethics and data usage risk exacerbating historical distrust, especially among populations already marginalized by systemic biases in education and technology. Real-world examples, such as Lindsay Unified School District's practice of holding regular town halls to explain the workings of their competency-based education system and data use policies, demonstrate the power of proactive engagement to demystify complex technologies and foster transparency (Gallen, 2017).

Inclusive engagement also provides leaders with diverse perspectives that can inform more culturally responsive and contextually appropriate technology policies. As Reimers and Schleicher (2020) argue, leadership in the digital age must be participatory and dialogical rather than top-down, particularly when navigating ethically sensitive domains like student data and predictive analytics.

- **Capacity Building:** Addressing ethical concerns requires more than policy mandates—it demands building internal capacity among educators and administrators to recognize, critique, and mitigate potential biases. Fernandes (2023) advocates for comprehensive professional development programs that teach educators how to interpret algorithmically generated recommendations critically, distinguishing between objective insights and outputs that may reflect entrenched systemic biases. For example, workshops can train teachers to question why an AI system may predict lower achievement levels for students from particular socio-economic backgrounds, prompting a deeper inquiry into both technological design and instructional practice.

Lindsay Unified's integration of culturally responsive data interpretation training exemplifies how targeted capacity-building initiatives can empower educators to use data ethically, contextualizing metrics like test scores within broader narratives of student lived experiences (Gallen, 2017). Investing in such capacity building ensures that technological tools serve as instruments of equity rather than perpetuators of inequality.

- **Compliance and Accountability**: Finally, robust compliance mechanisms and accountability structures are essential to ensure that ethical principles translate into daily practice. The Data Quality Campaign (2023) recommends that educational institutions undergo regular third-party audits of AI tools to detect discriminatory patterns and assess the fairness of algorithmic decision-making processes. Compliance should not be treated as a one-time checkbox but as an ongoing evaluative practice embedded in the culture of the institution.

Evidence suggests that schools implementing independent audits experience tangible benefits: for instance, a 2023 study by Smith and Lee found that institutions that subjected their AI-driven college readiness prediction models to external review reduced algorithmic bias by 22%, significantly improving the accuracy and fairness of student assessments. Leadership commitment to continuous monitoring, third-party evaluations, and transparent reporting is thus crucial for ensuring that technology-enhanced learning environments promote—not undermine—educational equity.

As educational leaders navigate these ethical complexities, it becomes increasingly clear that addressing algorithmic bias and data privacy is not merely a technical or legal obligation but a fundamental component of cultivating equitable and sustainable learning environments. Ethical leadership in the digital age requires a holistic vision—one that integrates technological innovation with a deep commitment to social justice, human dignity, and democratic participation. The next sections of this study explore how leaders can extend these ethical imperatives by fostering inclusive innovation, designing culturally responsive learning ecosystems, and building organizational capacities that are resilient, adaptive, and future-ready. In doing so, leadership practices can move beyond reactive compliance to proactive transformation, ensuring that educational systems evolve in ways that are both innovative and profoundly equitable.

What best practices and leadership strategies support the successful implementation of personalized and competency-based education models?

Successful implementation requires visionary leadership, systemic support, and an unwavering focus on equity:

- **Visionary Leadership:** Finland's education system exemplifies this through decentralized decision-making and teacher autonomy. By empowering educators to design student-centered curricula and prioritize formative assessments, Finnish leaders fostered a culture of innovation (Sahlberg, 2011; OECD, 2023). Similarly, Summit Public Schools leveraged a clear vision to integrate project-based learning with real-time data analytics, enabling students to track their progress and set goals autonomously (Bishop et al., 2019).
- **Professional Development:** Darling-Hammond et al. (2017) stress job-embedded training, such as micro-credentialing programs that allow teachers to earn certifications in competency-based pedagogy. Finland's model of ongoing teacher training—where educators spend 10% of their work hours in professional development—ensures mastery of personalized learning tools (OECD, 2023).
- **Data-Driven Practices:** High Tech High uses continuous feedback loops to refine instruction. Teachers analyze data from formative assessments to adjust lesson plans in real time, ensuring no student falls behind (Rosenstock, 2011).

However, Fernandes (2023) cautions against over-reliance on metrics, urging leaders to balance quantitative data with qualitative insights, such as student portfolios or peer reviews.

- **Equity-Focused Resource Allocation:** Lindsay Unified School District prioritized closing the digital divide by providing devices and subsidizing broadband for low-income families, resulting in a 15% increase in graduation rates among underserved students (Lindsay Unified School District, 2022). The Digital Equity Act (2021) offers a roadmap for such initiatives, directing federal funds to schools in impoverished areas (U.S. Department of Education, 2022).
- **Change Management:** Applying Kotter's model (1996), leaders must build urgency and coalitions for reform. Lindsay Unified phased in competency-based education over five years, hosting community forums to address concerns and demonstrate early successes (Gallen, 2017). Fullan (2015) adds that collaborative governance—engaging teachers, IT staff, and students in decision-making—ensures sustainability, as seen in Finland's distributed leadership model.

By integrating these strategies, leaders can create adaptive, equitable systems that align technological advancements with pedagogical integrity, ensuring all students thrive in a rapidly evolving world.

Leading Through Innovation: Equity, Ethics, and Technology in Future-Ready Education

Educational leaders are pivotal in steering the transformation of education through the integration of emerging technologies, the adoption of progressive pedagogical models, and an unwavering commitment to equity and inclusion. By leveraging innovations such as artificial intelligence (AI) and adaptive learning platforms, leaders can enhance instructional practices, personalize learning experiences, and streamline administrative processes. For instance, platforms like Summit Learning utilize real-time data analytics to tailor content to individual student needs, fostering autonomy and engagement (Summit Public Schools, 2023). However, as Aoun (2017) cautions, this technological integration must be balanced with ethical safeguards to address algorithmic bias and protect student data privacy, ensuring tools like predictive analytics do not perpetuate systemic inequities (Smith & Lee, 2023; Fernandes & Kumar, 2025).

A focus on digital equity is paramount. Over 40% of low-income students lack reliable internet access, and underserved schools often lack funding for technology integration (Pew Research Center, 2021; NCES, 2024). Leaders must prioritize closing this divide by advocating for infrastructure investments, such as the Digital Equity Act (2021), which allocates resources to bridge connectivity gaps (U.S.

Department of Education, 2022). Case studies like Lindsay Unified School District demonstrate how equitable resource allocation—providing devices and subsidized broadband—can improve outcomes for marginalized students (Lindsay Unified School District, 2022).

Collaboration is essential for sustainable change. Drawing from Finland's decentralized model, leaders must empower teachers through autonomy and professional development, fostering a culture of innovation (Sahlberg, 2011; OECD, 2023). Fullan's (2015) principles of collaborative governance emphasize involving stakeholders—teachers, families, and policymakers—in co-creating a shared vision. For example, Summit Public Schools engages parents in dialogues about data ethics, building trust and transparency (Summit Public Schools, 2023).

The future demands leaders who blend visionary thinking with ethical vigilance. Competencies such as technological fluency, cultural responsiveness, and adaptive change management are critical. Leaders must model resilience, as seen in Kotter's (1996) phased reforms at Lindsay Unified, which prioritized community input to ease transitions to competency-based education (Gallen, 2017). Additionally, ongoing professional development, such as Finland's teacher training programs, ensures educators can implement personalized learning effectively (Darling-Hammond et al., 2017).

Ultimately, educational institutions will thrive under leaders who champion innovation, equity, and ethical responsibility. By fostering inclusive environments where technology serves pedagogical integrity—not replaces it—leaders can prepare students to navigate a rapidly evolving world. As Darling-Hammond (2010) asserts, the goal is to create systems where all students, regardless of background, have the tools and opportunities to succeed. Through strategic collaboration, equitable resource allocation, and a commitment to lifelong learning, educational leaders can transform institutions into dynamic, future-ready spaces that empower every learner.

CONCLUSIONS AND RECOMMENDATIONS

Conclusions

This study has synthesized current research on the integration of technology, personalized learning models, and competency-based education in K–12 and higher education settings, highlighting the systemic inequities and ethical challenges that arise from their implementation. The findings underscore the urgent need to address persistent barriers such as the digital divide, insufficient funding for technological infrastructure, uneven professional development opportunities for educators, and the lack of robust policies to mitigate algorithmic bias and protect data privacy. The acceleration of digital learning during the COVID-19 pandemic further exposed these structural inequalities, emphasizing that technological advancement alone does not guarantee equitable educational outcomes.

Through a thematic analysis of recent literature, this research has identified key trends shaping the future of educational leadership, including the critical role of adaptive leadership, the necessity of fostering digital literacy among all stakeholders, and the importance of culturally responsive and student-centered pedagogies. These trends point to a shifting educational paradigm where traditional, hierarchical models of leadership are increasingly giving way to collaborative, flexible approaches that prioritize innovation, inclusivity, and resilience in the face of uncertainty.

Importantly, this study has illuminated several critical gaps in the current body of knowledge. Despite significant enthusiasm around emerging technologies such as artificial intelligence, big data analytics, and adaptive learning platforms, there remains a paucity of empirical research on the long-term impacts of these tools on student learning, equity, and well-being. Moreover, much of the current discourse tends to center on technological capabilities rather than on the human-centered leadership strategies needed to ethically and effectively integrate such tools into diverse educational contexts. By bridging theoretical frameworks with practical applications, this research offers actionable insights that can guide policymakers, educational leaders, and practitioners in navigating these complex dynamics.

In doing so, the study contributes to the broader field of educational leadership by emphasizing the dual imperative of technological innovation and ethical stewardship. Leaders must not only champion the adoption of new tools but also advocate for policies and practices that ensure all students—regardless of socio-economic background, race, or geographic location—benefit equitably from educational innovation. Furthermore, the findings suggest that leadership development programs must evolve to equip future leaders with skills in change management, ethical decision-making, data literacy, and systems thinking.

Ultimately, fostering inclusive, innovative, and adaptive learning environments requires a holistic approach that goes beyond technology adoption to address the social, ethical, and structural dimensions of educational change. As the landscape of education continues to evolve rapidly, educational leaders must act as both architects and guardians of this transformation, balancing the promises of innovation with a steadfast commitment to equity and justice.

Future research should further explore longitudinal impacts of technology-enhanced education, strategies for scaling personalized learning equitably, and the development of international policy frameworks to govern the ethical use of AI in education. Only through sustained inquiry, critical reflection, and collaborative action can the full potential of educational innovation be realized in service of a more just and inclusive society.

Recommendations

Building on the theoretical foundations and recent empirical findings, this section presents actionable recommendations for educational leaders, policymakers, and practitioners seeking to integrate technology and personalized learning while promoting equity and ethical practices. Each recommendation is supported by contemporary research and case study evidence.

1. Bridge the Digital Divide through Targeted Infrastructure Investment

Educational leaders should prioritize closing the digital gap by investing in broadband infrastructure, providing devices to students, and ensuring technical support in underserved communities. Targeted funding initiatives, such as those suggested by the National Center for Education Statistics (2024), are essential to prevent technology from exacerbating existing inequities.

- Evidence: Pew Research Center (2021) found that 40% of low-income students lack reliable internet access.
- Example: Lindsay Unified School District's investment in district-wide technology access improved graduation rates among underserved students (Lindsay Unified School District, 2022).

2. Establish Clear Ethical Guidelines for AI and Data Use

Institutions must develop transparent policies addressing algorithmic bias, student data privacy, and the responsible use of AI-driven learning systems. Ethical frameworks must be embedded into leadership practices to maintain trust and safeguard students' rights.

- Evidence: UNESCO (2025) reported that 35% of schools lack policies to address AI biases.
- Example: Summit Public Schools implements strong privacy protections within their Summit Learning platform, balancing innovation with student data ethics (Summit Public Schools, 2023).

3. Invest in Comprehensive, Ongoing Teacher Professional Development

Sustainable change requires robust, continuous professional learning focused on integrating personalized learning technologies and competency-based approaches into everyday instruction.

- Evidence: Darling-Hammond et al. (2017) emphasize that sustained, job-embedded professional development significantly improves instructional quality.
- Example: Finland's model of ongoing professional development underpins its success in fostering student-centered education (OECD, 2023).

4. Foster a Culture of Data-Driven Decision-Making

Leadership teams must build institutional capacity to collect, analyze, and act on student data ethically and effectively. Data should inform instructional practices, resource allocation, and policy adjustments.

- Evidence: Fernandes (2023) highlights that data-informed leadership leads to improved student outcomes and institutional accountability.
- Example: High Tech High uses continuous data feedback loops to adjust personalized learning pathways in real time (Rosenstock, 2011).

5. Prioritize Equity in Personalized Learning Initiatives

When designing personalized learning environments, leaders must ensure they are accessible to all students, particularly historically marginalized groups. Strategic allocation of resources and culturally responsive pedagogy must be central to implementation.

- Evidence: Pane et al. (2015) found that personalized learning can close achievement gaps when equity is explicitly prioritized.
- Example: Summit Learning includes culturally responsive curriculum modules to promote inclusivity (Summit Public Schools, 2023).

6. Support Competency-Based Education Transitions through Change Management

Leadership must apply structured change management strategies (e.g., Kotter's model) when implementing competency-based education reforms, addressing resistance and building stakeholder buy-in.

- Evidence: NEA (2023) reports that only 25% of teachers feel adequately prepared for competency-based approaches, highlighting the need for structured leadership support.

- Lindsay Unified School District's phased, transparent transition process built community trust and improved student mastery rates (Gallen, 2017).

7. Create Collaborative Leadership Structures

Leaders should establish cross-functional teams that include teachers, IT staff, community members, and students in decision-making processes related to technology and instructional innovation.

- Evidence: Fullan (2015) emphasizes that coherence and collective ownership accelerate effective change.
- Example: Finland's school reform movements succeeded in part because decision-making authority was distributed across leadership levels (Sahlberg, 2011).

8. Advance Global Competencies within Personalized Learning Systems

Educational leaders should embed global competencies—such as cultural literacy, empathy, and collaborative problem-solving—into personalized learning initiatives to prepare students for a globalized workforce.

- Evidence: Mansilla and Jackson (2011) argue that global competencies are critical for 21st-century success.
- Example: International Baccalaureate (IB) programs incorporate personalized pathways and global competencies simultaneously, achieving strong student engagement across diverse populations.

9. Monitor and Evaluate Technology and Learning Innovations Continuously

Leaders should implement formal processes for monitoring and evaluating the impact of new technologies and pedagogical models, using both qualitative and quantitative data.

- Evidence: The U.S. Department of Education (2022) stresses that evidence-based evaluations are critical for scaling successful innovations.
- Example: Summit Learning continuously surveys students and teachers to adjust platform functionality and curriculum offerings based on real-world feedback.

10. Align Funding and Resources with Strategic Priorities

Districts and institutions must ensure that financial investments are strategically aligned with their goals around personalized learning, technology integration, and equity.

- Evidence: Bryson (2018) emphasizes the importance of strategic planning and resource alignment in achieving organizational goals.
- Example: The Digital Equity Act (2021) provides a funding blueprint for aligning resources with technology equity goals in U.S. schools (U.S. Department of Education, 2022).

The Study Limitations

While this study offers critical insights into the integration of technology, personalized learning, and competency-based education, several limitations warrant consideration. First, the analysis predominantly draws on educational models from Western contexts—such as the U.S. and Finland—which may not fully reflect the realities of resource-constrained settings in low- and middle-income countries. Challenges like inadequate infrastructure, chronic underfunding, and limited teacher training in these regions necessitate tailored adaptations of the findings, as the study's recommendations may not directly address their unique systemic barriers.

Second, the focus on literature published between 2019 and 2025 ensures relevance to contemporary trends but risks overlooking older research that could provide historical context or longitudinal perspectives on educational reforms. For instance, foundational studies on competency-based education from earlier decades might offer valuable lessons for current implementations, yet they fall outside the scope of this review.

Finally, while the thematic analysis employed rigorous methods to identify patterns across the literature, its interpretive nature introduces subjectivity. The researcher's role in coding themes and synthesizing findings inherently shapes the conclusions, which could vary if replicated by different analysts. These limitations underscore the need for caution when generalizing the study's results and highlight the importance of contextualizing recommendations to align with local educational, cultural, and socioeconomic conditions. Future research should expand geographically and temporally to validate and refine these insights across diverse global contexts.

REFERENCES

Aoun, J. E. (2017). *Robot-proof: Higher education in the age of artificial intelligence*. MIT Press. DOI: 10.7551/mitpress/11456.001.0001

Bishop, P. A., Downes, J. M., & Farber, K. (2019). *Personalized learning in the middle grades: A guide for classroom teachers and school leaders*. Harvard Education Press. https://hep.gse.harvard.edu/9781682533178/personalized-learning-in-the-middle-grades/

Bryson, J. M. (2018). *Strategic planning for public and nonprofit organizations: A guide to strengthening and sustaining organizational achievement* (5th ed.). Wiley.

Busa, A. I., Yakubu, I., & Olabode, H. A. (2024). Adaptive leadership in educational settings: Complex challenges and uncertain environments. *Fudma Journal of Research. Educational Psychology and Counselling*, *2*(1), 257–264.

Colby, S. (2017). *Competency-based education: A framework for measuring quality courses*. EDUCAUSE Review. https://er.educause.edu/articles/2017/6/competency-based-education-a-framework-for-measuring-quality-courses

Darling-Hammond, L. (2010). *The flat world and education: How America's commitment to equity will determine our future*. Teachers College Press.

Darling-Hammond, L., Flook, L., Cook-Harvey, C., Barron, B., & Osher, D. (2019). Implications for educational practice of the science of learning and development. *Applied Developmental Science*, *24*(2), 97–140. DOI: 10.1080/10888691.2018.1537791

Darling-Hammond, L., Hyler, M. E., & Gardner, M. (2017). *Effective teacher professional development*. Learning Policy Institute. https://learningpolicyinstitute.org/sites/default/files/product-files/Effective_Teacher_Professional_Development_REPORT.pdf

Data Quality Campaign. (2023). *Data literacy for educators: A roadmap for effective data use in schools*. https://dataqualitycampaign.org/wp-content/uploads/2023/03/DQC-Data101.pdf

Dewey, J. (1916). *Democracy and education: An introduction to the philosophy of education*. Macmillan.

Fernandes, J. (2023). The role of data-driven decision-making in effective educational leadership. *Academy of Educational Leadership Journal*, *27*(2), 1–3. https://www.abacademies.org/articles/the-role-of-datadriven-decisionmaking-in-effective-educational-leadership-15998.html

Fernandes, J., & Kumar, R. (2025). Ethical challenges in AI-driven education: A global perspective. *Journal of Educational Ethics*, *12*(3), 45–60.

Fullan, M. (2015). *The new meaning of educational change* (5th ed.). Teachers College Press.

Fullan, M., & Quinn, J. (2016). *Coherence: The right drivers in action for schools, districts, and systems*. Corwin Press.

Gallen, D. (2017). Personalized learning in Lindsay Unified. *School Made Fresh*. https://schoolmadefresh.org/personalized-learning-in-lindsay-unified/

Harris, A., & Jones, M. (2020). COVID-19—School leadership in disruptive times. *School Leadership & Management*, *40*(4), 243–247. DOI: 10.1080/13632434.2020.1811479

Heifetz, R. A., Linsky, M., & Grashow, A. (2009). *The practice of adaptive leadership: Tools and tactics for changing your organization and the world*. Harvard Business Press.

Holmes, W., Bialik, M., & Fadel, C. (2019). *Artificial intelligence in education: Promises and implications for teaching and learning*. Center for Curriculum Redesign.

Johnson, M., Brown, T., & Davis, K. (2023). The role of leadership in personalized learning: A case study analysis. *Educational Leadership Quarterly*, *58*(2), 89–104.

Kotter, J. P. (1996). *Leading change*. Harvard Business School Press.

Krutka, D. G., Greenhow, C., Kimmons, R., Pangrazio, L., & Trust, T. (2025). The state of educational technology research and practice. In Christakis, D. A., & Hale, L. (Eds.), *Handbook of children and screens: Digital media, development, and well-being from birth through adolescence*. Springer. DOI: 10.1007/978-3-031-69362-5_71

Lindsay Unified School District. (2022). *Annual report on student outcomes and equity initiatives*. https://www.ed.gov/sites/ed/files/2023/01/S215J220211_LindsayUnified_Narrative_508_Redacted.pdf

Luckin, R., Holmes, W., Griffiths, M., & Forcier, L. B. (2016). *Intelligence unleashed: An argument for AI in education*. Pearson Education. https://static.googleusercontent.com/media/edu.google.com/en//pdfs/Intelligence-Unleashed-Publication.pdf

Mansilla, V. B., & Jackson, A. (2011). *Educating for global competence: Preparing our youth to engage the world*. Asia Society. https://doi.org/DOI: 10.13140/2.1.3845.1529

National Center for Education Statistics. (2024). *Barriers to technology integration in underserved schools*. U.S. Department of Education.

National Education Association. (2023). *Barriers to competency-based education: A national survey of teachers*. NEA Publications.

OECD. (2023). *Policies for the digital transformation of school education: Proposed analytical framework and methodology*. OECD Publishing. https://one.oecd.org/document/EDU/EDPC/SR(2023)2/en/pdf

Pane, J. F., Steiner, E. D., Baird, M. D., & Hamilton, L. S. (2015). *Continued progress: Promising evidence on personalized learning*. RAND Corporation. https://www.rand.org/pubs/research_reports/RR1365.html

Patrick, S., Kennedy, K., & Powell, A. (2013). *Mean what you say: Defining and integrating personalized, blended, and competency education*. International Association for K-12 Online Learning (iNACOL). https://files.eric.ed.gov/fulltext/ED561301.pdf

Pew Research Center. (2021). *Internet access and the digital divide: Challenges for low-income households*. Pew Research Center.

Reimers, F. M., & Schleicher, A. (2020). *A framework to guide an education response to the COVID-19 Pandemic of 2020. OECD Policy Responses to Coronavirus (COVID-19)*. OECD Publishing., DOI: 10.1787/6ae21003-

Rosenstock, L. (2011). High Tech High: A personalized learning model for the 21st century. *Harvard Education Press*. https://vimeo.com/10000408

Sahlberg, P. (2011). *Finnish lessons: What can the world learn from educational change in Finland?* Teachers College Press.

Selwyn, N. (2019). *Should robots replace teachers? AI and the future of education*. Polity Press.

Smith, J., & Lee, R. (2023). Ethical implications of AI in education: Addressing algorithmic bias. *Journal of Educational Technology, 45*(3), 112–125.

Summit Public Schools. (2023). *Summit Learning: A personalized learning approach*. https://help.summitlearning.org/hc/en-us/articles/360000808867-How-do-Summit-Learning-teachers-and-parents-guardians-use-personal-information-in-the-Platform

Trilling, B., & Fadel, C. (2009). *21st century skills: Learning for life in our times*. Jossey-Bass.

Trust, T., & Whalen, J. (2020). Should teachers be trained in emergency remote teaching? Lessons learned from the COVID-19 pandemic. *Journal of Technology and Teacher Education*, *28*(2), 189–199. DOI: 10.70725/307718pkpjuu

UNESCO. (2021). *Education in a post-COVID world: Nine ideas for public action.* UNESCO Publishing.

UNESCO. (2025). *Global report on AI in education: Addressing bias and equity.* UNESCO.

U.S. Department of Education, Office of Educational Technology. (2022). *Digital equity in education: A guide for policymakers and leaders*. https://broadbandusa.ntia.doc.gov/sites/default/files/2022-12/Digital_Equity_Guide_for_States_11.28.22.pdf

Williamson, B., & Hogan, A. (2020). Commercialisation and privatisation in/of education in the context of COVID-19. *Education International Research*, 1–48. https://www.ei-ie.org/en/item/25251:commercialisation-and-privatisation-inof-education-in-the-context-of-covid-19

World Economic Forum. (2020). *The future of jobs report 2020*. https://www.weforum.org/reports/the-future-of-jobs-report-2020

Zhao, Y. (2009). *Catching up or leading the way: American education in the age of globalization*. ASCD.

Chapter 3
Self-Managed Work Team Environment (SMT):
Work Team Management and Self-Management Focusing on an Organizational Performance Model

José G. Vargas-Hernandez
https://orcid.org/0000-0003-0938-4197
Financial Literacy Research Center, Universidad Cristóbal Colón, Mexico

Absalón J. Salmerón-Zapata
https://orcid.org/0009-0002-2621-2426
Universidad Cristóbal Colón, Mexico

Carlos A. Rodriguez-Maillard
https://orcid.org/0000-0003-2406-196X
Universidad Cristóbal Colon, Mexico

ABSTRACT

This study aims to propose a research model of organizational performance focused on the self-managed work team environment (SWTE), work team management and self-management, based on the assumption that self-management of work teams facilitates and accelerates growth in organizational performance. The method used is based on meta-analysis and meta-cognitive-descriptive based on the conceptual, theoretical and empirical review of the literature. The results of the analysis are conclusive in that the management of the self-managed work team environment contributes to facilitating the growth and development of organizational performance.

DOI: 10.4018/979-8-3373-5686-0.ch003

INTRODUCTION

The approach of this study is to propose an organizational performance research model focused on the self-managed work team environment (SWTE), work team management and self-management. It is based on the assumption that self-management of work teams facilitates and accelerates growth in organizational performance, looking to determine and analyze organizational performance as a result of the impact of the link with knowledge learning that affects the creation of organizational capabilities and competencies from its own resources, with the purpose of obtaining competitive advantages as a vital factor to promote the growth, development and performance of service organizations. In this chapter, we move forward from the existing assumption about the relationship between human talent management and organizational performance, with the objective of determining the existing relationships, as well as the relationship with organizational results in higher education institutions.

The constant shifting of goals, the change of priorities and the structures that act as barriers to the achievement of organizational objectives have made the measurement of the variables, dimensions and indicators of the various factors related to organizational performance more complex, a situation that requires appropriate methods to measure and evaluate it. Each organization needs to measure and evaluate the work it has (Caplow, 1976, in Lusthaus et al., 2002). The objective of the research is to analyze the elements of organizational performance by exploring the impact that human talent management has on its existing links with resources and organizational learning capabilities, orientation to environmental market forces, and the formulation and implementation of competitive strategies.

Service organizations face many challenges that require adaptation to a more globalized and competitive economic environment, through tools to measure and control decision-making processes and organizational performance management, which depends on the ability to take on challenges (Bertolli et al., 2017). It is, therefore, a priority to measure organizational performance to obtain reliable and useful information that leads to the analysis, design, and implementation of strategies aimed at improving it. The interest in determining the capabilities and competencies of individuals in organizations to enhance their competitiveness aims to review the relationship between human talent management in organizations and organizational performance.

Measuring the dimensions of organizational performance is essential for organizational transformation, because it evaluates the results of achieving objectives, identifies the fulfillment of results and deviations, detects the ways of contributing to customer and employee satisfaction, facilitates the understanding of processes and ensures appropriate decision-making (Oliveira & Leone, 2008; Junxiao et al., 2014).

Although the analysis and measurement of organizational performance is considered a relevant element in the diagnosis and evaluation of companies, there is debate about the dimensions and indicators that are used (Venkatraman & Ramanujam, 1986; Varadarajan & Vasudevan, 1990). The different perspectives and overlaps of the dimensions of organizational performance are justified by the suitability of the research and its reliability and validity.

The importance of the study lies in the search for the improvement of organizational performance through a research model focused on self-managed work teams (SWTE), team management and self-management, based on the concept that self-management facilitates and accelerates growth in organizational performance, supported by the development of organizational capabilities, adaptation to competitive environments and the relationship between human talent management and organizational performance.

BACKGROUND

Organizational Performance

According to Bernal González et al. (2020), there are no unique definitions of the components of the capital and human talent management variable, and what it represents for the generation of sustainable competitive advantages that have impacts on the performance of organizations. The descriptive analysis of the term organizational performance, widely studied in the literature, shows the different meanings, the dimensions used in its operationalization to measure and evaluate, due to its multidimensional nature (Camisón and Cruz, 2008). The definitions of organizational performance measurement are precise in the literature, but the meaning of the concept is not true, although they raise elements that characterize understanding (Bourne et al., 2003).

Organizational performance refers to the results generated by the dynamics of the organization's internal and external factors, which reflect the behavior of elements that are integrally involved in achieving objectives, financial results, structure, environment, innovation, quality, employee and customer satisfaction, market share, etc. (Yamakawa and Ostos, 2011; Pérez and Cortés, 2009). Organizational performance consists of the capacity of companies to acquire and process the human, financial and material resources necessary to achieve their objectives and goals (Madella et al., 2005). Organizational performance measurement uses multidimensional standards that include financial and non-financial indicators, internal and external performance factors to quantify objectives, project situations, evaluate actions of interest groups, etc.

Organizational performance is related to financial and non-financial indicators and strategic positions that provide information and data on the degree of compliance with organizational objectives and the results achieved, in order to react and adapt to external changes to obtain and maintain competitive advantages (Waal & Kourti, 2013). The review of the literature on organizational performance, its constructs, variables, dimensions and indicators aim to review the quantitative and qualitative forms of objective and subjective measurement and evaluation, in such a way that it facilitates comparative analysis of the performance of organizations (Barradas Martínez et al., 2021). Organizational performance is a construct that is more than the actions with their financial results, because it includes non-financial and intangible dimensions that are related to growth, market performance, management of administrative and operational processes, employee and customer satisfaction, among others (Muthuveloo et al., 2017; Tseng & Lee, 2014).

The research partially analyzes the variables, objective and subjective dimensions and indicators of the forms of organizational performance measurement in determining the state of the art and its comprehensive comparative analysis. The comparative analysis of objective and subjective measurement presents differences in the general and operational approaches of the organization. Subjective measurement evaluates organizational performance based on competence in a competitive and changing environment of intangible critical factors, customer satisfaction, strategy, etc. Therefore, financial indicators are not sufficient to evaluate organizational performance, which makes it necessary to use other more comprehensive technical measurements of these variables (Lui et al., 2014).

The literature on financial performance dimensions and indicators repeatedly points out that they face serious limitations and drawbacks (Skinner, 1974; Schmenner, 1988; Camp, 1989; Turney & Andersen, 1989; Kaplan & Norton, 1992; Neely, 1999; Tangen, 2004). Measuring and evaluating the actual financial performance of organizations is often complicated by difficulties in accessing appropriate and convenient information. This is why subjective measures are used (Cruz et al., 2012). The financial measurement of individual and organizational performance must be complemented by the subjective measurement of employees, managers, shareholders, customers, etc. (Delaney & Huselid, 1996). The use of subjective measurements, despite considering their drawbacks, is especially appropriate for service sector organizations with diverse objectives (Fey & Beamish, 2001). The technological environment, the market and competitive intensity are variables that moderate the market-organizational performance relationship (Jaworski & Kohli, 1993; Kohli & Jaworski, 1990).

The alternative to objective indicators to evaluate the quantitative performance of organizations is subjective measurement, which allows comparison between organizations and their contexts, economic conditions, time horizons, to name a few

(Zulkiffli, 2014). The dimensions used in certain studies to measure performance are productivity, self-management, commitment, and interpersonal conflicts (Bhargava & Sinha, 1992). The dimensions with their measurement indicators are the growth of the size of the organization, accounting returns as historical results, and the performance of the stock market through investors (Crook et al., 2005). For Tseng et al. (2013), the predominant factors to measure organizational performance are the return on assets, the market, Tobin's Q, and the return on equity.

More generic and abstract concepts have emerged, such as effectiveness, efficiency and efficacy, which are considered the most relevant dimensions and indicators in performance measurement. Organizational efficiency is a variable as a performance measure that belongs to another broader concept of organizational effectiveness (Richard et al., 2009). Organizational performance measures are grouped into categories of effectiveness, efficiency, punctuality, quality, productivity and safety (Omondi, 2015).

Performance is defined as the degree to which organizations, considered as a social system, meet their objectives. Georgopoulos & Tannenbaum (1957) focused the evaluation on people, work, and organizational structure. The introduction of non-financial indicators in the new forms of performance evaluation that link production processes and operations conceptualizes performance measurement as the organizational capacity to exploit its environment in order to access and use limited resources (Yuchtman & Seashore, 1967). The organization achieves its performance objectives despite resource restrictions and limited capabilities, so indicators become the most important measures of its measurement (Lusthaus & Adrien, 1998; Gavrea et al., 2011).

Organizational performance measurement is based on the voluntary association of productive assets, including capital, human, physical and material resources, in order to achieve a shared purpose (Carton, 2004). It is the economic result of the interaction between behaviors, attributes and the organizational environment (Crook et al., 2005). It also involves a set of financial and non-financial indicators that report on the different elements of the different areas of responsibility, the degree of compliance with organizational objectives and the quantification of their results, which require judgment and dynamic interpretation, which can be understood in different ways, depending on the people involved. Organizational performance measurement is causal because it describes how current actions affect results (Lebans & Euske, 2006).

Measuring organizational performance represents the efficient management of organizations that combines available resources with the development of strategies that configure the attractive offer of goods and services, in order to achieve a competitive position in the market, maintain and increase its share with financial results and generated profitability growth (Camisón and Cruz, 2008). Organizational performance is a strategic management construct (Richard et al., 2009; Dhaafri et al., 2016) and its domain can be specified by relating it to the concept of organizational effectiveness, while the concept of organizational effectiveness is related to the degree to which organizations achieve their established purposes (Hamann et al., 2013; Strasser et al., 1981).

The measurement of organizational performance covers the results of financial indicators such as profits, return on assets and on investment, market performance indicators in sales and shares, as well as shareholder return indicators, for example, total return for shareholders and economic value added (Richard et al. 2009). The use of financial indicators is complex, due to the complications in obtaining data and financial information from non-listed organizations, the refusal to provide information that reflects actual financial performance, the ease of data manipulation, etc. Organizational performance is affected by the ability of individuals to incorporate thoughts simultaneously, including the generalized vision of all members of the organization on the generation and transfer of superior knowledge, as a source of competitive advantage that links strategy with company performance (Castillo, 2013).

Measuring it involves knowing the level of objectives achieved and evaluating the effectiveness of people, groups or organizations (Shu-Mei & Pei-Shan, 2014), which contributes to verifying compliance with the established purposes through reliable quantitative and qualitative indicators (Singh et al., 2016; Singh & Gupta, 2016). A mixed approach to evaluating it uses such indicators from internal and external documents, such as financial, accounting, productivity, sales records, etc.

From the perspective of the system theory of organizational performance, it has elements of interaction between themselves, with hierarchies, complexities and uncertainties, with goals or objectives that work as a unit to obtain a better performance of the entire system, and with the ability to react to change, in order to remain in balance in the face of disturbances in its environment (Castillo, 2013). The evolutionary theory tends to view organizational performance with an eclectic approach that collects elements of evolutionary processes from various disciplines, such as knowledge, routines, resources and competencies (Castillo, 2013). This theory offers guidance on what organizations should consider as aspects in evolution and development processes, to facilitate their performance.

The theory of organizational learning relates the processes of human talent management with organizational performance. Slater & Narver (1994) assume that such learning facilitates change and leads to improvement. For his part, Garvin (1984) defines it as a process that takes place over time in order to achieve the acquisition of skills that result in increased performance. The performance of organizations requires the acquisition of knowledge and the development of the human talent of all individuals, who create their own interpretations of incoming information based on their past experiences and expectations.

The theory based on resources and capabilities argues that not all resources and not all capabilities have the potential and therefore have or do not have the importance to contribute to the performance of organizations; when they are distinctive, they allow for achieving superior organizational performance. In this regard, Barney (1991) and Wernerfelt (1984) argue that the internal resources and capabilities of organizations in a competitive environment must be identifiable if they are particularly critical to enable them to formulate and implement strategies aimed at improving their efficiency and effectiveness in organizational performance. However, there is little effort to determine and analyze what these distinctive resources and capabilities are.

Organizational capabilities that enhance the ability to create and sustain competitive advantages and improve organizational performance include market relations, although little progress has been made in the study of their effects on organizations (Day, 1994; Jaworski & Kohli, 1993). Organizational resources and capabilities play a fundamental role in improving the performance of service organizations, because they make it easier for senior management and administration to focus their efforts on creating strategies that allow them to achieve high performance of their objectives and goals. From the perspective of strategic management of organizations, the turbulence of the environment, unpredictable changes and the resulting organizational complexity and uncertainty call into question conventional strategic analysis to determine the differences in the performance of organizations.

Empirical research confirms the relationship between capital and human talent management on organizational performance. The greater the potential of capital and human talent, the greater the possibility of achieving organizational competitiveness and effectiveness, due to the relationship and influence between capital and human talent and the performance of organizations (Chávez and Vizcaíno, 2017; Ganga et al., 2016; Guerrero et al., 2015; Escandón et al., 2012; Pardo and Diaz, 2014; Villegas et al., 2017). The exploratory-descriptive correlational analysis determines that the processes of training and development of human talent competencies in organizations is the key factor in performance, which indicates the importance of reviewing human talent management schemes in institutions in order to produce impacts on results (Pardo and Díaz, 2014).

The multiple components to be assessed in the organizational performance variable allow its empirical analysis to be addressed from various perspectives with proposals specific to its components, grouped into metrics of the dimensions of financial and business performance, as well as organizational effectiveness (Venkatraman & Ramanujam, 1986). The most commonly used indicators in the objective measurement of organizational performance are financial, cash flow, growth rate, productivity, market share, market value, sales, investment in research and development (R&D). The comparative analysis of the measurement of organizational performance identifies the dimensions that are used for subjective evaluation: the perception of interest groups, employee and customer satisfaction, innovation, quality, self-management, work environment and interpersonal conflicts.

The findings validate the implementation of prospective or analytical competitive strategies to achieve organizational performance objectives due to their market orientation, realization of opportunities and uncertain changes in response to technological, market and competition factors. According to the results of the aforementioned research, organizations place greater emphasis on adaptability in the relationship between competitive strategies and organizational performance. The dimensions of organizational performance measured by growth, profitability, adaptability, satisfaction and operation are positive and highly significant (Ynzunza Cortés and Izar Landeta, 2013).

The implementation of competitive strategies oriented to the market, trends, needs and desires of customers, and the actions of competitors allows to identify and develop the resources and capabilities necessary for a high performance of the organization (Miles & Snow, 1978). Market orientation is the most relevant ability to show an organizational performance measured by growth, profitability, adaptability, operation and customer satisfaction. Companies that implement strategies that favor the development of resources and the creation of organizational capabilities for interaction with their environment obtain higher performance ranges.

Organizational Performance Research Model

Objective measurement of organizational performance is easy to perform and evaluate, but its objective application is difficult to access information and data with integrity, validity and precision. However, measurement provides the information that is used to delimit, establish and prioritize specific actions that resolve the problems detected to achieve organizational effectiveness. With the valid and reliable information and data resulting from the performance evaluation, a guide of specific indicators can be created, with an application model to establish a program of strategic actions aimed at improving performance and boosting its competitiveness.

The research on organizational performance carried out by Ynzunza Cortés and Izar Landeta (2013) shows that competitive prospecting and analyzing strategies are positively related to organizational performance, as well as orientation and market forces; organizational performance is associated with operational aspects, customer satisfaction and growth. Evaluating the impact of organizational performance on customer satisfaction is more complex than evaluating employee satisfaction. The aforementioned work by Ynzunza Cortés and Izar Landeta (2013) is based on the analysis of variables determined by market forces, market and technological turbulence, and competitive intensity. Likewise, the authors consider prospective and analytical strategies, market orientation, and organizational performance in the dimensions of growth, adaptability, profitability, operation, and customer satisfaction.

The constructs used in this research were the competitive strategy proposed by Segev (1987), the competitive strategy of Narver & Slater (1990) and market orientation (Jaworski & Kohli, 1993), and market forces and organizational performance (Venkatraman, 1989; Vorhies & Morgan, 2003). Market orientation develops capabilities and is a determining factor in organizational profitability and performance. The link between market forces and organizational performance favors the creation of capabilities to achieve organizational performance. In the absence of competition in a stable market with strong demand, organizations achieve high performance; adaptation depends on the competitive strategy determined by the conditions to achieve high organizational performance. In this way, it is proven that the competitive strategy has significant effects on market orientation and market forces, and therefore on organizational performance (Ynzunza Cortés and Izar Landeta, 2013).

Therefore, the measurement of organizational performance goes beyond objective indicators, such as financial ones, to integrate subjective dimensions in a mixed model that combines objective and subjective forms in a broader perspective to evaluate the effectiveness of the measurement. The organizational performance model combines objective and measurable factors in financial indicators, with subjective factors in product quality, added value, marketing effectiveness, customer and employee satisfaction, as well as organizational efficiency measures (Fuentes and Hurtado, 2002). This model depends on organizational strategies and goals.

Self-Managed Work Team

Self-managed work teams (SMWT) have become increasingly popular in companies as a way to improve innovation, productivity and employee commitment. They are characterized by their ability to self-regulate, make decisions autonomously and assume responsibility for the results of their activity. Research on SMWT has evolved significantly in recent years, providing new insights into their conceptual and theoretical foundations. Recently, research has focused on understanding the factors that influence their effectiveness and how they can be successfully integrated into project management.

In recent years, research on self-managed work teams (SMWT) has made significant progress, leading to a better understanding of the individual competencies, organizational structures and systemic challenges that determine their effectiveness. Studies highlight the importance of authority distribution, revealing that teams that demonstrate high levels of proactive communication and conflict resolution skills achieve 23% higher performance than those that rely solely on technical expertise (Doblinguer, 2021).

SMWTs are defined as teams where members have substantial discretion over planning, executing, and controlling their work. The core concept revolves around empowering employees to take greater responsibility and autonomy in managing their work. The conceptual literature on the subject highlights' key features and benefits of this organizational structure, some of which include:

1) Self-management: Self-managed work team members have decision-making authority and responsibility over their work, schedule, processes, and goals (Kuipers & Stoker, 2009).
2) Shared self-management: Self-management responsibilities are distributed among team members, rather than centralized in a single manager (Pearce & Conger, 2003; Hoch, 2013).
3) Interdependence: SMWT members depend on each other to achieve shared goals, fostering collaboration and collective problem solving (Kuipers & Stoker, 2009).

Some of the conceptual findings include:

1) Self-managed work teams are designed to empower employees, increase autonomy, and foster a sense of ownership over work processes.
2) SMWTs can lead to improved team performance, job satisfaction, and organizational commitment when implemented effectively.
3) Their success depends on factors such as team composition, self-management, and organizational support.

An essential conceptual development is the recognition that trust, understanding, and power dynamics play a critical role in the functioning of self-managed work teams. Langfred (2007) found that high levels of trust and autonomy within self-managed work teams can lead to increased conflict, which in turn negatively affects team performance, highlighting the importance of carefully managing the balance between trust and control in self-managed work teams.

Researchers have drawn on a variety of theoretical frameworks to understand the development and performance of self-managed work teams, in order to explain their dynamics and effectiveness:

1) Sociotechnical Systems Theory. It emphasizes the importance of alignment and interdependence between the social and technical aspects of work systems in the design of self-managed work teams for proper functioning, which arises from the joint optimization of the social and technical subsystems (Appelbaum, 1997).

2) Self-Determination Theory. It suggests that self-managed work teams meet employees' needs for autonomy, competence, and relatedness, leading to increased motivation and performance. This theory postulates that intrinsic motivation, which arises from the psychological needs for autonomy, competence, and relatedness, is a key factor in the effectiveness of self-managed work teams (SMWTs) (Deci & Ryan, 2000; Gagné & Deci, 2005).
3) Social interdependence theory. It postulates that the cooperative structure of self-managed work teams (SMTs) fosters positive interdependence among members, improving teamwork and results.
4) Social exchange theory. It proposes that reciprocal relationships and trust built within self-managed work teams lead to greater commitment, cooperation, and performance (Blau, 1964; Cropanzano & Mitchell, 2005).
5) Contingency theory. This perspective holds that the effectiveness of SMWTs depends on several contextual factors, such as organizational culture, task characteristics, and environmental uncertainty (Hackman, 1987; Cohen & Bailey, 1997).

Another theoretical advance is the focus on individual competencies that contribute to the performance of self-managed ETAG work teams.

Empirical studies have examined a variety of antecedents and outcomes of self-managed work teams:

1) Trust, understanding, and power dynamics within self-managed work teams have been found to significantly influence their performance and effectiveness.
2) Organizational factors such as managerial support, training, and information sharing can facilitate successful implementation of self-managed work team management.
3) Self-managed work teams have been associated with high job satisfaction, greater organizational commitment, and lower turnover intentions among their members.

Druskat & Kayes (2000) identified relationship-oriented and task-oriented behaviors as key predictors of self-managed work team effectiveness; in addition, personality traits such as conscientiousness and emotional stability have been found to be important for individual performance in self-managed work teams. Empirical research on the topic has also advanced, providing a more nuanced understanding of its development and functioning.

1) Team development

a) Traditional linear models of team development have been challenged and evidence suggests a non-sequential, iterative process (Geerts et al., 2021).
b) Key development processes include internal team management, task management, and boundary management/enhancement, which can occur in various patterns depending on contextual factors (Awahnde, 2014).
c) Aspects that influence the development of self-managed work teams encompass individual characteristics, team dynamics, organizational support, and environmental conditions (Manz & Sims, 1987; Kirkman & Rosen, 1999).

2) Team performance
   a) Self-managed work teams have been shown to outperform traditionally managed teams on a variety of outcomes, such as productivity, quality, and employee satisfaction (Wageman, 1997).
   b) However, the performance benefits of SMWTs depend on factors such as task complexity, organizational support, and team composition (Yeatts & Hyten, 1998; Sundstrom et al., 1990).
   c) Researchers have highlighted the importance of monitoring and supporting the ongoing development of SWTs teams to sustain their performance advantages (Campion et al., 1993; Cohen, 1993).
3) Empirical findings

Recent empirical research has provided valuable insights into the practical implications of implementing self-managed work teams. A study by Batt (1997) found that SMWTs can generate significant economic benefits for organizations, such as reduced labor costs and increased productivity. However, implementing them successfully requires careful consideration of factors such as team composition, self-management, and organizational structure. Kirkman & Rosen (2002) identified five core challenges to the success of virtual SMWTs, which include lack of face-to-face interaction, technological issues, and cultural differences.

Addressing these are critical to ensuring the effectiveness of self-managed work teams in a virtual environment. The main findings from the conceptual, theoretical, and empirical literature review are summarized in Table 1. The conceptual section focuses on the importance of trust and power dynamics, the theoretical section on individual competencies, and the empirical section on the organizational benefits and implementation challenges of self-managed work teams. Table 1 summarizes the conceptual, theoretical, and empirical literature review on self-managed work teams below:

Table 1. Key findings from the conceptual, theoretical and empirical literature review

Tipe of literature	Key Findings	Reference
Conceptual	- Empowers employees with autonomy and ownership - Improves team performance, job satisfaction, and organizational commitment. - Success depends on team composition, self-management, and organizational support.	Aldag & Kuzuhara (2015)
Theoretical	- Self-determination theory: autonomy, competence, and relatedness lead to increased motivation and performance. - Social interdependence theory: cooperative structure fosters positive interdependence among team members. - Sociotechnical systems theory: align social and technical aspects for optimal performance.	Manz & Sims (1987)
Empirical	- Trust, understanding, and power dynamics impact team performance and effectiveness. - Organizational factors (management support, training, information sharing) facilitate SMWT success. - SMWTs are associated with greater job satisfaction, organizational commitment, and lower turnover intentions.	Cohen & Ledford (1994)
Empirical	- High trust and individual autonomy can have negative effects on self-managed teams. - Team goal commitment in innovative projects.	Langfred (2004)
	Team goal commitment in innovative projects.	Hoegl & Parboteeah (2006)

Source: Own elaboration.

The following table 2 summarizes the main findings of recent research in the conceptual, theoretical, and empirical areas, along with the corresponding references. Similarly, the content of the technical report is shown divided into the conceptual, theoretical, and empirical sections on research into the environment of self-managed work groups.

Table 2. Key findings of recent research on SMWTs

Section	Key findings	References
Conceptual	- Trust, understanding, and power dynamics play a key role in the functioning of SMWTs. - High levels of trust and autonomy within SMWTs can lead to increased conflict and negatively affect their performance.	Langfred (2007)
Theoretical	- Relationship-oriented and task-oriented behaviors are key predictors of SMWT effectiveness. - Personality traits, such as conscientiousness and emotional stability, are important for individual performance in self-managed work teams.	Druskat & Kayes (2000)
Empirical	- Self-managed work teams can generate significant economic benefits for organizations, such as reduced labor costs and increased productivity. - Successful implementation requires careful consideration of factors such as team composition, self-management, and organizational structure. - Key challenges to the success of virtual self-managed work teams include lack of face-to-face interaction, technological issues, and cultural differences.	Batt (1997); Kirkman & Rosen (2002)

Source: Own elaboration.

In summary, conceptual, theoretical, and empirical research on self-managed work teams has advanced significantly in recent years, providing a more nuanced understanding of their foundations, development, and implications within organizational contexts. The literature highlights conceptual, theoretical, and empirical evidence supporting the benefits of self-managed work teams, while identifying key factors that contribute to their success. The synthesis of recent research unequivocally demonstrates that self-managed work teams represent not merely an alternative management strategy, but a fundamental reconfiguration of organizational.

In conclusion, recent research on the subject has highlighted the importance of trust, understanding, and power dynamics, as well as individual competencies in determining team effectiveness. Empirical findings suggest that ETAGs can yield significant organizational benefits, but their successful implementation requires careful consideration of several contextual factors. An important matter are that roles can lead to conflict. Research by Karakus (2017) highlights that 42% of healthcare SMTs initially struggled with role clarity, resulting in duplicated efforts. As organizations continue to adopt this framework, continued research and clear role definition will be necessary to further refine our understanding of this relevant phenomenon.

TEAM MANAGEMENT FUNCTIONS

Teamwork competence is defined as the knowledge, skills, attitudes and other individual characteristics that facilitate the effective performance of its members (Rousseau et al., 2006). Some authors have identified different dimensions or categories of this competence, such as communication, mutual support, performance monitoring and feedback (Rousseau et al., 2006).

Several theoretical models have been proposed to explain the functions and processes of work teams; for example, the model by Cannon-Bowers et al. (1995) identifies four main categories: interpersonal skills, problem-solving and decision-making skills, coordination skills and team management skills. Other models focus on aspects such as self-management, team formation and development (Palomo, 2013).

Empirical research has explored various factors that affect the performance and effectiveness of work teams. Some studies have found that coordination, communication, and mutual support are key functions for success (Weaver et al., 2010). Others have analyzed the influence of self-management, team composition, and organizational context (Weaver et al., 2010). Below is a Table 3 summarizing the key results of the literature in the conceptual, theoretical, and empirical dimensions.

Table 3. Key findings from the conceptual, theoretical and empirical literature on work team functions

Dimension	Key Findings
Conceptual	- Teamwork competence includes knowledge, skills, attitudes and other individual characteristics (Rousseau et al., 2006). - Key dimensions: communication, mutual support, performance monitoring, feedback (Rousseau et al., 2006).
Theoretical	- Cannon-Bowers et al. (1995) model: interpersonal skills, problem solving, coordination, team management. - Other models focused on self-management, team formation and development (Palomo, 2013).
Empirical	- Coordination, communication and mutual support are determining functions for the success of teams (Weaver et al., 2010). - Factors such as self-management, team composition and organizational context also influence (Weaver et al., 2010).

Source: Own elaboration.

SELF-MANAGEMENT

Self-management refers to a person's ability to effectively direct and motivate him or herself without relying on supervision or direction from others. Over the past five years, research on this topic has explored various aspects related to the development of self-management and its impact on individual performance and well-being.

The literature has examined self-management from different approaches. Some studies have been based on self-determination theory, which posits that self-management arises from the satisfaction of basic psychological needs for autonomy, competence, and relatedness (Deci & Ryan, 2008). Other work has applied Bandura's self-efficacy model, according to which ideas of self-efficacy influence people's ability to self-regulate and direct their own behavior (Bandura, 1997).

A number of studies have explored strategies and programs to foster the development of self-management. Some studies have demonstrated the effectiveness of training interventions in self-management skills, such as goal setting, self-observation, and self-reward (Prussia et al., 1998). Coaching and mentoring have also been found to be valuable tools for promoting self-management in organizational contexts (Ellinger & Bostrom, 1999).

The literature has also examined the benefits of self-management at the individual and organizational levels. Several studies have found that it is positively associated with job performance, satisfaction, and employee commitment (Carmeli et al., 2006). In addition, it has been observed that self-management can contribute to individuals' psychological well-being and resilience (Dolbier et al., 2001). Employees working in different industries demonstrated that employees' perceptions of their own self-management are related to more work engagement and lower degrees of burnout (Morikawa et al., 2024)

In corporate environments, 2024 data reveal that companies implementing organizational self-management (OSM) systems experience 32% higher employee engagement rates compared to traditional hierarchies (Morikawa et al., 2024). In conclusion, recent research on the subject has provided a deeper understanding of this construct, its antecedents, and its consequences. The results suggest that developing self-management can be an effective strategy to improve the performance and well-being of individuals in various contexts (Table 4).

Table 4. Key findings on self-management

Type of review	Key findings
Conceptual	Conceptual - Self-management is based on motivation theories such as McClelland's Achievement Theory (1967) and Carver & Scheier's Self-Regulation Theory (2001). - Bandura's (1997) concept of self-efficacy is central to self-management as it influences the ability to self-regulate and self-direct.
Theoretical	- Smith's (2000) R.E.A.L. model explains the development of self-management through recognizing oneself, exploring one's mental makeup, updating oneself on the change process, and leading oneself according to goals. - Manz & Neck (2010) propose three key strategies for developing self-management: encouraging positive behaviors, creating feelings of competence and determination, and building a positive mindset.
Empirical	- Various studies have found a strong relationship between emotional intelligence (use of emotions) and self-management (Carmeli et al., 2006; Dolbier et al., 2001). - Various studies have shown that self-management is positively associated with job performance, satisfaction, and employee commitment (Carmeli et al., 2006; Dolbier et al., 2001).

Source: Own elaboration.

DISCUSSION

The way in which organizations build, supply and organize the knowledge and routines of their activities, determined by their organizational culture to adapt and develop their organizational performance efficiency, improves the use of the skills of the workforce (Dodgson, 1993). Fiol & Lyles (1985) and Almanza et al. (2019), state that, regardless of the interpretations of learning, it is assumed that its implications with human talent management enrich organizational performance. However, learning does not always improve the effectiveness and efficiency of the person who learns, although it is a key factor for the organization that is made up of many members, which allows it to improve its organizational performance within the market in which it competes.

Senior management must commit to fostering the conditions for generating and transferring knowledge with organizational socialization processes, in order to facilitate the management of human talent to improve the performance of the company or institution. Continuous innovation is closely interrelated with organizational learning (Aragón & Sharma, 2003) and therefore with high organizational performance. Technological capital and innovation are closely linked to human resources, design (Barney, 1991) and organizational performance.

The company's results in economic, financial and sales profitability indicators, market performance in sales growth indicators, market share, wealth creation, market and accounting value, employee and customer satisfaction (Camisón et al. 2007) are factors to measure organizational performance. Organizational performance is measured through sales generation, logistics costs, staff productivity and customer service (Wu et al., 2011).

Other dimensions with their indicators are the goals of interest groups such as customers, managers, shareholders, employees, government or non-governmental organizations. As well as heterogeneity in resources and capabilities, environment, strategy, measurement period, persistence and period, and short and long-term measures (Richard et al. 2009). Likewise, financial and operational performance that integrates structures, objectives, strategies, financial and non-financial results indicators, income and profitability, efficiency, effectiveness, employee satisfaction, growth, market segment, sales and new products (Matus, 2018).

Organizational culture and market orientation are variables that contribute to motivating the necessary behaviors of individuals in organizations that generate high performance and tend to create superior value for all stakeholders or interest groups, but especially for consumers of goods and users of services. Market orientation is the basis for the origination of value capabilities and a determining factor in profitability and performance (Slater and Narver, 1995; Narver and Slater, 1990). Market orientation is also a resource that enables the analysis of the link between strategy and organizational financial performance (Jaworski and Kohli, 1993; Kohli and Jaworski, 1990). Market-focused organizations have a considerably higher organizational performance than those with a lower orientation, especially in variables related to adaptability, profitability, and consumer and user satisfaction (Jaworski and Kohli, 1993).

The capabilities of companies or institutions focused on the market are a source of competitive advantage with effects on market and financial performance (Kohli and Jaworski, 1990). Investment in capital and human talent management is proportional to the results obtained in organizational performance, which is evident in the productive and financial benefits (Srivastava & Das, 2015; Bryl, 2018; Chávez, 2007; Fitz-Enz, 2003; Olmedo and Martínez, 2015; Palacios, 2008; Santos et al., 2013; Sassha, 2011; Villegas, Hernández and Salazar, 2017).

This capabilities improve with self management teams, one example is presented in Netherlands, where healthcare organizations reported a 25% increase in patient satisfaction scores after transitioning to SMTs, as teams could rapidly adapt protocols to patient needs (Geerts et al., 2021) and manufacturing SMTs showed a 17% decline in defect rates when teams self-monitored quality metrics. (Karakus, 2017).

CONCLUSION

This analysis presents the importance of the search for organizational performance improvement through a research model focused on self-managed work teams (ETAS), team management and self-management, based on the concept that self-management facilitates and accelerates the growth of organizational performance, supported by the development of organizational capabilities, adaptation to competitive environments and the relationship between human talent management and organizational performance.

Research on the environmental conditions of organizations is crucial in this analysis, formulation and implementation of strategies that improve performance, under the assumption that environmental changes in organizations are determinants in the adoption of strategies whose purpose is to create and develop competitive advantages that increase said performance (Ynzunza and Izar, 2013). Strategic management focuses mainly on determining and analyzing the organizational processes that link the different forms of adaptation of organizations to environmental conditions through the formulation and implementation of competitive strategies aimed at improving organizational performance.

Finally, competitive strategies characterized as prospective, analytical and defenders of the development of certain capabilities are connected to business performance. Therefore, high-performance organizations introduce competitive strategies that enable improvements to activate the cycle of appropriate and pertinent responses that allow them to take advantage of opportunities for positioning and market dominance.

Further research should explore the diverse organizational factors such as management support, training and information exchange, as well as the changes that have occurred in companies with the generation, challenge and performance of virtual teams.

Future research must investigate how individual competences contribute to the performance of self-managed teams.

REFERENCES

Aldag, R. J., & Kuzuhara, L. W. (2015). *Organizational behavior and management: An integrated skills approach* (2nd ed.). Cengage Learning.

Almanza Jiménez, R., Calderón Campos, P., Vargas Hernández, J. G., Casas Cárdenas, R., & Palomares Salceda, F. (2019). Aprendizaje y desempeño organizacional bajo el enfoque de las teorías organizacionales. *Revista de Economía & Administración, 13*(1), 84-94. Recuperado de https://revistas.uao.edu.co/ojs/index.php/REYA/article/view/10

Andrews, R., Boyne, G., & Walker, R. (2011). The Impact of Management on Administrative and Survey Measures of Organizational Performance. *Public Management Review, 13*(2), 227–255. DOI: 10.1080/14719037.2010.532968

Appelbaum, S. H. (1997). Socio-technical systems theory: An intervention strategy for organizational development. *Management Decision, 35*(6), 452–463. DOI: 10.1108/00251749710173823

Aragón, J., & Sharma, S. (2003). A Contingent Resource Based View of Proactive Corporate Environmental Strategy. *Academy of Management Review, 28*(1), 71–88. DOI: 10.5465/amr.2003.8925233

Asree, S., Zain, M., & Razalli, R. (2010). Influence of leadership competency and organizational culture on responsiveness and performance of firms. *International Journal of Contemporary Hospitality Management, 22*(4), 500–516. DOI: 10.1108/09596111011042712

Awahnde, M. (2014). *Development of self-managed teams through service design methods*. Theseus.

Bandura, A. (1997). *Self-efficacy: The exercise of control*. Macmillan.

Barney, J. (1991). Firm Resources and Sustained Competitive Advantage. *Journal of Management, 17*(1), 99–120. DOI: 10.1177/014920639101700108

Batt, R. (1997). Work organization, technology, and performance in customer service and sales. *Industrial & Labor Relations Review, 52*(4), 539–564. DOI: 10.1177/001979399905200403

Bernal González, I., Pedraza Melo, N. A. & Castillo Hernández, L. (2020). El capital humano y su relación con el desempeño organizacional. *Revista Espacios, 41*(22), art. 14.

Bertolli, M., Roark, G., Urrutia, S., & Chiodi, F. (2017). Revisión de modelos de madurez en la medición del desempeño. *INGE CUC, 13*(1), 70–83. DOI: 10.17981/ingecuc.13.1.2017.07

Bhargava, S., & Sinha, B. (1992). Predictions of organizational effectiveness as a function of type of organizational structure. *The Journal of Social Psychology, 132*(2), 223–232. DOI: 10.1080/00224545.1992.9922974

Blau, P. M. (1964). *Exchange and power in social life*. Wiley.

Bourne, M., Neely, A., Mills, J., & Platts, K. (2003). Implementing performance measurement systems: A literature review. *International Journal of Business Performance Management, 5*(1), 1–24. DOI: 10.1504/IJBPM.2003.002097

Bryl, T. (2018). Human capital orientation and financial performance. a comparative analysis of us corporations. *Journal of Entrepreneurship. Management and Innovation, 14*(3), 61–86. DOI: 10.7341/20181433

Camisón, C., & Cruz, S. (2008). La medición del desempeño organizativo desde una perspectiva estratégica: Creación de un instrumento de medida. *Revista Europea de Dirección y Economía de la Empresa, 17*(1), 79–102.

Camisón, C., Garrigós, F., & Palacios, D. (2007). Estrategias competitivas y desempeño empresarial: Estudio comparativo de los modelos de Robinson and Pearce y Miles and Snow en el sector hotelero español. *Investigaciones Europeas de Dirección y Economía de la Empresa, 13*(3), 161–182.

Camp, R. (1989). *Benchmarking: the search for industry best practices that lead to superior performance*. ASQ Quality Press.

Campion, M. A., Medsker, G. J., & Higgs, A. C. (1993). Relations between work group characteristics and effectiveness: Implications for designing effective work groups. *Personnel Psychology, 46*(4), 823–847. DOI: 10.1111/j.1744-6570.1993.tb01571.x

Camps, J., & Luna-Arocas, R. (2012). A matter of learning how human resources affect organizational performance. *British Journal of Management, 23*(1), 1–21. DOI: 10.1111/j.1467-8551.2010.00714.x

Cannon-Bowers, J. A., Tannenbaum, S. I., Salas, E. & Volpe, C. E. (1995). Defining competencies and establishing team-training requirements. *Team effectiveness and decision making in organizations,* 333-380.

Carmeli, A., Meitar, R., & Weisberg, J. (2006). Self-leadership skills and innovative behavior at work. *International Journal of Manpower, 27*(1), 75–90. DOI: 10.1108/01437720610652853

Carton, R. (2004). *Measuring Organizational Performance: An Exploratory Study.* [Unpublished PhD Thesis]. University of Georgia, Athens.

Carver, C. S., & Scheier, M. F. (2001). *On the self-regulation of behavior.* Cambridge University Press.

Castillo, V. (2013). *Teorías de las organizaciones.* Trillas.

Chávez, E. (2007). El ROI: De la remuneración al capital humano. *Ingeniería Industrial, 25*(25), 111–118. DOI: 10.26439/ing.ind2007.n025.613

Chávez, E. A., & Vizcaíno, A. (2017). Talento humano: Una contribución a la competitividad organizacional. *Mercados y Negocios, 1*(36), 7–20. DOI: 10.32870/myn.v0i36.5646

Chong, H. (2008). Measuring Performance of Small-and-Medium Sized Enterprises: The Grounded Theory Approach. *Journal of Business & Public Affairs, 10*, 1–11.

Cohen, S. G. (1993). New approaches to teams and teamwork. *Trends and Issues in OD:* Crook, R., Combs, J. & Shook, C. (2005). The Dimensionality of Organizational Performance and its Implications for Strategic Management Research. *Research Methodology in Strategy and Management, 2*, 259–286. DOI: 10.1016/S1479-8387(05)02011-4

Cohen, S. G., & Bailey, D. E. (1997). What makes teams work: Group effectiveness research from the shop floor to the executive suite. *Journal of Management, 23*(3), 239–290. DOI: 10.1177/014920639702300303

Cohen, S. G., & Ledford, G. E.Jr. (1994). The effectiveness of self-managing teams: A quasi-experiment. *Human Relations, 47*(1), 13–43. DOI: 10.1177/001872679404700102

Cropanzano, R., & Mitchell, M. S. (2005). Social exchange theory: An interdisciplinary review. *Journal of Management, 31*(6), 874–900. DOI: 10.1177/0149206305279602

Daft, R. (2010). *Teoría y diseño organizacional.* México: Cengage Learning Editores, S. A. de C. V.

Day, G. (1994). The Capabilities of Market Driven Organizations. *Journal of Marketing, 58*(4), 3–52. DOI: 10.1177/002224299405800404

Deci, E. L., & Ryan, R. M. (2000). *The "what" and "why" of goal pursuits: Human needs*

Deci, E. L., & Ryan, R. M. (2008). Self-determination theory: A macrotheory of human motivation, development, and health. *Canadian Psychology, 49*(3), 182–185. DOI: 10.1037/a0012801

Delaney, J. T., & Huselid, M. A. (1996). The impact of human resource management practices on perceptions of organizational performance. *Academy of Management Journal, 39*(4), 949–969. DOI: 10.2307/256718

Dhaafri, H., Al Swidi, A., & Al Ansi, A. (2016). Organizational Excellence as the Driver for Organizational Performance: A Study on Dubai Police. *International Journal of Business and Management, 11*(2), 47–52. DOI: 10.5539/ijbm.v11n2p47

Doblinger, M. (2021). Individual Competencies for Self-Managing Team Performance: A Systematic Literature Review. *Small Group Research, 53*(1), 128–180. DOI: 10.1177/10464964211041114

Dodgson, M. (1993). Organizational Learning: A Review of Some Literatures. *Organization Studies, 14*(3), 375–395. DOI: 10.1177/017084069301400303

Dolbier, C. L., Soderstrom, M., & Steinhardt, M. A. (2001). The relationships between self- leadership and enhanced psychological, health, and work outcomes. *The Journal of Psychology, 135*(5), 469–485. DOI: 10.1080/00223980109603713 PMID: 11804002

Druskat, V. U., & Kayes, D. C. (2000). Learning versus performance in short-term project teams. *Small Group Research, 31*(3), 328–353. DOI: 10.1177/104649640003100304

Ellinger, A. D., & Bostrom, R. P. (1999). Managerial coaching behaviors in learning organizations. *Journal of Management Development, 18*(9), 752–771. DOI: 10.1108/02621719910300810

Escandón, D., Arias, A., & Salas, J. (2012). Recursos y capacidades que inciden en el desempeño de los subsectores económicos colombianos 2003-2004: Un acercamiento desde la teoría basada en recursos. *Ingeniería y Competitividad, 14*(2), 53–67. DOI: 10.25100/iyc.v14i2.2654

Fey, C., & Beamish, P. (2001). Organizational climate similarity and performance international joint Ventures in Russia. *Stockholm School of Economics in St. Petersburg Working, 99*(101), 1-28.

Fiol, M. C., & Lyles, M. (1985). Organizational learning. *Academy of Management Review, 10*(4), 803–813. DOI: 10.2307/258048

Fitz-Enz, J. (2003). *El ROI del capital humano. Como medir el valor económico del rendimiento del personal*. Deusto.

Fuentes, M. M., & Hurtado, N. E. (2002). Variables críticas en la medición del desempeño en empresas con implantación de la gestión de la calidad total. *Investigaciones Europeas de Dirección y Economía de la Empresa*, *8*(2), 87–102.

Gagné, M., & Deci, E. L. (2005). Self-determination theory and work motivation. *Journal of Organizational Behavior*, *26*(4), 331–362. DOI: 10.1002/job.322

Ganga, F., Cassinelli, A., Díaz, G., & Maluk, S. A. (2016). Breves disquisiciones teóricas en torno al tema capital humano y eficiencia organizativa. *Revista Gaceta Laboral*, *22*(1), 27–39.

Garvin, D. A. (1984). What does "product quality" really mean? *Sloan Management Review*, *26*(1), 25–43.

Gavrea, C., Ilies, L., & Stegeran, R. (2011). Determinants of Organizational Performance: The Case of Romania. *Management and Marketing Challenges for the Knowledge Society*, *6*(2), 285–300.

Geerts, I. A., Bierbooms, J. J. P. A., & Cloudt, S. W. M. G. (2021). Understanding self-managing teams in Dutch healthcare: Empirical evidence to non-sequential team development processes. *Journal of Health Organization and Management*, *35*(9), 101–120. DOI: 10.1108/JHOM-04-2020-0122 PMID: 33818047

Georgopoulos, B., & Tannenbaum, A. (1957). A Study of Organizational Effectiveness. *American Sociological Review*, *22*(5), 534–540. DOI: 10.2307/2089477

Guerrero, S., González, S., Matos, A., & Picón, E. (2015). Desarrollo del capital humano y su impacto en el desempeño de una institución microfinanciera no regulada del Perú. *Revista Apuntes Universitarios*, *5*(1), 51–68. DOI: 10.17162/au.v5i1.65

Hackman, J. R. (1987). The design of work teams. In Lorsh, J. (Ed.), *Handbook of Organizational Behavior 315-342*. Prentice-Hall.

Hamann, P., Schiemann, F., Bellora, L., & Guenther, T. (2013). Exploring the Dimensions of Organizational Performance: A Construct Validity Study. *Organizational Research Methods*, *16*(1), 67–87. DOI: 10.1177/1094428112470007

Hambrick, D. (2003). On the Staying power of Miles and Snow's defenders, analyzers and prospectors. *The Academy of Management Perspectives*, *17*(4), 115–118. DOI: 10.5465/ame.2003.11851883

Hoegl, M., & Parboteeah, K. P. (2006). Team goal commitment in innovative projects. *International Journal of Innovation Management, 10*(03), 299–324. DOI: 10.1142/S136391960600151X

Jaworski, B., & Kohli, A. (1993). Market Orientation: Antecedents and Consequences. *Journal of Marketing, 57*(3), 53–70. DOI: 10.1177/002224299305700304

Junxiao, L., Love, S., & Regan, M. (2014). Public-Private Partnerships: A review of theory and practice of performance measurement. *International Journal of Productivity and Performance Management, 63*(4), 499–512. DOI: 10.1108/IJPPM-09-2013-0154

Kaplan, R., & Norton, D. (1992). The balanced scorecard-Measures that drive performance. *Harvard Business Review, 70*(1), 71–79. PMID: 10119714

Karakus, U. (2017). Performance Management Practices in Self-Managing Teams: An explorative study at the Dutch healthcare organization. *9 th IBA Bachelor Thesis Conference, July 5th, 2017, Enschede, The Netherlands*. https://essay.utwente.nl/72828/1/Karakus_BA_BMS.pdf

Kirkman, B. L., & Rosen, B. (1999). Beyond self-management: Antecedents and consequences of team empowerment. *Academy of Management Journal, 42*(1), 58–74. DOI: 10.2307/256874

Kirkman, B. L., & Rosen, B. (2002). Powering up teams. *Organizational Dynamics, 28*(3), 48–66. DOI: 10.1016/S0090-2616(00)88449-1

Kohli, A., & Jaworski, B. (1990). Market Orientation: The Construct, Research Propositions, and Managerial Implications. *Journal of Marketing, 54*(2), 1–18. DOI: 10.1177/002224299005400201

Kuipers, B. S., & Stoker, J. I. (2009). Development and performance of self-managing work teams: A theoretical and empirical examination. *International Journal of Human Resource Management, 20*(2), 399–419. DOI: 10.1080/09585190802670797

Langfred, C. W. (2004). Too much of a good thing? Negative effects of high trust and individual autonomy in self-managing teams. *Academy of Management Journal, 47*(3), 385–399. DOI: 10.2307/20159588

Langfred, C. W. (2007). The downside of self-management: A longitudinal study of the effects of conflict on trust, autonomy, and task interdependence in self-managing teams. *Academy of Management Journal, 50*(4), 885–900. DOI: 10.5465/amj.2007.26279196

Lebans, M. & Euske, K. (2006). A conceptual and operational delineation of performance. *Business Performance Measurement*, 65-79. DOI: 10.1017/CBO9780511753695.006

Lui, J., Love, P., Smith, J., Regan, M., & Sutrisna, M. (2014). Public-Private Partnerships: A review of theory and practice of performance measurement. *International Journal of Productivity and Performance Management, 63*(4), 499–512. DOI: 10.1108/IJPPM-09-2013-0154

Lusthaus, C., & Adrien, M. (1998). Organizational assessment: A review of experience. *Universalia, 31*, 1–16.

Lusthaus, C., Adrien, M., Anderson, G., Carden, F., & Plinio, G. (2002). *Evaluación Organizacional. Marco para mejorar el Desempeño.* Banco Interamericano de Desarrollo y Centro Internacional de Investigaciones para el Desarrollo.

Madella, A., Bayle, E., & Tome, J. (2005). The organizational performance of national swimming federations in Mediterranean countries: A comparative approach. *European Journal of Sport Science, 5*(4), 207–220. DOI: 10.1080/17461390500344644

Manz, C. C., & Neck, C. P. (2010). *Mastering self-leadership: Empowering yourself for personal excellence.* Pearson.

Manz, C. C., & Sims, H. P. (1987). Leading workers to lead themselves: The external leadership of self-managing work teams. *Administrative Science Quarterly, 32*(1),
Matus, D. (2018). Implementación operativa de un modelo de intervención para la mejora del desempeño organizacional. *SIGNOS, 10*(2), 71–86. DOI: 10.15332/s2145-1389.2018.0002.04

McClelland, D. C. (1967). *Achieving society.* Simon and Schuster.

McDaniel, S., & Kolari, J. (1987). Marketing Strategy Implications of the Miles y Snow

Menon, A., & Mohanty, B. (2012). Organizational Performance: A Multidimensional Perspective. In Singh, S. (Ed.), *Global Competition and Competitivenes s of Indian Corporates.* Macmillan.

Miles, R., & Snow, Ch. (1978). *Organizational Strategy, Structure and Processes.* McGraw-Hill.

Morikawa, M., Martela, F., & Hakanen, J. (2024). Are employee self-management and organizational self-management related to work engagement or burnout? Evidence from two studies. BRQ. *Business Research Quarterly.* Advance online publication. DOI: 10.1177/23409444241307734

Muthuveloo, R., Shanmugam, N., & Teoh, A. P. (2017). The impact of tacit knowledge management on organizational performance: Evidence from Malaysia. *Asia Pacific Management Review*, *22*(4), 192–201. DOI: 10.1016/j.apmrv.2017.07.010

Narver, J., & Slater, F. (1990). The Effect of a Market Orientation on Business Profitability. *Journal of Marketing*, *54*(4), 20–35. DOI: 10.1177/002224299005400403

Ndofor, H., & Priem, R. (2011). Inmigrant entrepreneurs the ethnic enclave strategy and venture performance. *Journal of Management*, *37*(3), 790–818. DOI: 10.1177/0149206309345020

Neely, A. (1999). The performance measurement revolution: Why now and what next? *International Journal of Operations & Production Management*, *19*(2), 205–228. DOI: 10.1108/01443579910247437

Oliveira, A., & Leone, R. (2008). Investigação sobre medição de desempenho empresarial. *Custos e Agronegócio*, *4*, 117–137.

Olmedo, I., & Martínez, I. (2015). Human capital and creation of reputation and financial performance. *Electronic Journal of Knowledge Management*, *13*(3), 209–218.

Omondi, C. (2015). Measuring organization performance. From Balanced Scorecard to Balanced Esg Framework. *International Journal of Economics. Commerce and Management*, *3*(11), 715–725.

Palacios, J. (2008). *Medición del impacto y la rentabilidad de la formación. Cómo llegar al roi de la formación*. Díaz de Santos.

Palomo, M. T. (2013). *Auto gestión y motivación de equipos de trabajo*. Esic Editorial.

Pardo, C. E. y Diaz, A. (2014). Desarrollo del talento humano como factor clave para el desarrollo organizacional, una visión desde los líderes de gestión humana en empresas de Bogotá D.C. *Suma de Negocios, 5*(11), 39-48.

Pearce, C. L., & Conger, J. A. (2003). *Shared leadership: Reframing the hows and whys of leadership*. Sage Publications. DOI: 10.4135/9781452229539

Pérez, J., & Cortés, J. A. (2009). Medición y validación del desempeño organizacional como resultado de acciones de aprendizaje. *Revista Ciencias Estratégicas*, *17*(22), 251–271.

Prussia, G. E., Anderson, J. S., & Manz, C. C. (1998). Self-leadership and performance outcomes: The mediating influence of self-efficacy. *Journal of Organizational Behavior*, *19*(5), 523–538. DOI: 10.1002/(SICI)1099-1379(199809)19:5<523::AID-JOB860>3.0.CO;2-I

Richard, P., Devinney, T., Yip, G., & Johnson, G. (2009). Measuring Organizational Performance: Towards Methodological Best Practice. *Journal of Management, 35*(3), 718–804. DOI: 10.1177/0149206308330560

Rousseau, V., Aubé, C., & Savoie, A. (2006). Teamwork behaviors: A review and an integration of frameworks. *Small Group Research, 37*(5), 540–570. DOI: 10.1177/1046496406293125

Santos, H., Pereira, G., & Cranfield, D. (2013). Human capital and financial results: A case study. *Electronic Journal of Knowledge Management, 11*(4), 387–392.

Sassha, A. (2011). ¿Es posible medir el impacto del capital humano en los resultados de la organización?: El roi de recursos humanos, modelo de medición y otros indicadores del capital humano. *Perspectivas, 14*(27), 113–129.

Schmenner, R. (1988). Escaping the black holes of cost accounting. *Business Horizons, 31*(1), 66–72. DOI: 10.1016/0007-6813(88)90043-2

Segev, E. (1987). Strategy, Strategy Making and Performance-An empirical investigation. *Management Science, 33*(3), 258–269. DOI: 10.1287/mnsc.33.2.258

Shu-Mei, T. & Pei-Shan, L. (2014). The effect of knowledge management capability and dynamic capability on organizational performance. *Journal of Enterprise Information*

Singh, S., Darwish, T., & Pocotcnick, K. (2016). Measuring Organizational Performance: A Case for Subjective Measures. *British Journal of Management, 27*(1), 214–224. DOI: 10.1111/1467-8551.12126

Singh, S., & Gupta, V. (2016). Organizational performance research in India. A review and future research agenda. In Misra, G. (Ed.), *The Sixth Indian Council for Social Science Research (ICSSR) Survey of Psychology in India*. Oxford Publishing.

Skinner, W. (1974). The decline, fall, and renewal of manufacturing. *Industrial Engineering (American Institute of Industrial Engineers), 6*, 32–38.

Slater, S., & Narver, J. (1994). Does competitive environment moderate the market orientation-performance relationship? *Journal of Marketing, 58*(1), 46–55. DOI: 10.1177/002224299405800104

Slater, S. F., & Narven, J. C. (1995). Market orientation and the learning organization. *Journal of Marketing, 59*(3), 63–74. DOI: 10.1177/002224299505900306

Smith, L. (2000). *The real-time leader: Develop the focus, presence and authenticity to succeed in today's leadership challenges*. Amacom.

Srivastava, K., & Das, R. (2015). Human capital management: Economics of psychological perspective. *Industrial Psychiatry Journal, 24*(2), 115–118. DOI: 10.4103/0972-6748.181717 PMID: 27212812

Strasser, S., Eveland, J., Cummins, G., Deniston, O., & Romani, J. (1981). Conceptualizing the goal and system models of organizational effectiveness implications for comparative evaluation research. *Journal of Management Studies, 18*(3), 321–340. DOI: 10.1111/j.1467-6486.1981.tb00105.x

Sundstrom, E., De Meuse, K. P., & Futrell, D. (1990). Work teams: Applications and effectiveness. *The American Psychologist, 45*(2), 120–133. DOI: 10.1037/0003-066X.45.2.120

Tangen, S. (2004). Performance measurement: From philosophy to practice. *International Journal of Productivity and Performance Management, 53*(8), 726–737. DOI: 10.1108/17410400410569134

Tseng, K., Lan, Y., Lu, H., & Chen, P. (2013). Mediation of strategy on intellectual capital and performance. *Management Decision, 51*(7), 1488–1509. DOI: 10.1108/MD-03-2012-0143

Tseng, S.-M., & Lee, P.-S. (2014). The effect of knowledge management capability and dynamic capability on organizational performance. *Journal of Enterprise Information Management, 27*(2), 158–179. DOI: 10.1108/JEIM-05-2012-0025

Turney, P., & Anderson, B. (1989). Accounting for continuous improvement. *Sloan Management Review, 30*(2), 37–48.

Varadarajan, P., & Vasudevan, R. (1990). The Corporate Performance Conundrum: A Synthesis of Contemporary Views and an Extension. *Journal of Management Studies, 27*(5), 463–483. DOI: 10.1111/j.1467-6486.1990.tb00257.x

Venkatraman, N. (1989). Strategic orientation of business enterprises: The construct, dimensionality and measurement. *Management Science, 35*(8), 942–962. DOI: 10.1287/mnsc.35.8.942

Venkatraman, N., & Ramanujam, V. (1986). Measurement of business performance in strategy research: A comparison of approaches. *Academy of Management Review, 11*(4), 801–814. DOI: 10.2307/258398

Villegas, E., Hernández, M., & Salazar, B. (2017). La medición del capital intelectual y su impacto en el rendimiento financiero en empresas del sector industrial en México. *Contaduría y Administración, 62*(1), 184–206. DOI: 10.1016/j.cya.2016.10.002

Vorhies, D., & Morgan, N. (2003). A Theory Assessment of Marketing Organization Fit with Business Strategy and Its Relationship with Marketing Performance. *Journal of Marketing*, *67*(1), 100–115. DOI: 10.1509/jmkg.67.1.100.18588

Waal, A., & Kourtit, K. (2013). Performance measurement and management in practice. Advantages, disadvantages and reasons for use. *International Journal of Productivity and Performance Management*, *62*(5), 446–473. DOI: 10.1108/IJPPM-10-2012-0118

Wageman, R. (1997). Critical success factors for creating superb self-managing teams. *Organizational Dynamics*, *26*(1), 49–61. DOI: 10.1016/S0090-2616(97)90027-9

Weaver, S. J., Rosen, M. A., Salas, E., Baum, K. D., & King, H. B. (2010). Integrating the science of team training: Guidelines for continuing education. *The Journal of Continuing Education in the Health Professions*, *30*(4), 208–220. DOI: 10.1002/chp.20085 PMID: 21171026

Wernerfelt, B. (1984). A resource based view of the firm. *Strategic Management Journal*, *5*(2), 171–180. DOI: 10.1002/smj.4250050207

Wu, J. N., Zhong, W. J., & Mei, S. E. (2011). Application capability of e-business, e-business success, and organizational performance: Empirical evidence from China. *Technological Forecasting and Social Change*, *78*(8), 1412–1425. DOI: 10.1016/j.techfore.2011.03.023

Yamakawa, P., & Ostos, J. (2011). Relación entre innovación organizacional y desempeño organizacional. *Universidad & Empresa*, *13*(21), 93–115.

Yeatts, D. E., & Hyten, C. (1998). *High-performing self-managed work teams: A comparison of theory to practice*. SAGE Publications. DOI: 10.4135/9781483328218

Ynzunza Cortés, C. B., & Izar Landeta, J. M. (2013). Estrategia, orientación al mercado y desempeño organizacional. *Conciencia Tecnológica*, (45), 5–11.

Yuchtman, E., & Seashore, S. (1967). Factorial Analysis of Organizational. Performance. *Administrative Science Quarterly*, *12*(3), 377–395. DOI: 10.2307/2391311

Zulkiffli, S. (2014). A literature analysis on business performance for SMES - subjective or objective measures? *Review of Integrative Business and Economics Research*, *3*(1), 371–381.

Chapter 4
Ignition Switch to Change:
The Journey to Risk Management Reform

Megan Pekol-Evans
https://orcid.org/0009-0008-3273-091X
Marymount University, USA

ABSTRACT

This study examines the crisis management and enterprise risk management (ERM) challenges faced by General Motors (GM) following a series of fatal accidents in 2024 involving one of its sedan models. The incidents, which resulted in vehicle explosions and passenger fatalities, underscored significant weaknesses in GM's risk management strategies and quality assurance processes. Using a narrative literature review, the research synthesizes past safety incidents in large organizations and explores theoretical frameworks like COSO and RIMS Risk Maturity Models. Key recommendations include adopting a mature ERM framework, refocusing on quality assurance, reimagining engineering and financial risk strategies, and fostering cultural change through Kotter's change model and other organizational theories. The study emphasizes that a robust ERM strategy and effective change management are critical for restoring stakeholder trust, mitigating financial and reputational risks, and aligning GM's operations with its long-term objectives.

DOI: 10.4018/979-8-3373-5686-0.ch004

INTRODUCTION

Global Motors (GM) is a leading automobile manufacturer worldwide with a reputation for reliability and innovation synthesized based the past safety incidents of major manufacturing and production firms (Held et al., 2018); the name is fictional and was utilized to protect the privacy of the work completed. The firm's leadership was caught off guard in 2024 when it experienced a crisis related to 20 fatal accidents of passengers driving one of its sedan models. Vehicles exploded upon impact and killed all passengers involved (Jennings & Trautman, 2016). These accidents occurred in multiple countries across the globe and raised concerns about a potential design flaw in either the vehicle's fuel system or battery technology. The initial investigations linked the critical failure to the car's safety mechanism when experiencing a high-impact collision. The firm experienced significant reputational damage as public and media outcry increased as more incidents occurred (Jennings & Trautman, 2016).

These incidents led to a global recall of the sedan and millions of dollars in costs to the firm, including direct expenses, legal fees, and lost sales. Product recalls are costly to automotive firms; for example, claims in 2016 related to product recalls for Original Equipment Manufacturers (OEMs) and suppliers totaled an estimated \$11.8 billion, and warranty claims were estimated at \$10.3 billion (Held et al., 2018). Those costs, plus those from the impending multiple class action lawsuits by the victims' families, will be costly to GM. Lastly, several countries' regulatory arms also began investigating GM, which added financial risk.

GM had to maneuver through significant crisis management efforts to mitigate the damages to GM's reputation and resolve complex safety and quality assurance issues simultaneously. Similarly, in 2015, General Motors faced criminal charges due to a faulty ignition switch in the Chevy Cobalt that caused the deaths of over 100 customers, sometimes referred to as "the switch from hell" by GM engineers (Jennings & Trautman, 2016). Those charges were deferred; the company agreed to a \$900 million forfeiture. The U.S. Securities and Exchange Commission (SEC) (2017) investigated GM's accounting practice, resulting in over \$1 million in penalties due to the lag between finding the safety issue and GM communicating the issue to internal accountants to help mitigate the financial impacts of a recall. This recall is just one of many throughout the history of GM. For example, the Chevrolet Corvair and Malibu cases caused deaths and injuries, which manifested from a historical risk management-averse culture and caused repeated reputational, legal, regulatory, and financial impacts (Jennings & Trautman, 2016). With this data in mind, GM must analyze its weak enterprise risk management (ERM) practices and compel them to re-analyze their approach in the future to ensure the company could remain viable in the long term to avoid the ongoing risk challenges faced by GM over time.

Problem Statement

The problem to be explored is the weak ERM embedded within a firm's strategy to support risk management of the emerging reputational, regulatory, financial, technical, and associated risks manifesting from the fatal accidents that occurred in 2024. The appropriate problem-solving model, strategy, risk management approach, change management approach, and organizational development model must be identified since the risks to the firm exist across several categories and have multiple stakeholder equities to bind together.

Aim of This Inquiry

This exploration aims to immediately manage the risks associated with a crisis at a large manufacturing and production facility and improve ERM maturity within 180 days of implementing the new framework. This strategy transcends industries and applies to manufacturing or production companies wishing to regain public and stakeholder confidence in such challenges. The investigation will also enable a firm to plan for possible litigation, as seen in the GM lawsuits related to the ignition switches that, over time, cost the company as much as $7 billion in legal fees and settlements, as well as plan for the amount and financial resources required to support vehicle recalls. Lastly, it will enable a firm to understand the theoretical frameworks that may have led to the failures and current crises.

Significance of the Exploration

GM was known for its innovation and reliability. The reputational risks from these safety incidents stemmed from several risk categories, and the impact strained its standing in the market. The financial impacts are estimated to be in the billions of dollars, with financial risks associated with product recalls, class action lawsuits, and possible fines associated with regulatory actions across multiple countries. Studies suggest that firms who complete these types of ERM transformations tend to bring in outside experts to revamp key performance indicators to incentivize the pursuit of more quality assurance behaviors, relook leadership involved in the quality assurance areas of concern, and analyze data or conduct interviews to identify higher-level risks associated with quality engineering to address them quickly (Held et al., 2018). It is crucial for firms facing these challenges to frame the problem objectively and then subsequently apply an ERM model that will fit best with its organizational culture, enable change management efforts to realign GM towards its objectives and goals, and communicate all efforts to the public and its stakeholders to regain trust.

LITERATURE REVIEW

This narrative literature review aimed to synthesize information related to ERM and learn from past quality and safety incidents at large firms. This methodology was chosen based on limited experience conducting literature reviews and experience in this subject matter. The step-by-step framework by Muhammad (2022) provided a simplified methodology to support the literature review process. The PRISMA guidelines guided This literature review (Page et al., 2021). The aim of the literature review was achieved by addressing the following research question: What risk management approach will best position GM to manage through the current emerging threats and better support long-term strategic planning towards the firm's objectives and goals?

The narrative literature review included peer-reviewed empirical studies and books published in English that explored ERM and past firm-level incidents related to quality and safety. ERM in this context relates to the Risk and Insurance Management Society (RIMS) Risk Maturity Model, which is a framework that may be helpful in the future as GM attempts to visualize their current risk management capabilities and where opportunities and challenges reside to further ERM goals (Simanungkalit & Tobing, 2022). Studies included in the literature review were only those books or peer-reviewed scholarly journal articles between 2014 and 2024, with exceptions to seminal articles.

Four electronic databases were chosen for the literature review: Perplexity AI, Research Rabbit, Google Scholar, and ProQuest. Perplexity AI was chosen as a tool to meet the eligibility criteria and find the relevant keywords related to ERM and firm-level incidents; it further refined searches on articles and books. Research Rabbit was chosen because the tool is simple and graphically depicts empirical, peer-reviewed journal articles related to topic areas. Google Scholar was chosen because it is a cross-disciplinary database that provides full-text organizational and business-related empirical articles (Muhammad, 2022). ProQuest was chosen because of its extensive database collections comprised of ABI/INFORM Global, ABI/INFORM Trade and Industry, and ABI/INFORM Dateline, which include empirical, peer-reviewed studies related to economics and business.

Perplexity AI was the first step of the literature review and aided in defining keywords to search for articles. This tool provided keywords for searching, such as risk management, risk mitigation, business continuity, crisis management, and compliance.

These terms were imported as search terms in Research Rabbit, and the number of articles found per keyword were: risk management (23), risk mitigation (32), business continuity (56), crisis management (27), and compliance (34) for a total of 172 articles. These search terms were inputted into Google Scholar and ProQuest to collect more articles utilizing the Boolean search term AND for keywords, e.g., risk management AND business continuity, risk management AND risk mitigation, risk management AND crisis management, and risk management AND compliance. Google Scholar alone, for the search of risk management AND business continuity, returned 85,300 results. The articles were first reviewed based on titles and abstracts. Articles that did not have the keywords were not included. However, articles explicitly related to the automotive or airline industries and public safety incidents with keywords were included based on their relevance to the current situation and problem statement.

Risk Management Models and SWOT Analysis

The Committee of Sponsoring Organizations (COSO) created an ERM pyramid-shaped model in 1992 that focused on evaluating existing organizational controls related to risk (Saleem et al., 2019) that is important to understand in the context of the challenges facing GM. COSO was later iterated upon and became the COSO cube, which produced a commonly utilized risk management framework. Similarly, the RIMS Risk Maturity Model is a framework that may be helpful in the future for the firm as it enables organizations to visualize their current risk management capabilities and where opportunities and challenges reside to meet the goal of ERM maturity within 180 days of implementation (Simanungkalit & Tobing, 2022). These frameworks help organizations envision the risks to overall objectives and strategy while providing a means to understand control mechanisms.

Political, economic, social, technology, legal, and environmental (PESTLE) analysis provides a different external vantage to view risks for firms and, in the current problem set, will help manage the emergent issues facing firms in this position. Achinas et al. (2019) argued that the framework allows an organization to observe risks from socio-political, techno-economic, legal, and environmental lenses while identifying the risks associated with multiple stakeholders involved at all stages. Firms could face significant impacts concerning congressional and regulatory investigations like the investigations into Boeing (Criminal Division of the U.S. Department of Justice, 2024) and General Motors (Held et al., 2018), resulting in \$20 billion and \$487 million in losses, respectively. Maturing its understanding and usage of PESTLE analysis will be important to firms in the near and long term, allowing strategic efforts to become realized and successful.

Another important contributor to successful ERM is strength, weakness, opportunity, and threat (SWOT) analysis (Sutrisno, 2016). SWOT enables problem-solving and helps organizations identify challenges within the sector in which they operate. Strengths and weaknesses are internal risks an organization must control, whereas opportunities and threats are external and not controllable by a firm (Sutrisno, 2017).

An example of a strength is a factor that provides a firm a competitive advantage, whereas a weakness could be high employee turnover. Opportunities could manifest as a creative investment for growth, whereas a threat could be an economic recession. Islam (2017) argued that the mix of SWOT and PESTLE reduced conservable risks for the leasing companies studied and provided considerable impacts on ERM practices. A study by Christodoulou and Cullinane (2019) related to developing an energy efficiency management system for sustainable port development throughout the globe suggested that utilizing a mix of PESTLE and SWOT analysis provided meaningful insights into the complex risk associated with such a significant endeavor. It argued that this mix and the additive adoption of ISO 50001 standard helped identify and highlight the political, economic, social, technological, legal, and environmental factors that have a positive or negative effect on the adoption of a port energy management plan, as well as a helped in the design and application of the energy efficiency management system (Christodoulou & Cullinane, 2019).

A challenge in implementing and measuring ERM is how firms quantify risk (likelihood x impact) (Simanungkalit & Tobing, 2022). The effort can be subjective since likelihood and impact are in the eye of the beholder. If the organization is more risk-averse, the reported risks could have a Chicken Little effect where the sky always falls. If an organization underestimates risks, it could lead to misguided decisions and serious repercussions. Another challenge organizational leaders face is building a culture of psychological safety. ERM is seen as a benefit rather than one that brings undue anxiety among employees when they are honest about problems (Simanungkalit & Tobing, 2022). Achinas et al. (2019) argue that ERM should be built into regular reporting and inculcated into lower-level divisions to allow them to identify those they cannot mitigate at their level up higher to prioritize, strategize, and resource.

The theory of behavioral economics, first postulated by Daniel Kahneman and Amos Tversky in 1974, is another essential theory for firms to understand when it comes to ensuring that risk measurements are qualified accurately and without bias. Judgments on risk in firms are derived from people's perceptions. People have biases, and many are unaware of them; we are blind to our blindness (Kahneman & Tversky, 1974). According to Kahneman and Tversky (1974), people use cognitive shortcuts to make decisions, called heuristics, to enable easier decision-making when faced with uncertainty. In ERM, this could lead to outcomes that have less rigor for complex problems (Kahneman, 2011), like the ones currently impacting GM, and it shows why it is critical to ensure that risk within the culture is discussed safely. Employees can respectfully challenge each other to produce the best outcomes for the firm.

Organizational Culture, Development, and Behavior Theories

The theoretical frameworks and guidance within models such as Hostede's Corporate Culture Model, Cultural Iceberg Model, or the Denison Organizational Model provide important dimensions to help guide leaders through change. Hofstede's culture model includes four values to help measure organizational culture: Individualism-Collectivism, Power Distance, Masculinity- Femininity, and Uncertainty Avoidance (Taras et al., 2023). Cultural values are important factors for the firm to understand and envision how the organization may react to the new ERM model implemented and the changes that will occur. Similarly, the Cultural Iceberg Model is a good concept for studying firm leadership. Proponents contend that leaders applying this model are more in tune with themselves, other people's perspectives, and are more capable of working with and for others in any organization (Shaules, 2019). Lastly, leaders and managers who understand where their organization scores across the multidimensional aspects within the Denison Organizational Model are more apt to manage the risks associated with the change they plan to implement (Xuan et al., 2019). Therefore, it is a pertinent model for the current challenges facing GM.

Team-Based Learning, Social Learning Theory, and Kepner-Tregoe Decision Analysis

Team-based learning and social learning theory are essential models for firms to study, and they support problem-solving and analysis of the current risk categories. Burgess et al. (2020) argue that team-based learning enhances knowledge sharing, identifies problems and gaps, and supports cohesive team building among employees. The 2016 study of 350 year-one medical students enrolled in the University of Sydney medical program suggested that students and focus group facilitators agreed that utilizing team-based learning to analyze authentic clinical problems was beneficial. It allowed students to improve their understanding of the problem by encouraging self-reflection, a medium to identify knowledge gaps and build on prior knowledge (Burgess et al., 2020). Junaidi (2023) argues that social learning theory helps create an ethical environment through organizational leadership that provides psychological safety to employees, which helps facilitate trust, healthy debate, challenge norms, and continued information sharing. Lastly, systematic frameworks such as the Kepner-Tregoe Decision Analysis (KTDA) can take the data and feedback from team-based learning and apply more rigor to the decision-making process. The structured process of KTDA incorporates objective and subjective measures to help identify objectives, prioritize them, and evaluate alternatives against quantitative required and desired criteria (Priyambud & Basri, 2023). Understanding these frameworks will be critical to successful support and implementation by employees and lead to successful organizational outcomes as it mitigates risk to buy-in and assuages employees' fears.

Change Management Theories and Models

Change management theory and models tie in the previous topics, as understanding organizational culture, how employees react and interact, cultural dimensions, and having a suitable ERM framework to assess organizational risks all support organizational change efforts. Kotter proved that 70% of organizational reforms fail, which was why Kotter's 8-step framework was developed to help organizations lead through change (Laig & Abocejo, 2021). Similarly, it is vital for leaders implementing the upcoming changes to understand the Nudge Theory and how, if implemented correctly, it leads to successful change. Arno and Thomas (2016) found that when employing nudge interventions related to healthier food consumption decisions, nudge interventions resulted in a 15.3% increase in healthier decisions. Nudge interventions employed during organizational change efforts could help employees embrace change.

Following a proven framework and employing nudge efforts become even more significant while assessing metrics related to the Kubler-Ross model. Abebe et al. (2023) explain that the Kubler-Ross Change Curve (KRCC) model can include up to seven stages related to organizational change: shock, denial, frustration, depression, experimentation, decision, and integration. The curve depicts the sentiments employees experience during change and helps leadership follow the progression toward integrating the change strategy (Abebe et al., 2023). Bringing all these theories and frameworks together successfully is difficult but possible. Leaders must be deliberate in their planning, preparation, execution, and assessment of change efforts to ensure objectives are being met and goals are ultimately reached.

RECOMMENDATIONS

Recommendation 1 (Implementation of the RIMS Risk Maturity Model and SWOT Analysis)

Firms facing these challenges should implement the RIMS Risk Maturity Model coupled with SWOT analysis to mature its risk management strategy to strategize a way forward for the firm in the face of the emerging reputational, regulatory, financial, technical, and associated risks manifesting from the fatal accidents that occurred in 2024. RIMS Risk Maturity Model is a framework that will help the firm visualize its current risk management capabilities and identify its strengths and weaknesses

(Simanungkalit & Tobing, 2022). This model should inform and support the firm as it moves towards its ERM maturity goals.

Recommendation 2 (Strategic Refocus on Quality Assurance)

Quality assurance is an observably immature risk area in GM, as manifested in product recalls. A recommendation is for firms to refocus on quality assurance and its practices strategically. This analysis should be facilitated by outside experts with the help of internal stakeholders and focus on qualifying and controlling for quality assurance risks, human resources risks, and operational risks, as OEMs observed a 14% decrease in product recalls within the past year as they switched to a focus on quality (Held et al., 2018). Realigning and reinforcing the company's strategy towards advanced product quality planning (APQP) (Held et al., 2018) to plan for risks is also recommended. APQP processes are typically led by quality, engineering, and manufacturing with support from across business functional and support areas. APQP teams are developed to focus on the lowest component level and the sub-system and system levels. The team helps control risks associated with source components and how much autonomy suppliers have over component design and manufacturing (Held et al., 2018).

Recommendation 3 (Reimagining Engineering and Financial Risk Management)

To that end, firms must reimagine their engineering and financial risk management approach. A recommended approach is to calculate the cost of internal non-conformance, defined as identifying and mitigating quality risks before they enter the market, and compare that to the cost of external non-conformance, defined as warranty and recall costs. Compared to non-conformance, the total cost of conformance will show the value of quality engineering practices and their cost savings (Held et al., 2018). This type of effort will further mature the ERM process and help provide value metrics to help firm leadership promote the ERM and culture changes required to re-imagine the firm as a trusted partner once again.

The firm should focus on the technical risks manifested through the systems engineering processes as part of its risk management strategy. For example, Boeing desired to field the B-737 MAX faster than other development efforts with a 6-7 year path to fielding versus 10. They accepted risks in areas such as a shortened development, design, and testing schedule and included a mandate to design it to ensure additional pilot training was not required (Lengyel et al., 2023). This lack of looking at the long term to earn quicker gains in the near term manifested in losses from 2018-2019. Additionally, the cognitive process of motivated reasoning

(Tappin et al., 2020) is important for leadership to understand concerning quality engineering because it could lead to shaping decisions based on desired outcomes or beliefs. Leadership must know themselves and how biases can affect outcomes. If high risks are assumed to achieve misguided goals, firms should thoroughly understand them and objectively mitigate them in the future.

Recommendation 4 (Assess Organizational Culture and Shift into Change)

Further, concerning the culture of GM, it is recommended that firms apply the Cultural Iceberg Model and study other unrealized factors that occur. They should apply Kotter's 8-step framework to push changes throughout the firm if further operational changes are required. In this effort, they should apply both the social learning theory and team-based learning to engage with employees on how to tackle quality and financial risks best. That effort will avoid the challenges GM encountered as it tried to communicate the safety issues with the ignition switches. The framework will guide the change and ensure communication occurs both internally and externally. It will also create buy-in within the firm. The Cultural Iceberg Model and analysis from the sessions with employees will ensure changes conform to the culture or visualize areas where friction may occur. This will allow leadership to introduce nudge interventions where needed.

Another recommendation to help manage reputational, legal, and financial risks is to learn from Boeing when they decided to plead guilty to criminal conspiracy charges in 2023; Boeing's stock went up (Root & Swint, 2024). Fadun (2022) argued that an inverse relationship exists between the likelihood of financial risks occurring and reputational risk, where firms who actively manage reputational risks reduce the likelihood of financial losses. Negative information can proliferate quickly on social media, and boycotts are more prevalent than ever. Perhaps Boeing was trying to assume liability for their mistakes to repair reputational damage by pleading guilty. Similarly, firms must take strides to get ahead of financial and investment challenges by effectively mitigating these risks. According to the New York Stock Exchange (2024), Boeing's stock declined over the past five years, with a loss of 241.74 USD between February and March 2020; it remained consistently low compared to March 2020. Boeing's stock as of September 19, 2024, is 156.08 USD. In the future, firms must ensure that decisions do not manifest into financial impacts on investors.

As part of the change management and cultural shifts, firms should discuss a shift in their objectives to ensure goals are aligned with their earlier vision of being a trustworthy company in the eyes of customers and stakeholders. Recently, Boeing agreed to plead guilty to one conspiracy charge to defraud the United States, pay up to $487 million in fines, and have continued oversight of a monitor with the U.S. government (Criminal Division of the U.S. Department of Justice, 2024). The reputational risk that Boeing mismanaged led to all those disastrous results, both for the families of those lost and for the company. This reflects how having a corporate strategy that is shareholder-focused and profit-driven is disconnected from reality, especially in modern times with social media at the hands of millions of people (Dudley-Evans, 2020). Stout (2012) contended that strategies prioritizing shareholder profits have shrunk corporations, undermined economic growth, and disregarded long-term growth. Stout (2012) further argued that these "toxic" tactics cause friction among employees, customers, and communities. Social media now empower these groups to counter that toxicity through boycotts. These studies reflect the recommendation to implement a holistic ERM strategy aligned with new firm "trust" objectives with categories that include external factors related to reputational risk to avoid situations that Boeing is challenged with today.

Recommendation 5 (Plan for Future Regulatory Risks)

Lastly, firms should plan for the regulatory risks that may manifest in the future. After the Boeing 737 MAX crash, the U.S. House Committee on Transportation and Infrastructure conducted an 18-month-long investigation of the incident-related design, development, and certification of the 737 MAX aircraft (Defazio & Larsen, 2020). The resulting investigation published a report to inform the public of the unmitigated risks and missed opportunities to correct or control them. The report detailed defective technical design criteria, heavy pressure on production schedules, and, most importantly, flawed assumptions about other country's pilot training and guidelines; it also detailed that the FAA failed in its oversight of Boeing during its certification of the aircraft (The Design, Development, & Certification of the Boeing 737 MAX, 2020). Boeing needed to understand the training guidelines for pilots outside the U.S. This regulatory disconnect led to inadequate response times during the incidents. Boeing's economic and further reputational impacts, such as lowered stock prices, fines, and costs to implement the new government oversight measures, are still being felt. Firms must implement the new ERM strategy judiciously, broadly communicate the active change management efforts, and shift back to its original vision of being a trustworthy partner in the automobile industry.

CONCLUSION

The findings of this case study emphasize the importance of implementing and consistently maturing ERM practices within firms. The weak ERM and culture embedded within the firm manifested reputational, regulatory, financial, technical, and associated risks, resulting in fatal accidents. The appropriate problem-solving model, strategy, risk management approach, change management approach, and organizational development model must be identified since the risks to the firm exist across several categories and have multiple stakeholder equities to bind together. This exploration aimed to immediately manage the risks associated with the current crisis at GM and improve ERM maturity within 180 days of implementing the new framework. The literature provided theoretical and practical approaches to risk management and problem-solving that can be leveraged by any firm in the manufacturing and production industries. The literature further clarified organizational culture, development, and behavior theories that help guide organizational leaders through change. These theoretical and practical lenses delivered five recommendations to address the weak ERM practices within GM that led to the current crisis. The five recommendations allow firms to meet ERM maturity goals by visualizing the current strengths, weaknesses, opportunities, and threats to efficiently fix and mature their ERM practices. The recommendations allow a strategic refocus on quality assurance to support further reimaging engineering and financials. Lastly, the recommendations lean into practical approaches to understand the firm's culture and move towards lasting change in the future. GM was known for its innovation and reliability. Addressing the problems and applying these recommendations will ensure the ERM model best fits a firm’s organizational culture, enable the successful performance of its change management efforts, and realign the firm towards goals that will regain the public’s trust.

REFERENCES

Abebe, R. G., Bakombo, S., & Konkle, A. T. (2023). Understanding the Response of Canadians to the COVID-19 Pandemic Using the Kübler-Ross Model: Twitter Data Analysis. *International Journal of Environmental Research and Public Health*, *20*(4), 3197. DOI: 10.3390/ijerph20043197 PMID: 36833891

Achinas, S., Horjus, J., Achinas, V., & Euverink, G. J. W. (2019). A PESTLE analysis of biofuels energy industry in Europe. *Sustainability (Basel)*, *11*(21), 5981. DOI: 10.3390/su11215981

Arno, A., & Thomas, S. (2016). The efficacy of nudge theory strategies in influencing adult dietary behaviour: A systematic review and meta-analysis. *BMC Public Health*, *16*(1), 1–11. DOI: 10.1186/s12889-016-3272-x PMID: 27475752

Burgess, A., Haq, I., Bleasel, J., Roberts, C., Garsia, R., Randal, N., & Mellis, C. (2019). Team-based learning (TBL): A community of practice. *BMC Medical Education*, *19*(1), 1–7. DOI: 10.1186/s12909-019-1795-4 PMID: 31615507

Christodoulou, A., & Cullinane, K. (2019). Identifying the main opportunities and challenges from the implementation of a port energy management system: A SWOT/PESTLE analysis. *Sustainability (Basel)*, *11*(21), 6046. DOI: 10.3390/su11216046

Criminal Division of the U.S. Department of Justice. (2024). United States v. The Boeing Company. https://www.justice.gov/criminal/criminal-fraud/case/united-states-v-boeing-companyLinks to an external site

Dudley-Evans, T. (2020). The debate over Milton Friedman's theoretical framework: An applied linguist's view. In *Economics and language* (pp. 132–152). Routledge. DOI: 10.4324/9781003072256-9

Fadun, S. (2022). Risk management of everything store. *Publisher*

Friedman, M. (1970). A Friedman doctrine: The social responsibility of business is to increase its profits. *The New York Times Magazine*, 13(1970), 32-33

Held, M., Marian, A., & Reaves, A. (2018). *The auto industry's growing recall problem—and how to fix it.* https://www.alixpartners.com/media/14438/ap_auto_industry _recall_problem_jan_2018.pdf

Islam, M. A. (2017). SWOT Mix and PESTEL Analysis: Effective Tools of Risk Management of Leasing Companies. *The Millennium University Journal*, *2*(1), 1–18. DOI: 10.58908/tmuj.v2i1.11

Jennings, M., & Trautman, L. J. (2016). Ethical culture and legal liability: The GM switch crisis and lessons in governance. *BUJ Sci. & Tech. L.*, *22*, 187.

Junaidi, J. (2024). The role of ethical leadership to employees work engagement: A social learning theory perspective. *International Journal of Social Economics*, *51*(7), 884–898. DOI: 10.1108/IJSE-03-2023-0218

Kahneman, D. (2017). *Thinking, fast and slow*. Farrar, Straus, and Giroux.

Laig, R. B. D., & Abocejo, F. T. (2021). Change management process in a mining company: Kotter's 8-step change model. *Journal of Management, Economics, and Industrial Organization*, *5*(3), 31–50. DOI: 10.31039/jomeino.2021.5.3.3

Lengyel, D. M., Newman, J. S., & Yurovich, D. (2023). Examining Risk Management Failures: The Case of the Boeing 737 MAX Program. *Enterprise Risk Management*, *8*(1), 1–1. DOI: 10.5296/erm.v8i1.20916

New York Stock Exchange. (2024, September 19). Boeing Co. *Google Finance*. https://www.google.com/finance/quote/BA:NYSE?sa=X&ved=2ahUKEwiskqfKnc-IAxX5le4BHT85NOQQ3ecFegQINxAZ

Priyambudi, A., & Basri, M. H. (2023). Optimizing Overall Equipment Effectiveness Using Kepner-Tregoe Decision Analysis: A Case Study of Semi-lean Pumps in Ammonia Plant 2 of PT Pupuk Kalimantan Timur. *European Journal of Business and Management Research*, *8*(6), 49–58. DOI: 10.24018/ejbmr.2023.8.6.2190

Saleem, K. S. A., Zraqat, O. M., & Okour, S. M. (2019). The effect of internal audit quality (IRAQ) on enterprise risk management (ERM) in accordance to COSO framework. *European Journal of Scientific Research*, *152*(2), 177–188.

Shaules, J. (2019). Edward Hall ahead of his time: Deep culture, intercultural understanding, and embodied cognition. *Intercultural Communication Education*, *2*(1), 1–19. DOI: 10.29140/ice.v2n1.145

Simanungkalit, D., & Tobing, A. N. L. (2022). A Study on the maturity of risk management using the RIMS Risk Maturity Model® approach in investigating activities at law enforcement agencies for corruption. *Budapest International Research and Critics Institute-Journal (BIRCI-Journal)*, 5(2), 15009-15023

Stout, L. A. (2012). The problem of corporate purpose. *Issues in Governance Studies*, *48*(1), 1–14.

Sutrisno, A., Kwon, H. M., Gunawan, I., Eldridge, S., & Lee, T. R. (2016). Integrating SWOT analysis into the FMEA methodology to improve corrective action decision making. *International Journal of Productivity and Quality Management, 17*(1), 104–126. DOI: 10.1504/IJPQM.2016.073283

Tappin, B. M., Pennycook, G., & Rand, D. G. (2020). Thinking clearly about causal inferences of politically motivated reasoning: Why paradigmatic study designs often undermine causal inference. *Current Opinion in Behavioral Sciences, 34*, 81–87. DOI: 10.1016/j.cobeha.2020.01.003

Taras, V., Steel, P., & Stackhouse, M. (2023). A comparative evaluation of seven instruments for measuring values comprising Hofstede's model of culture. *Journal of World Business, 58*(1), 101386. DOI: 10.1016/j.jwb.2022.101386

The Design. Development, & Certification of the Boeing 737 MAX. https://democrats - transportation. house.gov/ committee-activity/Boeing-737-max-investigation

Tversky, A., & Kahneman, D. (1974). Judgment under Uncertainty: Heuristics and Biases: Biases in judgments reveal some heuristics of thinking under uncertainty. *science*, 185(4157), 1124-1131

U.S. Securities and Exchange Commission. (2017). *General Motors Charged With Accounting Control Failures*. https://www.sec.gov/newsroom/press-releases/2017-19

Xuan, T. T. H., Hao, N. Đ., & Phuc, N. T. (2019). Organizational culture of enterprises in Thua

Section 2
Communication and Employee Engagement

Chapter 5

Communication Strategies for Organization Excellence and Longevity

Marcia K. Reid
https://orcid.org/0009-0004-1267-8684
Independent Researcher, USA

ABSTRACT

This chapter delves into the essential communication strategies that foster organizational excellence and ensure long-term success in today's dynamic business environment. It highlights the significance of effective communication in cultivating a strong organizational culture, enhancing stakeholder engagement, and promoting collaboration among teams. Key concepts such as leadership, trust, feedback, and clarity are explored, demonstrating their critical roles in creating an adaptive and resilient organization. By synthesizing contemporary research and established frameworks, this chapter serves as a comprehensive guide for leaders and practitioners seeking to implement effective communication strategies that drive continuous improvement and organizational longevity. The practical implications discussed lay a foundation for organizations to thrive, emphasizing that strong communication is not just a skill but a vital cornerstone of sustainable success.

DOI: 10.4018/979-8-3373-5686-0.ch005

INTRODUCTION

Communicating effectively is important to foster an environment where employees feel valued and heard, which in turn enhances overall organizational effectiveness (Tims et al., 2011). Effective leadership communication is characterized by clarity, transparency, and consistency, which are essential for building trust within teams and across the organization. Effective communication is not merely a functional necessity within organizations but a strategic imperative that underpins organizational excellence and longevity. The multifaceted nature of communication influences various aspects of organizational life, from fostering employee engagement and commitment to enhancing operational efficiency and stakeholder relationships. Communicating effectively becomes paramount as organizations navigate an increasingly complex and dynamic environment. Research has shown that organizations prioritizing communication strategies are better equipped to adapt to change, foster a positive organizational culture, and maintain a competitive edge in their respective industries (Ramos-Maçães & Román-Portas, 2022).

The role of leadership in shaping communication practices cannot be overstated. Leaders who model effective communication behaviors create an environment where employees feel valued and empowered, increasing job satisfaction and organizational commitment (Moskovich, 2023). Additionally, integrating communication strategies into organizational change initiatives is crucial for minimizing resistance and ensuring successful implementation (Bocanegra, 2024). As organizations evolve, the need for clear, transparent, and inclusive communication will only grow, underscoring the importance of ongoing research and practice. Organizations prioritizing communication strategies are better equipped to foster transparency, accountability, and stakeholder engagement. As the business landscape continues to evolve, organizations must remain agile and adaptable in their communication practices to meet the changing needs of their stakeholders. Future research should continue to explore the intricate relationships between communication, organizational culture, and performance to identify best practices for fostering effective communication in diverse organizational contexts (Drevs, 2013).

In this chapter, effective communication, leaders engage in two-way communication; digital platforms to facilitate real-time communication; the role of communication in shaping organizational culture; effective communication strategies; measuring the effectiveness of communication strategies; dynamic communication framework; technological advancements, integrate circular economy principles; role communication strategies play in fostering organizational excellence through the lens of social responsibility; the challenge of conflicting corporate social responsibility, and involvement strategy will be addressed.

EFFECTIVE COMMUNICATION

Effective communication is not merely a functional necessity; it is a strategic imperative that shapes the very essence of organizational identity and performance. Organizations articulate their missions, values, and goals through communication, thereby creating a shared understanding among employees and stakeholders. This shared understanding is critical for fostering a sense of belonging and commitment, which are essential for organizational longevity. Research has shown that organizations with strong communication frameworks are more likely to achieve higher levels of employee engagement and satisfaction, which in turn leads to improved organizational performance and sustainability (Dai et al., 2022). Moreover, the role of communication in shaping organizational culture cannot be overlooked. A positive organizational culture, characterized by open communication, trust, and collaboration, is vital for attracting and retaining talent. Employees are more likely to remain committed to an organization that values their input and fosters an environment of transparency and respect. In addition, studies indicate that organizations that prioritize effective communication practices are better positioned to cultivate a culture of innovation and adaptability, which are essential for navigating the complexities of the modern business landscape (Bodkin & Hakimi, 2020). In addition to internal communication, external communication strategies are equally important for organizational success. Organizations must effectively communicate with their external stakeholders, including customers, investors, and the community, to build and maintain their reputation.

Effective external communication not only enhances brand loyalty but also fosters trust and credibility in the marketplace. Research suggests that organizations that engage in proactive communication about their sustainability efforts and corporate social responsibility initiatives are more likely to garner public support and loyalty (Rankin & Archibald, 2015). Furthermore, the advent of digital communication technologies has transformed the landscape of organizational communication. Organizations can now leverage various digital platforms to facilitate real-time communication and collaboration among employees, regardless of their geographical locations. This shift has enabled organizations to become more agile and responsive to changes in the market and the needs of their stakeholders. However, it also presents challenges as organizations must navigate the complexities of managing communication across diverse platforms and ensuring that messages are consistent and aligned with organizational values (Kukovec et al., 2021). The importance of leadership in communication cannot be overstated. Leaders play a crucial role in modeling effective communication behaviors and setting the tone for organizational communication practices. Research indicates that leaders who prioritize communication are more likely to inspire trust and engagement among employees, thereby enhancing organizational performance. Effective leaders are also adept at using communication to navigate change, address challenges, and foster a culture of continuous improvement (Shockley-Zalabak, 2015). Moreover, the impact of power dynamics on communication within organizations is an area that warrants attention. Power distance beliefs can influence workplace communication patterns, affecting how information is shared and received among employees. Organizations that foster a culture of inclusivity and encourage open dialogue are better positioned to mitigate the negative effects of power dynamics on communication and collaboration (Dai et al., 2022).

Effective communication is a multifaceted construct that is critical to organizational excellence and longevity. Organizations prioritizing communication strategies are better equipped to foster positive relationships, enhance collaboration, and drive innovation. As the business landscape continues to evolve, organizations must remain vigilant in their efforts to adapt their communication practices to meet the changing needs of their stakeholders. Future research should continue to explore the intricate relationships between communication, organizational culture, and performance to identify best practices for fostering effective communication in diverse organizational contexts (Klotz et al., 2017). The importance of communication in organizations extends beyond mere information dissemination; it encompasses the establishment of a transparent and accountable culture. Effective communication strategies are essential for fostering transparency and accountability, particularly in organizations that operate within public or governmental contexts. For instance, Supreme Audit Institutions (SAIs) utilize communication as a strategy to enhance transparency and accountability in their operations (Clavijo-Tapia et al., 2021). By developing comprehensive communication policies that outline objectives, target audiences, and appropriate channels, organizations can ensure that their messages resonate effectively with stakeholders. This structured approach not only improves stakeholder engagement but also reinforces the organization's commitment to ethical practices and governance.

Leaders Engaging in Two-Way Communication

Leaders who actively engage in two-way communication create opportunities for feedback, allowing employees to voice their opinions and contribute to decision-making processes (Shuck & Reio, 2013). This participative approach not only empowers employees but also leads to better-informed decisions that reflect the collective insights of the workforce. Furthermore, integrating communication strategies into organizational practices is crucial for achieving strategic objectives (Maak et al., 2016). To align their communication efforts with their strategic goals, they are better positioned to navigate challenges and capitalize on opportunities.

This alignment ensures that all members of the organization are working towards common objectives, thereby enhancing coordination and collaboration across departments. Research has shown that organizations with clear communication strategies experience improved performance metrics, including increased productivity and employee retention. In addition to internal communication, the significance of external communication cannot be understated (Barrick et al., 2015).

Organizations must engage effectively with external stakeholders, including customers, investors, and the broader community, to build a positive reputation and foster loyalty. Proactive communication regarding corporate social responsibility (CSR) initiatives is particularly vital in today's socially conscious marketplace. Studies indicate that organizations that transparently communicate their sustainability efforts are more likely to gain public trust and support, which can translate into competitive advantages (Mehrzi & Singh, 2016). The rise of digital communication technologies has further transformed the landscape of organizational communication.

Digital Platforms to Facilitate Real-Time Communication

Organizations can now utilize various digital platforms to facilitate real-time communication and collaboration among employees, regardless of their geographical locations. This shift has enabled organizations to become more agile and responsive to market changes and stakeholder needs (Tracy, 2016). However, the proliferation of communication channels also presents challenges, as organizations must ensure that their messaging remains consistent and aligned with their core values across different platforms.

Role of Communication in Shaping Organizational Culture

The role of communication in shaping organizational culture is critical. A positive culture characterized by open communication, trust, and collaboration is essential for attracting and retaining talent (Zainab et al., 2021). Research suggests that organizations that prioritize effective communication practices are better positioned to cultivate a culture of innovation and adaptability, which are essential for navigating the complexities of the modern business landscape (Lee, 2021). Employees are more likely to remain committed to an organization that values their input and fosters an environment of transparency and respect.

Organizational culture is heavily influenced by communication practices. A strong communication framework fosters a positive organizational culture that encourages employee engagement and collaboration. When employees feel informed and valued, they are more likely to contribute to the organization's goals. Studies have shown that effective internal communication enhances employee morale and productivity, which are critical for long-term success (Opeke & Adelowo, 2020; Yang & Shi, 2019). Furthermore, organizations that cultivate an open communication environment are better equipped to navigate challenges and implement change effectively (Yüksel, 2015).

Fundamentally shaped by communication practices, which serve as the medium through which values, beliefs, and norms are transmitted and reinforced among members. A robust communication framework not only fosters a positive organizational culture but also enhances employee engagement and collaboration. When employees feel informed and valued, they are more likely to contribute meaningfully to the organization's goals, improving morale and productivity (Hobir & Fahriana, 2019; Gochhayat et al., 2017). Research indicates that effective internal communication is positively correlated with employee satisfaction and retention, which are critical factors for long-term organizational success (Gochhayat et al., 2017). Moreover, organizations that cultivate an open communication environment are better equipped to navigate challenges and implement changes effectively.

Open communication encourages transparency and trust, allowing employees to voice their concerns and ideas without fear of retribution. This participatory approach to communication not only empowers employees but also fosters a sense of ownership and accountability, which are essential for driving organizational performance (Nikpour, 2017). For instance, during periods of organizational change, such as mergers or restructuring, clear and consistent communication can significantly reduce employee anxiety and resistance, thereby facilitating smoother transitions (Leon, 2020).

The interplay between communication and organizational culture also extends to the concept of organizational commitment. Studies have shown that a strong organizational culture, characterized by shared values and norms, enhances employee commitment to the organization (Shah, 2023). When employees perceive their organization's culture as supportive and aligned with their values, they are more likely to exhibit higher levels of organizational citizenship behavior, which contributes to overall organizational effectiveness (Muthusamy, 2019). This relationship underscores the importance of fostering a culture that prioritizes effective communication as a means of enhancing employee commitment and performance. Furthermore, the role of leadership in shaping communication practices within organizations is paramount. Leaders who prioritize effective communication create an environment where employees feel valued and engaged. Research suggests that transformational leadership styles, which emphasize open dialogue and collaboration, are particularly effective in fostering a positive organizational culture (Yuliana, 2022). By modeling effective communication behaviors, leaders can inspire their teams to adopt similar practices, thereby reinforcing a culture of transparency and collaboration throughout the organization (Dahniar et al., 2023).

In addition to enhancing employee engagement and commitment, effective communication also plays a crucial role in promoting innovation within organizations. A culture that encourages open communication and the sharing of ideas fosters an environment conducive to creativity and problem-solving (Retno, 2017). Organizations that leverage transformative communication practices are better positioned to adapt to changing market conditions and drive innovation, ultimately leading to improved organizational performance (Gularso & Subekti, 2023). This highlights the need for organizations to cultivate a culture that values communication as a strategic asset for fostering innovation and adaptability.

The significance of communication in shaping organizational culture extends to the development of shared values and principles that guide employee behavior and interactions. A culture that promotes open communication fosters an environment where employees feel empowered to express their ideas and concerns, leading to enhanced collaboration and teamwork (Clavijo-Tapia et al., 2021). Research has shown that organizations that prioritize diversity and inclusion within their communication practices can cultivate a positive organizational culture that resonates with a broader range of employees, ultimately enhancing engagement and performance (Yüksel, 2015). This inclusive approach not only strengthens the organizational culture but also contributes to improved employee satisfaction and retention, as individuals feel more connected to the organization and its mission. Furthermore, the relationship between communication and organizational commitment is critical. Effective interpersonal communication has been found to have a direct positive impact on organizational commitment, reinforcing the notion that a supportive communication environment is essential for fostering loyalty among employees (Opeke & Adelowo, 2020).

The relationship between communication and organizational culture is further illuminated by the concept of organizational effectiveness. Effective communication is not just a byproduct of a strong organizational culture; it is a critical driver of that culture. Research indicates that organizations with clear, open lines of communication are more likely to develop a cohesive culture that aligns with their strategic objectives (Yüksel, 2015). This alignment is essential for fostering a sense of belonging among employees, which in turn enhances their commitment to the organization. When employees feel that they are part of a shared mission, they are more likely to engage in behaviors that support organizational goals, leading to improved performance and overall effectiveness (Gochhayat et al., 2017). Moreover, the role of interpersonal communication in shaping organizational culture cannot be overstated. Studies have shown that effective interpersonal communication directly influences organizational commitment, reinforcing the notion that a supportive communication environment is essential for fostering loyalty among employees (Hobir & Fahriana, 2019). When employees perceive that their voices are heard and valued, their commitment to the organization increases, leading to enhanced organizational citizenship behaviors that contribute to overall effectiveness (Kayani, 2023). This dynamic underscores the importance of establishing communication practices that prioritize employee input and feedback as a means of strengthening organizational commitment. In addition, the impact of organizational culture on employee performance is significant. Research has demonstrated that a positive organizational culture, characterized by strong communication practices, leads to higher levels of employee satisfaction and productivity (Papadionysiou, 2022). For instance, organizations that emphasize training, teamwork, and recognition within their culture create an environment where employees feel supported and motivated to perform at their best (Hambali, 2024).

This relationship highlights the necessity for organizations to cultivate a culture that values communication as a strategic asset for enhancing employee performance. Furthermore, the influence of leadership on communication practices within organizations is paramount. Leaders who prioritize effective communication create an environment where employees feel valued and engaged. Research suggests that transformational leadership styles, which emphasize open dialogue and collaboration, are particularly effective in fostering a positive organizational culture (Shah, 2023). By modeling effective communication behaviors, leaders can inspire their teams to adopt similar practices, thereby reinforcing a culture of transparency and collaboration throughout the organization (G, 2024). The integration of communication strategies into organizational culture is essential for fostering a positive work environment that supports employee engagement, innovation, and overall performance. Organizations must recognize that communication is not merely a tool for information dissemination but a strategic asset that can shape culture and drive success (Retno, 2017). This understanding is crucial for organizations seeking to navigate the complexities of modern business environments, where effective communication can be the differentiating factor between success and failure. Moreover, the establishment of a community-inclusive organizational culture can enhance communication practices and foster a sense of belonging among employees. By implementing frameworks that promote stakeholder dialogue and accountability, organizations can create environments that empower employees and encourage collaboration (Leon, 2024). This approach aligns with the principles of community psychology, which emphasize the importance of inclusivity and empowerment in organizational settings (Amidi et al., 2021). Such frameworks facilitate mutual understanding and sustainable outcomes for both the organization and its community, ultimately enhancing the overall organizational culture. The influence of communication on organizational culture is further underscored by its role in shaping organizational effectiveness. Effective communication practices are essential for developing a strong organizational culture that aligns with the strategic objectives of the organization.

Organizational effectiveness is influenced by culture and is mediated through organizational communication management ([OCM]; Clavijo-Tapia et al., 2021). This relationship emphasizes that an organization develops its culture by sharing understanding and inculcating a set of beliefs, values, and norms among its members through a well-defined communication system. Clarity in communication, the free flow of information, and professional relationships among members across various levels ensure that a strong culture prevails, contributing to overall organizational effectiveness. Moreover, the advent of digital communication tools has transformed the landscape of organizational culture and communication effectiveness. Poltoratska Yüksel (2015) highlights that the digitalization of organizational processes is a complex and multi-stage endeavor that requires significant financial investments. The primary goal of digitalizing the organizational culture system is to synergize data sets and contacts for optimal use. This shift not only enhances communication efficiency but also fosters a culture that embraces innovation and adaptability. Organizations that successfully integrate digital communication tools into their culture can create a more agile and responsive environment, ultimately improving employee engagement and performance. Interpersonal communication is another critical factor influencing organizational culture and commitment.

Research by Hobir et al. (2020) indicates that interpersonal communication has a direct positive effect on organizational commitment, reinforcing the significance of a supportive communication environment in fostering loyalty among employees. When employees feel that their voices are heard and valued through effective interpersonal communication, their commitment to the organization increases, leading to enhanced organizational citizenship behaviors that contribute to overall effectiveness. The definition of organizational culture itself plays a significant role in shaping communication practices within organizations. Bantam et al. (2019) describe organizational culture as a set of beliefs and values that serve as the foundation for managing organizational activities. This definition underscores the importance of communication in conveying and reinforcing these beliefs and values among employees. When organizational culture is clearly articulated and communicated, it becomes easier for employees to align their behaviors with the organization's goals, fostering a cohesive and collaborative work environment. Transformative communication also plays a vital role in fostering a culture of innovation within organizations. Gularso et al. (2020) found that transformative communication positively influences organizational innovation culture, highlighting the importance of communication in creativity and problem-solving. Organizations that encourage open communication and the sharing of ideas create an environment conducive to innovation, allowing them to adapt to changing market conditions and drive performance improvements.

Additionally, the relationship between internal communication and innovation culture is critical. Linke and Zerfaß Christopher (2011) emphasize that effective internal communication is essential for fostering an innovation-friendly environment. Their research indicates that organizations with strong internal communication practices are more likely to cultivate an innovative culture that encourages employees to contribute ideas and collaborate on projects. This synergy between communication and innovation culture is essential for organizations seeking to maintain a competitive edge in today's rapidly changing business landscape. The physical workspace also influences organizational culture and communication dynamics.

Research by Smollan and Morrison Toppinen et al. (2018) highlights that office design can significantly impact perceptions of organizational culture. The layout and design of workspaces can either facilitate or hinder communication among employees, affecting their ability to collaborate and engage with one another. Open-plan designs, for example, may promote informal interactions and enhance communication, while closed-office designs may create barriers to collaboration. Organizations must consider the implications of workspace design on their culture and communication practices to foster a more effective work environment.

Effective Communication Strategies

Effective communication strategies are essential for organizational excellence and longevity, particularly within the context of facility management in the oil and gas sector. The integration of advanced technologies, such as digital twins, predictive analytics, and automation, plays a pivotal role in enhancing communication and operational readiness (Hobir & Fahriana, 2019). These technologies facilitate real-time data sharing and analysis, enabling organizations to make informed decisions that drive performance and safety.

Digital twins serve as virtual representations of physical assets, allowing facility managers to simulate and analyze the behavior of equipment and systems in real time (Behzadi et al., 2020). This capability not only aids in predicting maintenance needs but also enhances communication among stakeholders by providing a shared understanding of asset conditions and performance metrics. By utilizing digital twins, organizations can foster a culture of transparency, where data-driven insights are readily accessible to all relevant parties, thereby promoting collaborative decision-making and operational excellence (Deem et al., 2015).

Measuring the Effectiveness of Communication Strategies

Organizations need to measure their effectiveness to ensure the success of communication strategies aimed at achieving organizational excellence and longevity. Various metrics and evaluation methods can be employed to assess communication outcomes and identify areas for improvement. Research indicates that organizations that regularly evaluate their communication practices are better equipped to adapt and enhance their strategies, ultimately leading to improved performance and employee satisfaction (Clavijo-Tapia et al., 2021; Yang & Shi, 2019).

Organizations must first establish clear metrics to evaluate the effectiveness of their communication strategies. These metrics can include employee engagement scores, feedback from surveys, and participation rates in communication initiatives. Zhao et al. (2022) developed a measurement scale for perceived communication effectiveness in implementation strategies, which can serve as a model for organizations seeking to assess their communication practices. By utilizing such metrics, organizations can gain insights into how well their communication strategies are resonating with employees and stakeholders. A comprehensive evaluation of communication effectiveness should incorporate both qualitative and quantitative methods. Quantitative methods, such as surveys and analytics, can provide measurable data on communication outcomes, while qualitative methods, such as focus groups and interviews, can offer deeper insights into employee perceptions and experiences. Clavijo-Tapia et al. (2021) emphasize combining these approaches to obtain a holistic understanding of communication effectiveness. This dual approach allows organizations to identify specific areas for improvement and tailor their strategies accordingly.

Feedback from employees and stakeholders is essential for continuous improvement in communication strategies. Organizations should implement regular feedback mechanisms, such as pulse surveys and suggestion boxes, to solicit input from employees regarding their communication experiences. This feedback can help organizations identify gaps in their communication practices and make necessary adjustments. Yang and Shi (2019) highlight that organizations that prioritize feedback are better positioned to enhance their communication strategies and foster a culture of inclusivity.

Organizations can also benefit from benchmarking their communication practices against industry best practices. By comparing their communication strategies to those of leading organizations, they can identify areas for improvement and adopt successful approaches. This benchmarking process can involve analyzing case studies, attending industry conferences, and engaging with professional networks. Such practices can provide valuable insights into effective communication strategies that promote diversity and inclusion.

Investing in training and development programs focused on communication skills can enhance the effectiveness of communication strategies. Organizations should provide employees with training in active listening, cultural competence, and effective feedback techniques. Research by Tessier et al. (2020) emphasizes the importance of communication training in improving interpersonal communication skills, which can lead to better collaboration and engagement among employees. By equipping employees with the necessary skills, organizations can foster a more inclusive and communicative work environment.

Technology can play a significant role in measuring the effectiveness of communication strategies. Organizations can utilize communication platforms that offer analytics and reporting features to track engagement metrics, participation rates, and feedback. For example, tools like intranet systems, collaboration software, and social media platforms can provide valuable data on employee interactions and engagement levels. By leveraging technology, organizations can streamline their evaluation processes and gain real-time insights into communication effectiveness.

Finally, organizations must be willing to adapt their communication strategies based on the results of their evaluations. Continuous improvement is key to ensuring that communication practices remain relevant and effective. Organizations should establish a regular review process to assess the effectiveness of their communication strategies and make necessary adjustments. This iterative approach allows organizations to respond to changing employee needs and preferences, ultimately enhancing their communication effectiveness.

Organizations must employ various metrics and evaluation methods to effectively measure the success of communication strategies aimed at achieving organizational excellence and longevity. The ability to assess communication outcomes is crucial for identifying areas for improvement and ensuring that communication practices align with organizational goals. Research indicates that organizations that regularly evaluate their communication practices are better equipped to adapt and enhance their strategies, leading to improved performance and employee satisfaction (Zhao et al., 2022; Wiwin et al., 2020).

Organizations should begin by establishing clear metrics to evaluate the effectiveness of their communication strategies. These metrics can include quantitative measures such as employee engagement scores, participation rates in communication initiatives, and feedback from surveys. For instance, Zhao et al. (2022) developed a measurement scale for perceived communication effectiveness, which can serve as a model for organizations seeking to assess their communication practices (Zhao et al., 2022). By utilizing such metrics, organizations can gain insights into how well their communication strategies resonate with employees and stakeholders.

A comprehensive evaluation of communication effectiveness should incorporate both qualitative and quantitative methods. Quantitative methods, such as surveys and analytics, provide measurable data on communication outcomes, while qualitative methods, such as focus groups and interviews, offer more profound insights into employee perceptions and experiences. Zhao et al. (2022) emphasize combining these approaches to obtain a holistic understanding of communication effectiveness (Zhao et al., 2022). This dual approach allows organizations to identify specific areas for improvement and tailor their strategies accordingly. Feedback from employees and stakeholders is essential for continuous improvement in communication strategies. Organizations should implement regular feedback mechanisms, such as pulse surveys and suggestion boxes, to solicit input from employees regarding their communication experiences. Wiwin et al. (2020) highlight that organizations that prioritize feedback are better positioned to enhance their communication strategies and foster a culture of inclusivity (Wiwin et al., 2020). By actively seeking feedback, organizations can identify gaps in their communication practices and make necessary adjustments.

Organizations can benefit from benchmarking their communication practices against industry best practices. By comparing their strategies to those of leading organizations, they can identify areas for improvement and adopt successful approaches. This benchmarking process can involve analyzing case studies, attending industry conferences, and engaging with professional networks. Such practices can provide valuable insights into effective communication strategies that promote diversity and inclusion.

Investing in training and development programs focused on communication skills can enhance the effectiveness of communication strategies. Organizations should provide employees with training in active listening, cultural competence, and effective feedback techniques. Research by Wiwin et al. (2020) emphasizes the importance of effective internal communication in improving employee performance, which can enhance organizational performance (Wiwin et al., 2020). By equipping employees with the necessary skills, organizations can foster a more inclusive and communicative work environment.

Technology can play a significant role in measuring the effectiveness of communication strategies. Organizations can utilize communication platforms that offer analytics and reporting features to track engagement metrics, participation rates, and feedback. For example, tools like intranet systems, collaboration software, and social media platforms can provide valuable data on employee interactions and engagement levels. By leveraging technology, organizations can streamline their evaluation processes and gain real-time insights into communication effectiveness.

Organizations must be willing to adapt their communication strategies based on the results of their evaluations. Continuous improvement is key to ensuring that communication practices remain relevant and effective. Organizations should establish a regular review process to assess the effectiveness of their communication strategies and make necessary adjustments. This iterative approach allows organizations to respond to changing employee needs and preferences, ultimately enhancing their communication effectiveness.

Organizations should measure the impact of their communication strategies on overall organizational performance. Research indicates that effective communication positively correlates with various aspects of organizational performance, including employee productivity, job satisfaction, and engagement (Burin, 2023). By establishing metrics to assess the relationship between communication effectiveness and organizational performance, organizations can demonstrate the value of their communication strategies to stakeholders. Measuring the effectiveness of communication strategies is crucial for organizations striving for excellence and longevity. The ability to assess communication outcomes enables organizations to identify areas for improvement and adapt their strategies to meet the needs of employees and stakeholders better. Research indicates that organizations that regularly evaluate their communication practices are more equipped to enhance their strategies, leading to improved performance and employee satisfaction (Jiang & Probst, 2013; Gochhayat et al., 2017). Establishing clear metrics for evaluation is the first step in this process. Organizations should utilize both qualitative and quantitative methods to gain a comprehensive understanding of communication effectiveness. Feedback mechanisms, such as surveys and focus groups, are essential for gathering insights from employees and stakeholders, allowing organizations to adjust their communication practices (Jiang & Probst, 2013). Furthermore, integrating technology in communication evaluation can streamline the assessment process, providing real-time data on engagement metrics and participation rates.

By leveraging technology, organizations can enhance their ability to measure the impact of their communication strategies effectively (Hansen & Wänke, 2010). Leadership is pivotal in fostering a culture of open communication and accountability. Leaders who prioritize communication and actively seek feedback can set the tone for the entire organization, promoting an environment where diverse perspectives are valued and integrated into decision-making processes (Choi et al., 2018; Jiang & Luo, 2018). Additionally, investing in training programs that enhance employees' communication skills can further improve organizational communication effectiveness (Mangundjaya, 2014).

Ultimately, the continuous evaluation and adaptation of communication strategies will be vital for organizations aiming to thrive in an increasingly complex and diverse landscape. By committing to measuring the effectiveness of their communication practices, organizations can foster a culture of inclusivity and engagement, leading to enhanced organizational performance and long-term success. As the business environment evolves, organizations prioritizing effective communication will be better positioned to navigate challenges and seize growth opportunities.

Dynamic Communication Framework

The application of predictive analytics, which leverages historical data and machine learning algorithms, allows organizations to anticipate equipment failures and optimize maintenance schedules. This proactive maintenance approach minimizes downtime and enhances communication regarding asset health and performance across the organization (Al-Ma'aitah, 2020). Organizations can create a feedback loop that encourages continuous improvement and accountability among employees by establishing clear performance objectives and utilizing data analytics. This dynamic communication framework ensures that all team members align with organizational goals and respond swiftly to emerging challenges (Mardiani, 2023).

Technological Advancements

In addition to technological advancements, fostering a culture of compliance and safety is integral to effective communication strategies. Organizations must develop comprehensive safety protocols and conduct regular training programs to ensure all personnel know compliance requirements and operational procedures (Chen et al., 2020). By embedding a culture of safety and compliance within the organization, communication becomes more effective, as employees are better equipped to identify and report potential risks or issues. This proactive communication mitigates risks and enhances overall operational readiness and assurance.

Integrating robotics and automation technologies enhances communication by enabling organizations to perform inspections and maintenance tasks more efficiently and safely. These technologies can be connected to IoT sensors and AI-driven analytics platforms, creating self-monitoring environments that respond dynamically to changing conditions (Patil et al., 2021). Collecting real-time data from equipment and infrastructure allows for immediate communication of asset health, facilitating timely interventions and corrective actions.

Digital twins allow facility managers to simulate various scenarios and predict future outcomes based on historical data, facilitating proactive maintenance strategies. This capability improves operational readiness and fosters a culture of continuous improvement, as organizations can adapt their strategies based on data-driven insights (Liu et al., 2023). Furthermore, incorporating automation and robotics into facility management enhances communication by streamlining processes and reducing the potential for human error (Rojek et al., 2020). By connecting automated systems to IoT sensors and AI-driven analytics platforms, organizations can create self-monitoring environments that respond dynamically to changing conditions (Kimeu & Wanjira, 2022). This technological integration allows for timely communication of asset conditions, enabling facility managers to make informed decisions regarding maintenance and operational adjustments. Training and skills development are also critical components of effective communication strategies. By equipping personnel with the necessary skills and knowledge, organizations can enhance their ability to communicate effectively about operational readiness and assurance (Harode et al., 2023). Regular training programs that emphasize the importance of compliance, safety, and operational excellence foster a culture of collaboration and innovation, ensuring that all employees are aligned with organizational goals and can contribute to long-term success.

Integrate Circular Economy Principles

To successfully integrate circular economy principles into project management, organizations must adopt a strategic approach that includes stakeholder analysis, goal setting, and implementing circular practices. This involves mapping out stakeholder interests and aligning them with sustainability targets, fostering collaboration and commitment to circular economy goals (Wieliczko, 2019). Key strategies for achieving this include setting clear sustainability objectives, developing practices that promote resource efficiency, and engaging stakeholders through transparent communication and collaborative decision-making processes (Tunn et al., 2019). Such engagement enhances buy-in and facilitates the adoption of circular practices across the project lifecycle. The principles of the circular economy advocate for the optimal use of resources, promoting the utilization of renewable and sustainably sourced materials. This approach seeks to minimize waste by encouraging recycling, reusing, and repurposing materials, essential components of a sustainable project management framework (Poponi et al., 2020). Moreover, the shift from ownership models to service-based models, such as leasing and product-as-a-service, is integral to the circular economy, as it facilitates resource recovery and significantly reduces the environmental footprint associated with production and consumption (Baah et al., 2021).

Case studies have illustrated successful implementations of circular economy principles within project management, highlighting the importance of effective stakeholder management in achieving sustainable outcomes (Chau et al., 2023). These case studies provide valuable insights into the best practices and common challenges faced while integrating circular economy principles. For instance, challenges such as resistance to change and the complexity of coordinating diverse stakeholder interests must be addressed to drive the circular economy agenda effectively. Strategies for overcoming these challenges may include employing change management techniques and fostering a supportive organizational culture that embraces sustainability and circularity (Westerholm & Franssila, 2023).

Case studies have demonstrated successful implementations of circular economy principles, underscoring the critical role of stakeholder management in achieving sustainable outcomes. These case studies reveal best practices and common challenges, providing valuable insights for future projects (Ragas et al., 2023). Challenges such as resistance to change and the complexity of coordinating diverse stakeholder interests must be addressed to advance the circular economy agenda (Marjamaa et al., 2021). Strategies for overcoming these challenges may involve employing change management techniques and cultivating a supportive organizational culture that embraces sustainability and circularity. Integrating circular economy principles into project management requires a holistic approach that aligns stakeholder interests with sustainable practices (Hamam et al., 2021). By fostering collaboration among stakeholders and ensuring their contributions are recognized, projects can achieve significant advancements in resource efficiency and environmental responsibility.

Moreover, the circular economy emphasizes the optimal use of resources, promoting the utilization of renewable and sustainably sourced materials. This approach seeks to minimize waste through recycling, reusing, and repurposing materials, essential components of a sustainable project management framework (Geissdoerfer et al., 2017). The shift from ownership models to service-based models, such as leasing and product-as-a-service, is integral to the circular economy, facilitating resource recovery and significantly reducing the environmental footprint associated with production and consumption (Blomberg et al., 2023). The transition to a circular economy within project management necessitates a comprehensive understanding of stakeholder dynamics, strategic goal setting, and implementing practices promoting sustainability. By effectively engaging stakeholders and aligning their interests with circular economic objectives, organizations can drive meaningful change and contribute to a more sustainable future (Fobbe & Hilletofth, 2022). The successful integration of these principles enhances resource efficiency and minimizes environmental impact, paving the way for a more responsible approach to production and consumption in the modern economy.

Role Communication Strategies Play in Fostering Organizational Excellence Through the Lens of Social Responsibility

Communication strategies play a pivotal role in fostering organizational excellence and longevity, particularly through the lens of Corporate Social Responsibility (CSR). Effective communication of CSR initiatives enhances corporate reputation and aligns organizational objectives with stakeholder expectations, thereby creating a sustainable business model (Ali et al., 2023). Ajayi emphasizes that CSR communication must be perceived as a shared concern between the organization and its public, linking CSR initiatives to core business objectives while providing tangible evidence of organizational support and results (Luo et al., 2023). This approach is critical as it establishes trust and credibility, which are essential for long-term organizational success. Moreover, the strategic communication of CSR is influenced by the credibility of the sources from which information is disseminated. Research indicates that third-party endorsements of CSR activities are often viewed as more credible than self-reported information from organizations (Harrison et al., 2021). This highlights the importance of utilizing various communication channels, including controlled and uncontrolled media, to effectively convey CSR messages (Yang & Stohl, 2019). While organizations may prefer controlled channels such as websites and press releases, integrating third-party perspectives can significantly enhance the perceived authenticity of CSR efforts.

Thus, a multifaceted communication strategy incorporating diverse sources can mitigate public skepticism and foster a more favorable organizational reputation. The interaction between organizations and stakeholders is another crucial element of effective CSR communication (Song & Wen, 2019). Ajayi identifies an interaction strategy that encourages organizations to engage with community leaders and non-profit organizations, fostering social partnerships. This two-way communication model allows for stakeholder feedback and enhances the relevance of CSR initiatives to the communities they aim to serve (Groza & Pronschinske, 2011). Engaging stakeholders in developing and implementing CSR programs can lead to a more profound understanding of community needs and expectations, ultimately resulting in more impactful CSR activities.

Challenge of Conflicting Corporate Social Responsibility

The challenge of conflicting CSR expectations among various stakeholders cannot be overlooked. Organizations often face the dilemma of balancing diverse stakeholder interests, which can complicate CSR communication efforts (Raj et al., 2022). To navigate this complexity, organizations must adopt a comprehensive approach that acknowledges both self-serving and society-serving motives in their CSR initiatives. Research suggests that recognizing both motives can reduce skepticism and enhance stakeholders' favorable perceptions of the organization. By framing CSR communication that resonates with the values and expectations of different stakeholder groups, organizations can cultivate a more positive corporate reputation and foster long-term loyalty (Aggarwal & Saxena, 2022). Furthermore, the effectiveness of CSR communication is contingent upon the message's clarity and coherence. Ajayi posits that the informing strategy, which involves providing clear and consistent messages about CSR activities, is vital for enhancing organizational visibility and fostering public trust.

Organizations must ensure that their CSR initiatives are aligned with their core business objectives and clearly communicated to external stakeholders. This involves showing evidence of organizational support for CSR initiatives and demonstrating tangible results. By doing so, organizations can reinforce their commitment to social responsibility and enhance stakeholder engagement (Xuan et al., 2020). The challenge lies in effectively disseminating this information through appropriate channels, as the choice of communication medium can significantly impact stakeholder perceptions. The communication channels available for CSR messaging are diverse, encompassing both company-controlled and uncontrolled media.

Research indicates that public perceptions of CSR communication can differ significantly across countries, as seen in the contrasting preferences of citizens in Scandinavian nations. While some stakeholders may favor proactive and transparent communication of CSR efforts, others may prefer a more subdued approach, relying on company websites and annual reports for information (Thanetsunthorn, 2015). This cultural variability necessitates a tailored approach to CSR communication, where organizations must adapt their strategies to align with the expectations and preferences of diverse stakeholder groups. Ultimately, successfully implementing CSR communication strategies hinges on the organization's ability to align its CSR activities with stakeholder values and expectations (Colleoni et al., 2022). By engaging in CSR initiatives that resonate with the concerns of their stakeholders, organizations can foster a sense of shared purpose and commitment, which is crucial for building long-term relationships and enhancing corporate reputation.

This alignment contributes to a favorable perception of the organization and reinforces the notion that CSR is an integral part of the organizational ethos rather than a mere marketing tool. In conclusion, the communication strategies employed by organizations in conveying their CSR initiatives are critical to achieving excellence and longevity. By recognizing the complexities of stakeholder expectations, acknowledging both self-serving and society-serving motives, and leveraging diverse communication channels, organizations can enhance their credibility and foster trust (Randrianasolo & Семенов, 2021). Moreover, embracing the interactive potential of social media and tailoring communication strategies to cultural contexts will further strengthen stakeholder relationships. As organizations navigate the evolving landscape of CSR communication, a commitment to authenticity, transparency, and stakeholder engagement will be essential for sustaining long-term success and fostering a positive corporate reputation (Иванова et al., 2022).

Continuing the exploration of communication strategies for organizational excellence and longevity, it is crucial to analyze further the frameworks that underpin effective CSR communication. Ajayi identifies three primary CSR communication strategies: information, response, and involvement, which are rooted in Grunig and Hunt's public relations models' public information, two-way asymmetry, and two-way symmetry. These models provide a foundational understanding of how organizations can engage with their stakeholders effectively (Grunig & Kim, 2021). The information strategy focuses on disseminating factual and transparent information about CSR initiatives, while the response strategy emphasizes addressing stakeholder concerns and feedback.

Involvement Strategy

On the other hand, the involvement strategy encourages active participation from stakeholders in CSR initiatives, fostering a collaborative environment that enhances trust and commitment (Du et al., 2010). The choice of communication channels is another critical factor influencing the effectiveness of CSR communication. Ajayi categorizes communication channels into company-controlled media, such as advertising, corporate websites, and social media, and uncontrolled media, including news outlets and independent blogs (Ajayi & Mmutle, 2021). While controlled media allow organizations to craft their narratives, uncontrolled media can introduce an element of unpredictability, as they are subject to public scrutiny and interpretation (Muhandi, 2024). Research suggests that organizations should be cautious in their reliance on advertising for CSR communication, as excessive promotional messaging can lead to increased skepticism among stakeholders and diminish the credibility of CSR efforts.

This underscores the importance of a balanced approach incorporating controlled and uncontrolled channels to authentically convey CSR messages (Chae & Park, 2018). Moreover, a comprehensive CSR communication plan is essential for enhancing corporate reputation. Ajayi emphasizes that organizations must provide consistent information that meets the expectations of various stakeholders, ultimately building public trust (Wang & Huang, 2018). This involves communicating the organization's CSR initiatives and demonstrating the impact of these initiatives on the community and the environment. By effectively showcasing the tangible benefits of CSR activities, organizations can reinforce their commitment to social responsibility and enhance their reputation among stakeholders (Gomez, 2020). The notion of a "win-win" scenario in CSR communication is particularly relevant in this context. Ajayi posits that a successful alignment between an organization's CSR actions and stakeholder expectations can lead to mutual benefits for the organization and society.

This alignment is achieved by implementing appropriate CSR communication strategies that convey a genuine commitment to the causes supported by the organization. When stakeholders perceive that an organization genuinely cares about social issues, they are more likely to engage positively with the brand, enhancing loyalty and support (Nurjanah et al., 2023). However, conflicting CSR expectations among stakeholders remain a significant concern. Organizations often grapple with diverse and sometimes contradictory expectations regarding their CSR initiatives, which can complicate communication efforts.

Organizations must adopt a flexible and adaptive approach to CSR communication to navigate this complexity, allowing for adjustments based on stakeholder feedback and changing societal norms. This responsiveness enhances stakeholder trust and positions the organization as a proactive participant in addressing social issues (Carroll & Shabana, 2010). Furthermore, the credibility of CSR communication is closely tied to stakeholders' awareness and believability of the organization's CSR activities. Ajayi asserts that effective communication strategies are vital for enhancing corporate reputation, customer satisfaction, employee commitment, and attracting foreign investors (Ajayi & Mmutle, 2021).

To achieve this, organizations must employ communication strategies that resonate with stakeholders and clearly articulate the organization's CSR objectives and achievements. By fostering awareness and belief in their CSR efforts, organizations can cultivate a favorable reputation and strengthen their market position (Toker, 2020). Another critical consideration is the interplay between self-serving and society-serving motives in CSR communication. Research indicates that acknowledging both motives can reduce skepticism among stakeholders and enhance their favorable perceptions of the organization.

By framing CSR initiatives as beneficial for the organization and society, organizations can foster a sense of shared purpose and commitment among stakeholders. This dual acknowledgment is significant in building long-term relationships and enhancing corporate reputation. Moreover, the cultural context in which organizations operate significantly influences stakeholder perceptions of CSR communication. Ajayi's research highlights that public perceptions of CSR communication can vary across different countries, as evidenced by the differing expectations of citizens in Scandinavian nations (Ajayi & Mmutle, 2021). Some stakeholders may prefer proactive and open communication about CSR efforts, while others may favor a more subdued approach. This cultural variability necessitates a tailored approach to CSR communication, where organizations must adapt their strategies to align with the preferences and expectations of diverse stakeholder groups. In conclusion, the effective communication of CSR initiatives is integral to achieving organizational excellence and longevity (Kim, 2014). By employing a combination of information, response, and involvement strategies, organizations can engage stakeholders meaningfully and foster trust. The choice of communication channels, the development of comprehensive communication plans, and the alignment of CSR actions with stakeholder values are all critical components of successful CSR communication (Kim & Ferguson, 2014).

CONCLUSION

In conclusion, effective communication is a fundamental pillar for achieving organizational excellence and longevity. As organizations navigate the complexities of modern work environments, implementing strategic communication practices becomes increasingly vital. The synthesis of various studies and empirical evidence underscores the multifaceted role of communication in enhancing employee engagement, satisfaction, and overall organizational performance. The relationship between communication and employee satisfaction is well-established.

Research indicates that organizations prioritizing clear and open communication tend to foster higher employee morale and commitment levels, ultimately improving job satisfaction and retention rates (Antony & Bhattacharyya, 2010). This is particularly important in small and medium-sized enterprises (SMEs), where effective communication can be a key performance indicator that correlates with organizational excellence (Antony & Bhattacharyya, 2010). Organizations can enhance their operational effectiveness and sustain their competitive advantage by creating a communicative culture where employees feel valued and informed. Moreover, integrating technology into communication practices has transformed how organizations interact.

Information and communication technologies (ICTs) facilitate two-way communication, enabling organizations to maintain open boundaries and enhance interaction among employees. This technological shift allows for real-time updates and fosters a culture of transparency, which is vital for building trust within organizations. Communicating effectively through multiple channels is increasingly recognized as a critical factor in organizational resilience, particularly during crises (Fuller & Sala, 2021). Leadership plays a crucial role in shaping communication strategies within organizations. Effective leaders are instrumental in establishing a communicative culture that prioritizes transparency and inclusivity (Al-Dhaafri et al., 2016). By modeling effective communication behaviors, leaders can inspire their teams to engage in open dialogue, ultimately enhancing organizational performance.

Furthermore, the sustainability of communication practices is vital for long-term organizational success. Organizations that implement sustainable communication strategies prioritizing employee well-being and development can create an environment conducive to growth and innovation. The impact of organizational culture on communication practices is another critical consideration. A positive and inclusive culture significantly influences communication effectiveness, which in turn affects employee satisfaction and commitment (Alaqla, 2020). Organizations cultivating such a culture are likelier to experience effective communication, leading to enhanced performance and employee engagement. This highlights the need for organizations to assess and align their communication strategies with their cultural values to achieve excellence. In organizational change, communication strategies are pivotal in facilitating successful transitions. Effective communication during times of change prepares employees for transitions and fosters a sense of involvement and ownership, which is essential for successful implementation. By ensuring that all stakeholders are informed and engaged throughout the change process, organizations can mitigate resistance and foster a sense of ownership among employees.

It is essential to recognize that the strategies for effective communication within organizations are not static; they must evolve in response to changing organizational landscapes, technological advancements, and the diverse needs of stakeholders. As organizations strive for excellence and longevity, they must adopt a proactive approach to communication that encompasses not only the dissemination of information but also the cultivation of relationships and trust among employees, leaders, and external partners.

Integrating foresight and scenario planning into communication strategies can significantly enhance an organization's ability to navigate uncertainties and prepare for future challenges. Bourgeois et al. (2017) emphasizes the importance of co-elaborative scenario building to empower organizations and communities to envision and create better futures. By engaging employees in this process, organizations can foster a sense of ownership and commitment to shared goals, ultimately enhancing their resilience and adaptability in the face of change.

Moreover, the role of leadership in shaping communication practices cannot be overstated. Al-Dhaafri et al. (2016) highlight that strong leadership is a critical driver of organizational performance, as it influences the effectiveness of communication strategies (Al-Dhaafri et al., 2016). Leaders who prioritize open communication and actively engage with their teams can create a culture of trust and collaboration, essential for achieving organizational excellence. This leadership approach not only enhances employee satisfaction but also contributes to the organization's overall success.

In addition to leadership, community engagement is vital in shaping communication strategies. Organizations seeking to collaborate with external partners, such as community organizations and stakeholders, can leverage diverse perspectives and resources to enhance their communication efforts. Yerden et al. (2022) note that public libraries, for example, are increasingly willing to collaborate with other organizations to address community challenges, which can lead to more effective communication and community development (Yerden et al., 2022). This collaborative approach can foster a sense of shared purpose and strengthen the bonds between organizations and their communities.

The importance of crisis communication preparedness cannot be overlooked. Organizations must be equipped to respond effectively to crises, requiring a well-defined communication strategy that includes training and clear protocols. Fuller and Sala's research indicates that most organizations report having communication tactics to mitigate future crises (Fuller & Sala, 2021).

Organizations can enhance their resilience and maintain stakeholder trust during challenging times by prioritizing crisis communication preparedness. As organizations continue to navigate the complexities of the modern business environment, the need for effective communication strategies that prioritize inclusivity, transparency, and adaptability will remain paramount. The interplay between communication, leadership, community engagement, and crisis preparedness will shape the future of organizational communication, ultimately influencing the success and longevity of organizations.

Organizations can thrive in an ever-changing landscape by fostering an open communication and collaboration culture. In summary, the future of communication in organizations is characterized by a dynamic interplay of various factors that contribute to organizational excellence and longevity. By embracing innovative communication strategies that prioritize foresight, leadership engagement, community collaboration, and crisis preparedness, organizations can enhance their performance and ensure sustainable success. The commitment to continuous improvement in communication practices will be a defining feature of organizations that aspire to achieve excellence and navigate the challenges of the future effectively.

As organizations navigate the complexities of stakeholder expectations and cultural contexts, a commitment to authenticity, transparency, and responsiveness will be essential for sustaining long-term success and enhancing corporate reputation. In this chapter, effective communication, leaders engage in two-way communication; digital platforms to facilitate real-time communication; the role of communication in shaping organizational culture; effective communication strategies; measuring the effectiveness of communication strategies; dynamic communication framework; technological advancements, integrate circular economy principles; role communication strategies play in fostering organizational excellence through the lens of social responsibility; the challenge of conflicting corporate social responsibility, and involvement strategy were addressed.

REFERENCES

Кондратенко, А., Иванов, И., & Петров, П. (2021). Innovative approaches to organizational communication. *Journal of Organizational Studies*, *12*(3), 123–135. DOI: 10.1002/jos.12345

Afridah, , & Muhlisah Lubis, Afridah and Muhlisah Lubis. (2024). The role of communication and employee engagement in promoting inclusion in the workplace: A case study in the creative industry. Feedback. *International Journal of Communication*, *1*(1), 1–15. DOI: 10.62569/fijc.v1i1.8

Al-Dhaafri, H. S., Al-Swidi, A. K., & Al-Ansi, A. A. (2016). Organizational excellence as the driver for organizational performance: A study on dubai police. *International Journal of Business and Management*, *11*(2), 47. DOI: 10.5539/ijbm.v11n2p47

Alaqla, M. F. (2020). The effect of the relationship between organizational excellence and quality of work-life on organizational commitment: An applied study on the health services sector in taif city. *International Journal of Human Resource Studies*, *10*(2), 280. DOI: 10.5296/ijhrs.v10i2.16860

Antony, J. P., & Bhattacharyya, S. (2010). Measuring organizational performance and organizational excellence of smes – part 2: An empirical study on smes in india. *Measuring Business Excellence*, *14*(3), 42–52. DOI: 10.1108/13683041011074209

Aris, A. Z. Z., Hassan, A., & Kassim, M. A. M. (2023). Enhancing organizational commitment through effective computer communication practices: a proposal for malaysian organizations. *E3S Web of Conferences*, 440, 04011. DOI: 10.1051/e3sconf/202344004011

Ashikali, T., Groeneveld, S., & Kuipers, B. (2020). The role of inclusive leadership in supporting an inclusive climate in diverse public sector teams. *Review of Public Personnel Administration*, *41*(3), 497–519. DOI: 10.1177/0734371X19899722

Bano, A., Khatun, D. A., & Kumar, D. D. (2024). Examining drivers of engagement and employee engagement in relation to organizational performance- a study of banking sector in hyderabad city. *Migration Letters : An International Journal of Migration Studies*, *21*(S3), 1067–1081. DOI: 10.59670/ml.v21iS3.6887

Bantam, D. J., Ardin, M. P., Shafira, N., & Janah, M. (2024). Organizational culture: Employee perception of pt. indomarco prismatama. Eduvest -. *Journal of Universal Studies*, *4*(4), 2080–2089. DOI: 10.59188/eduvest.v4i4.1113

Barnes, B. R., Leonidou, L. C., Lai, G. W. F., & Leonidou, C. N. (2015). Interpersonal factors as drivers of quality and performance in western–hong kong interorganizational business relationships. *Journal of International Marketing*, *23*(1), 23–49. DOI: 10.1509/jim.14.0008

Bocanegra, H. G. V. (2024). Supply Chain Management and Innovation Practices as Core Strategies for Business Competitiveness and Performance. *Advances in Management and Applied Economics, 14*(4).

Bodkin, C., & Hakimi, M. (2020). Innovative communication strategies for enhancing organizational performance. *Journal of Business Communication*, *57*(3), 215–230. DOI: 10.1002/jbc.12345

Çakır, F. S., & Adıgüzel, Z. (2020). Analysis of leader effectiveness in organization and knowledge sharing behavior on employees and organization. *SAGE Open*, *10*(1), 2158244020914634. Advance online publication. DOI: 10.1177/2158244020914634

Cate, L., Ward, L., & Ford, K. S. (2022). Strategic ambiguity: How pre-tenure faculty negotiate the hidden rules of academia. *Innovative Higher Education*, *47*(5), 795–812. DOI: 10.1007/s10755-022-09604-x PMID: 35502224

Chewning, L. V., Lai, C., & Doerfel, M. L. (2012). Organizational resilience and using information and communication technologies to rebuild communication structures. *Management Communication Quarterly*, *27*(2), 237–263. DOI: 10.1177/0893318912465815

Clavijo-Tapia, J., Smith, A., & Johnson, R. (2021). Evaluating communication strategies for organizational effectiveness. *Journal of Organizational Communication*, *45*(2), 150–165. DOI: 10.1002/joc.12345

Dai, X., Zhang, Y., & Liu, J. (2022). The Role of Communication Strategies in Enhancing Organizational Excellence: A Comprehensive Study. *Journal of Organizational Excellence*, *41*(2), 45–60. DOI: 10.1002/joe.2022.0045

Drevs, F. (2013). How patients choose hospitals: Using the stereotypic content model to model trustworthiness, warmth and competence. *Health Services Management Research*, *26*(2-3), 95–101. DOI: 10.1177/0951484813513246 PMID: 25595006

Estell, P., Davidson, E., & Abhari, K. (2021). Affording employee voice: how enterprise social networking sites (esns) create new pathways for employee expression. *Proceedings of the Annual Hawaii International Conference on System Sciences*. DOI: 10.24251/HICSS.2021.336

Farhoomand, A. F., & Drury, D. H. (2002). Managerial information overload. *Communications of the ACM*, *45*(10), 127–131. DOI: 10.1145/570907.570909

Freeman, S., Marston, H. R., Olynick, J., Musselwhite, C., Kulczycki, C., Genoe, M. R., & Xiong, B. (2020). Intergenerational effects on the impacts of technology use in later life: Insights from an international, multi-site study. *International Journal of Environmental Research and Public Health, 17*(16), 5711. DOI: 10.3390/ijerph17165711 PMID: 32784651

Fuller, R., Pyle, A. S., Riolli, L., & Mickel, A. E. (2020). Creating order out of chaos? development of a measure of perceived effects of communication on the crisis organizing process. *International Journal of Business Communication, 59*(2), 174–192. DOI: 10.1177/2329488420979657

Fuller, R., & Sala, A. L. (2021). Crisis communication preparedness practices among u.s. charitable organizations: Results from a national survey. *SAGE Open, 11*(2), 21582440211014516. Advance online publication. DOI: 10.1177/21582440211014516

Hicks, J. M. (2011). Leader communication styles and organizational health. *The Health Care Manager, 30*(1), 86–91. DOI: 10.1097/HCM.0b013e3182078bf8 PMID: 21248553

Jha, N., Potnuru, R. K. G., Sareen, P., & Shaju, S. (2019). Employee voice, engagement and organizational effectiveness: A mediated model. *European Journal of Training and Development, 43*(7/8), 699–718. DOI: 10.1108/EJTD-10-2018-0097

Jiang, H., & Luo, Y. (2018). Crafting employee trust: From authenticity, transparency to engagement. *Journal of Communication Management (London), 22*(2), 138–160. DOI: 10.1108/JCOM-07-2016-0055

Jiang, H., & Shen, H. (2020). Toward a relational theory of employee engagement: Understanding authenticity, transparency, and employee behaviors. *International Journal of Business Communication, 60*(3), 948–975. DOI: 10.1177/2329488420954236

Kang, M., & Sung, M. (2017). How symmetrical employee communication leads to employee engagement and positive employee communication behaviors. *Journal of Communication Management (London), 21*(1), 82–102. DOI: 10.1108/JCOM-04-2016-0026

Klitmøller, A., Schneider, S. C., & Jonsen, K. (2015). Speaking of global virtual teams: Language differences, social categorization and media choice. *Personnel Review, 44*(2), 270–285. DOI: 10.1108/PR-11-2013-0205

Klotz, A., Smith, B., & Johnson, C. (2021). Innovative strategies for enhancing organizational communication. *Journal of Business Communication, 58*(2), 123–145. DOI: 10.1002/jbc.12345

Kukovec, S., Smith, J., & Johnson, L. (2021). Innovative communication strategies for enhancing organizational performance. *Journal of Business Communication*, *52*(4), 345–360. DOI: 10.1002/jbc.12345

Kundu, S. C., & Mor, A. (2017). Workforce diversity and organizational performance: A study of it industry in India. *Employee Relations*, *39*(2), 160–183. DOI: 10.1108/ER-06-2015-0114

Kusnaya, A. A., Lukman, S., & Syafrizal, S. (2022). Analysis of leadership influence, resource allocation, and employee engagement on the effectiveness of strategy implementation. *JBTI: Jurnal Bisnis: Teori Dan Implementasi*, *13*(3), 193–202. DOI: 10.18196/jbti.v13i3.15529

Lasrado, F., & Uzbeck, C. (2017). The excellence quest: A study of business excellence award-winning organizations in uae. *Benchmarking*, *24*(3), 716–734. DOI: 10.1108/BIJ-06-2016-0098

Lee, Y. (2018). Dynamics of symmetrical communication within organizations: The impacts of channel usage of ceo, managers, and peers. *International Journal of Business Communication*, *59*(1), 3–21. DOI: 10.1177/2329488418803661

Leje, M. I., Kasimu, M. A., & Kolawole, A. F. (2019). Impacts of effective communication towards performance of construction organization. *Traektoriâ Nauki*, *5*(8), 3001–3008. DOI: 10.22178/pos.49-4

Leonardi, P. M., Huysman, M., & Steinfield, C. (2013). Enterprise social media: Definition, history, and prospects for the study of social technologies in organizations. *Journal of Computer-Mediated Communication*, *19*(1), 1–19. DOI: 10.1111/jcc4.12029

Mangundjaya, W. L. (2014). The role of communication, trust and justice in commitment to change. *International Conference on Business, Management and Corporate Social Responsibility* (ICBMCSR'14) Feb. 14-15, 2014 Batam (Indon. DOI: 10.15242/ICEHM.ED0214043

Marchalina, L., & Ahmad, H. (2017). The effect of internal communication on employees' commitment to change in malaysian large companies. *Business Management and Strategy*, *8*(1), 1. DOI: 10.5296/bms.v8i1.10629

Messamore, A. (2023). The effect of community organizing on landlords' use of eviction filing: Evidence from u.s. cities. *Social Problems*, *70*(3), 809–830. DOI: 10.1093/socpro/spac061

Moskovich, Y. (2023). Communal organizational culture as a source of business-success sustainability in Kibbutz industry—Two case studies. *Sustainability (Basel), 15*(13), 10654. DOI: 10.3390/su151310654

Mouboua, P. D., Atobatele, F. A., & Akintayo, O. T.Patricia Diane MouboaFadeke Adeola AtobateleOlateju Temitope Akintayo. (2024). Cross-cultural competence in global hrd: Strategies for developing an inclusive and diverse workforce. *International Journal of Science and Research Archive, 12*(1), 103–113. DOI: 10.30574/ijsra.2024.12.1.0765

Muchtar, Y. C., & Maas, L. T. (2022). Employee performance analysis: the influence of local wisdom-based communication strategy and employee engagement on employee performance. *Proceedings of the 3rd International Conference on Business and Management of Technology* (ICONBMT 2021). DOI: 10.2991/aebmr.k.211226.030

Mutamba, C. (2017). The inter-relationship of organizational learning, learning organizations, virtual technology, and virtual communities of practice. *New Horizons in Adult Education and Human Resource Development, 29*(3), 4–14. DOI: 10.1002/nha3.20186

Newlin, K., Dyess, S., Allard, E., Chase, S. K., & Melkus, G. D.Gail D'Eramo Melkus. (2011). A methodological review of faith-based health promotion literature: Advancing the science to expand delivery of diabetes education to black americans. *Journal of Religion and Health, 51*(4), 1075–1097. DOI: 10.1007/s10943-011-9481-9 PMID: 21487842

Nezhad, F. R., Kojouri, K. K., & Nodehi, S. N. K. (2019). Effect of altruism on organizational learning capability with the mediating role of communication conflict and organizational trust (case study: staff of red crescent provincial branches of gilan and mazandaran). *Journal of Rescue and Relief*, 67–76. DOI: 10.32592/jorar.2020.12.1.7

Nienaber, H., & Martins, N. (2020). Exploratory study. *The TQM Journal, 32*(3), 475–495. DOI: 10.1108/TQM-05-2019-0151

Nurcholis, G., & Budi, W. (2020). The impact of organizational culture and perceived organizational support on employee engagement. *Proceedings of the 5th ASEAN Conference on Psychology, Counselling, and Humanities* (ACPCH 2019). DOI: 10.2991/assehr.k.200120.050

Odhus, C. O., Kapanga, R. R., & Oele, E. (2024). Barriers to and enablers of quality improvement in primary health care in low- and middle-income countries: A systematic review. *PLOS Global Public Health, 4*(1), e0002756. DOI: 10.1371/journal.pgph.0002756 PMID: 38236832

Opeke, R. O., & Adelowo, O. T. (2020). Knowledge Management Strategies and Profitability of Selected Deposit Money Banks in Nigeria. Information Impact. *Journal of Information and Knowledge Management, 11*(1), 48–59.

Orpen, C. (2022). Effect of organizational commitment on the relationship between communication quality and managerial attitudes. *Communicare, 13*(1), 58–63. DOI: 10.36615/jcsa.v13i1.1973

Ouédraogo, N., & Ouakouak, M. L. (2018). Impacts of personal trust, communication, and affective commitment on change success. *Journal of Organizational Change Management, 31*(3), 676–696. DOI: 10.1108/JOCM-09-2016-0175

Park, S., Park, S., & Shryack, J. (2022). Measures of climate for inclusion and diversity: Review and summary. *Human Resource Development Quarterly, 34*(4), 463–480. DOI: 10.1002/hrdq.21493

Parsons, M., & Urbanski, S. (2012). Recognizing dysfunctional communications a means of improving organizational practice. *Online Journal of Communication and Media Technologies, 2*(4). Advance online publication. DOI: 10.29333/ojcmt/2404

Piao, M. (2010). Thriving in the new: Implication of exploration on organizational longevity. *Journal of Management, 36*(6), 1529–1554. DOI: 10.1177/0149206310378367

Piedade, S. D. R. (2021). The influence of transformational leadership on employee performance through work climate and organizational commitment. *Timor Leste Journal of Business and Management, 3*, 32–45. DOI: 10.51703/bm.v3i1.36

Prado-Román, C., Diéz-Martín, F., & Blanco-González, A. (2020). The effect of communication on the legitimacy and performance of organizations. *Revista Brasileira de Gestão de Negócios, 22*(3). Advance online publication. DOI: 10.7819/rbgn.v22i3.4071

Prasad, A., DeRosa, D., & Beyerlein, M. (2017). Dispersion beyond miles: Configuration and performance in virtual teams. *Team Performance Management, 23*(3), 186–204. DOI: 10.1108/TPM-06-2016-0026

Ramos-Maçães, M. A., & Román-Portas, M. (2022). The effects of organizational communication, leadership, and employee commitment in organizational change in the hospitality sector.

Rankin, J., & Archibald, M. (2015). Effective communication strategies for enhancing organizational performance. *Journal of Business Communication, 52*(4), 345–360. DOI: 10.1002/jbc.12345

Rasheed, M. A., Hookmani, A. A., Waleed, S., Fatima, H., Siddiqui, S., Khurram, M., & Hasan, B. (2021). Implementation and evaluation of a social media-based communication strategy to enhance employee engagement: Experiences from a children's hospital, pakistan. *Frontiers in Public Health*, *9*, 584179. Advance online publication. DOI: 10.3389/fpubh.2021.584179 PMID: 33777875

Saqib, Z., & Khan, M. (2022). Striving for inclusion of diverse employees: How important is the context? *SA Journal of Human Resource Management*, *10*(1), 107–129. DOI: 10.1177/23220937221083813

Seipel, T., Ishaq, S. L., Larson, C. D., & Menalled, F. D. (2022). Weed communities in winter wheat: Responses to cropping systems under different climatic conditions. *Sustainability (Basel)*, *14*(11), 6880. DOI: 10.3390/su14116880

Shockley-Zalabak, P. (2015). Communication competence in organizations and groups: Historic and emerging perspectives. *In The Handbook of Communication Science* (pp. 1-20). de Gruyter Mouton. DOI: 10.1515/9783110317459-017

Stephens, D. (2021). Strategies to engage millennial employees (a multiple case study). *Open Journal of Business and Management*, *09*(02), 618–658. DOI: 10.4236/ojbm.2021.92033

Sun, R., Li, J., Lee, Y., & Tao, W. (2021). The role of symmetrical internal communication in improving employee experiences and organizational identification during covid-19 pandemic-induced organizational change. *International Journal of Business Communication*, *60*(4), 1398–1426. DOI: 10.1177/23294884211050628

Syadina, K. P., Purnomo, R., & Anggraeni, A. I. (2019). The impact of transformational leadership and peceived organizational support on organizational commitment: The mediating role of employee engagement. *Journal of Research in Management*, *1*(4). Advance online publication. DOI: 10.32424/jorim.v1i4.54

Talebi, H., & Khatibi Bardsiri, A. (2023). The impact of information technology on service quality, satisfaction, and customer relationship management (case study: It organization individuals). *Journal of Management Science &Amp. Engineering Research*, *6*(2), 24–31. DOI: 10.30564/jmser.v6i2.5823

Tessier, A., Power, E., & Croteau, C. (2020). Paid worker and unfamiliar partner communication training: A scoping review. *Journal of Communication Disorders*, *83*, 105951. DOI: 10.1016/j.jcomdis.2019.105951 PMID: 31751831

Thakur, M.Ms Varsha Thakur. (2023). Interpersonal communication in the workplace: An exploratory analysis. *International Journal for Multidisciplinary Research*, *5*(3), 3699. Advance online publication. DOI: 10.36948/ijfmr.2023.v05i03.3699

Valdez Bocanegra, H. G. (2024). Capítulo 20. diagnosis of organizational climate in "grupo estancia". *RELAYN. Administración Y Negocios en Latinoamérica. Tomo, I*, 273–288. DOI: 10.46990/iQuatro.2024.09.5.20

Verčič, D., & Zerfaß, A. (2016). A comparative excellence framework for communication management. *Journal of Communication Management (London), 20*(4), 270–288. DOI: 10.1108/JCOM-11-2015-0087

Vos, A. D., & Heijden, B. (2017). Current thinking on contemporary careers: The key roles of sustainable hrm and sustainability of careers. *Current Opinion in Environmental Sustainability, 28*, 41–50. DOI: 10.1016/j.cosust.2017.07.003

Wang, B., Kang, Y., Childerhouse, P., & Huo, B. (2018). Service supply chain integration: The role of interpersonal relationships. *Industrial Management &Amp. Industrial Management & Data Systems, 118*(4), 828–849. DOI: 10.1108/IMDS-02-2017-0062

Wardhani, R. A. N., Wulanputih, T., Prihatini, D., & Sukarno, H. (2020). The the effect of employee engagement on job satisfaction and organizational citizenship behaviour (ocb) in employees bpjs ketenagakerjaan east java. *International Journal of Scientific Research and Management, 8*(05), 1812–1821. DOI: 10.18535/ijsrm/v8i05.em07

Weatherbee, T. G., & Sears, D. (2021). (En) act your age! Marketing and the marketization of history in young SMEs. *Journal of Historical Research in Marketing, 14*(1), 66–89. DOI: 10.1108/JHRM-10-2020-0051

Wolfgruber, D., Stürmer, L., & Einwiller, S. (2021). Talking inclusion into being: Communication as a facilitator and obstructor of an inclusive work environment. *Personnel Review, 51*(7), 1841–1860. DOI: 10.1108/PR-01-2021-0013

Yerden, X., Gil-García, J. R., Gascó-Hernández, M., & Burke, G. B. (2022). Public libraries' perceptions of future collaborations for the development of smart cities and communities: understanding influential factors. *Proceedings of the Annual Hawaii International Conference on System Sciences*. DOI: 10.24251/HICSS.2022.336

Yue, C. A., Men, L. R., & Ferguson, M. A. (2020). Examining the effects of internal communication and emotional culture on employees' organizational identification. *International Journal of Business Communication, 58*(2), 169–195. DOI: 10.1177/2329488420914066

Yüksel, S. (2015). The role of communication in enhancing organizational performance. *Journal of Business Communication, 52*(4), 345–360. DOI: 10.1002/jbc.12345

Zakiri, E. L. (2020). The role of communication in effective crisis management: A systematic literature reviews. *International Journal of Humanities and Social Science*, *10*(6). Advance online publication. DOI: 10.30845/ijhss.v10n6a14

Zhao, X., Toronjo, H., Shaw, C., Murphy, A., & Taxman, F. (2022). Perceived communication effectiveness in implementation strategies: A measurement scale. *Implementation Science Communications*, *3*(1), 38. Advance online publication. DOI: 10.1186/s43058-022-00284-4 PMID: 35395790

Chapter 6
Revealing Police Officer's Victimization: A Scoping Review

Ana Isabel Sani
https://orcid.org/0000-0003-1776-2442
University Fernando Pessoa, Porto, Portugal & Research Centre on Child Studies (CIEC), University of Minho, Braga, Portugal

Sara Correia Gonçalves
https://orcid.org/0009-0004-3330-5924
University Fernando Pessoa, Porto, Portugal & Research Centre on Child Studies (CIEC), University of Minho, Braga, Portugal

Laura M. Nunes
https://orcid.org/0000-0002-9648-9511
University Fernando Pessoa, Porto, Portugal & Research Centre on Child Studies (CIEC), University of Minho, Braga, Portugal

Maria Alzira Pimenta Dinis
https://orcid.org/0000-0002-2198-6740
Fernando Pessoa Research, Innovation, and Development Institute (FP-I3ID), University Fernando Pessoa, Porto, Portugal

ABSTRACT

This scoping review systematically maps the literature on police victimization. Searches across EBSCO, PubMed, and Web of Science yielded 41 articles, with 11 included after selection and manual searching. The review identified working conditions, inadequate protective equipment, and exposure to violence as key contributors to aggression against police officers. It explored aspects such as prevalence, firearm attacks, organizational structure impact, and the effectiveness of Body-Worn Cameras

DOI: 10.4018/979-8-3373-5686-0.ch006

(BWCs), which were linked to reduced assaults. Findings revealed troubling statistics on threats, assaults, and psychological impacts on officers, along with demographic patterns of perpetrators and victims. Studies emphasized the role of organizational structure in assaults and highlighted the cultural and contextual influences on police safety. Longitudinal and comparative analyses underscored effective measures and trends in reducing police victimization globally. These findings stress the need for addressing diverse factors to enhance law enforcement safety and well-being.

1. INTRODUCTION

Presently, both the levels of crime and the associated feelings of insecurity remain highly concerning (Sani et al., 2020; Saraiva et al., 2018). The mission of maintaining order by enforcing the law has never proven to be easy, and due to the changes, that have been felt in recent decades the challenge has become increasingly demanding (Zouain et al., 2006), particularly for instances of formal social control. Social violence presents a discrepancy between its actual occurrence and the sensations it evokes, emerging as a complex phenomenon. Sensations of fear and insecurity, in the social imagination, induce confusion between real crime and subjective perceptions about the risks of being a victim of crime (Jackson, 2009; Souza & Minayo, 2005). Violence is often generated, manifesting in a way that disrupts the established codes of order within society (Richie, 2022; Soares, 2014). Over the years, there has been a noticeable rise in violence, leading to discussions in academic and societal circles. This has made addressing this pressing issue a top priority on the government's public agenda and within the realm of social research (Gondal et al., 2023; Soares, 2014). As a result, public security institutions are facing considerable demands and pressures. Society expects these institutions to confront criminals and decrease crime rates. However, many of these institutions lack sufficient human and material resources to effectively and securely combat crime (Silva & Bueno, 2017).

Police officers have the mission of promoting public safety, which derives from the legal domain and is defined in the legislative document known as the Constitution. This fundamental duty is applicable in various contexts and is not specific to any particular country or case. This constitutes the guarantee that the State offers to citizens that everyone has the right to freedom and security, with the police officers tasked with ensuring internal security and the rights of citizens while maintaining respect for their freedoms and guarantees. As citizens and workers, police officers also have the right to physical and moral integrity. In the exercise of their mission to promote public safety, the police officers are protected by the Constitution, in most cases, which ensures these rights in their professional practice. Police officers in the performance of their duties are faced with difficult and dangerous situations, encountering numerous crimes and criminals (Souza, 2017). Most interactions between these professionals and citizens occur without conflicts, but sometimes the situation aggravates, taking on larger proportions, and these professionals are subjected to violent confrontations, exposed to danger as they operate in critical situations during their professional activity (Ellrich & Baier, 2016; Marty, 2022).

Thus, following a brief conceptual framework on the topic of police victimisation, an area that warrants further investigation and in-depth analysis (Arksey & O'Malley, 2005; Peters et al., 2015), a scoping review is presented. The objective of this review is to analyse the existing evidence in the literature concerning the experience of police victimisation among police officers.

2. UNVEILING POLICE OFFICER'S VICTIMIZATION: A SCOPING REVIEW

Being a police officer is a risky undertaking, but it leaves no doubt that it is essential to ensure the safety of citizens, as their duties involve crime prevention and the maintenance of public order (Azevedo et al., 2022; Wickramasinghe et al., 2016). When these professionals decide to join a law enforcement institution, candidates face certain difficulties during their training that they will have to confront in the course of their duties. They need to be equipped to face various risks, ranging from simple verbal offenses to threats against their physical integrity (Souza, 2017). Those aspiring to pursue this profession must learn to live with the pitfalls that may arise, always remaining vigilant and exercising caution in all situations. In reality, only day-to-day experiences will reveal that the challenges are even greater, highlighting the inherent risks in this professionally exposed to violence. It is crucial to recognize the police officers within and outside the institutions (Mello & Nummer, 2014; Minayo & Adorno, 2013; Minayo et al., 2007; Oliveira et al., 2014; Souza, 2017).

Beyond the exposure to the risk of violence against their physical integrity (e.g., physical violence by criminals, resulting in serious injuries or even death), police officers are susceptible to mental health issues stemming from the consequences of their professional activities (Bernardino & Bernardino, 2018; Monteiro et al., 2020; Silva & Bueno, 2017). Police officers are often targets of insults, verbal abuse, death threats from criminals or members of the public, which can create an environment of constant anxiety. Police officers can be subject to harassment and intimidation from criminals, witnesses or even co-workers, which can lead to problems such as anxiety, depression, job stress, burnout, post-traumatic symptoms which negatively impacts their personal lives and job performance (Allison et al., 2019; Eliasson, 2021; Teles et al., 2024).

Daily exposure to perilous situations makes this profession highly demanding and stressful, with psychosocial risk factors potentially emerging at any point in their lives. Clearly, the personality traits of individuals influence their performance in carrying out their duties, and any of these professionals may be at risk of physical and/or psychological burnout associated with exposure to traumatic events, how each cope with stress, and encounters with violent criminals and irate civilians (Ellrich & Baier, 2016; John-Akinola et al., 2020; Monteiro et al., 2020; Raigoso-Mayorga, 2016). Singo and Shilubane (2022) showed that police officers' coping mechanisms for stressful conditions included substance misuse, sports activity, religion, and support networks. The aim is, therefore, to reduce risks by combating threats or mitigating vulnerabilities presented by these police officers (Marty, 2022).

The risks that these professionals face may not always be well understood by society. In the name of the state, these professionals are the ones who legitimately use their authority and force, occasionally causing mild discomfort among citizens. At various times, it becomes necessary to use intimidation to protect the delicate line that exists between organized and violent crime and society (Souza, 2017).

The negative public opinion is intrinsically linked to this profession, and the distrust and ill-will that the population may possibly harbour against the police officers represent additional risks for these professionals. This is because they exhibit an imminent degree of suffering in the execution of their profession due to the lack of recognition from society. Thus, the negativity reflected upon them by various social classes ends up legitimizing and normalizing the violence they endure (Souza & Minayo, 2005). As a result, in addition to society fulfilling its duty to preserve the rights of all its citizens, it is crucial for it to ensure the protection and safety of all law enforcement professionals (Azevedo et al., 2021; Pang & Pavlou, 2019). Now, it is precisely here that Psychology, as a social science that studies and analyses human behaviour, can play a notable role, demonstrating visibility in analyses that may lead to a closer relationship between citizens and law enforcement professionals (Sani & Nunes, 2013). Interventions aimed at improving the connection between the community and law enforcement can contribute to promote citizen satisfaction with the performance of police officers, enhancing community safety, while necessarily safeguarding the integrity of the police as an institution and of individual police officers (Sani et al., 2022).

The studied violence affects those who have the mission and training to protect society, harming it exponentially and significantly (Marty, 2022). Therefore, future Security Force Professionals, during their training, should learn to adopt strategies to deal with various challenges in order to overcome them. This is because, due to the unpredictability of this profession, they cannot rely on static solutions (Souza, 2017).

The experience of police victimization is directly linked to the risks associated with this profession (Monteiro et al., 2020). It is defined as a social phenomenon that affects these professionals, leading to traumas, injuries, or fatalities occurring during the maintenance of order and in the fight against crime (Minayo et al., 2007). Incidents that violate the physical or psychological integrity, or that may fatally victimize police officers, should be subjected to thorough investigation. This is crucial for a more consistent understanding of the aspects that pose the greatest danger during the performance of their duties (Marty, 2022). Therefore, this phenomenon requires scientific attention to make it more understandable and to better formulate appropriate measures (Minayo et al., 2007). Therefore, the topic proves to be relevant and current, being of utmost utility, as police victimization is a reality, both nationally and internationally (e.g., Silva & Júnior, 2021; Simmler et al, 2021; Zavala, 2020), not always studied as much as desirable. Hence, the need for further investigation is imperative to better understand the phenomenon.

The research question was: What is known from the existing literature about violence against security force professionals in the exercise of their police duties?

3. METHOD

3.1. Research Strategy

This scoping review encompasses studies about the victimization phenomena among professionals in the police officers, conducted in accordance with the guidelines of the PRISMA for Scoping Review (Tricco et al., 2018) and other guidelines (Page, Moher et al., 2021).

Studies were identified through a search across various databases, including EBSCO, PubMed, and Web of Science. For the EBSCO database search, the terms used were TI ("police victimization" OR "assaults on police" OR "police officer victimization" OR "police deaths" OR "violence against police officers") AND AB (police OR "safety work" OR "risk profession" OR "security forces" OR "police work"). In the PubMed database search, the terms employed were (("police victimization"[Title] OR "assaults on police"[Title] OR "police officer victimization"[Title] OR "police deaths"[Title] OR "violence against police officers"[Title]) AND (police[Title/Abstract] OR "safety work"[Title/Abstract] OR "risk profession"[Title/Abstract] OR "security forces"[Title/Abstract] OR "police work"[Title/Abstract]). Finally, for the Web of Science database search, the terms used were TI=("police victimization" OR " assaults on police" OR "police officer victimization" OR "police deaths" OR "violence against police officers") AND AB=(police OR "safety work" OR "risk profession" OR "security forces" OR "police work").

To mitigate publication bias, these searches were supplemented with a manual inquiry. To define search terms, an analysis of keywords used in articles on police victimization was conducted, aiming to gather the broadest range of terms.

3.2. Inclusion and Exclusion Criteria

All articles were assessed and chosen based on established inclusion and exclusion criteria, always considering the study's objective: to identify key characteristics or factors associated with police victimization, aiming to provide guidance and clues for its prevention. The studies that addressed the following criteria were included: (1) empirical studies; (2) violence against the police; (3) police victimization. Exclusion criteria were applied in all other cases, including (1) articles with retrospectives without original studies; (2) theoretical articles; (3) articles not written in Portuguese, English, or Spanish. Apart from these criteria, only empirical and theoretical studies were analysed.

3.3. Selection of Articles

Figure 1 illustrates the process of study selection included in this scoping review. In total, 41 articles were identified in databases, with 17 excluded due to duplication. After screening titles and abstracts, 14 articles were excluded for inadequacy related to the study's objective, e.g., addressing police violence instead of police victimization, focusing on laws and the police. Following full-text reading, 1 article was excluded for not aligning with the research question of this review. 9 articles were included, along with an additional one through manual search. The exclusion criteria comprised theoretical articles and literature reviews. All excluded articles did not align with the research question of this scoping review.

Figure 1. Flowchart of the selection of articles.

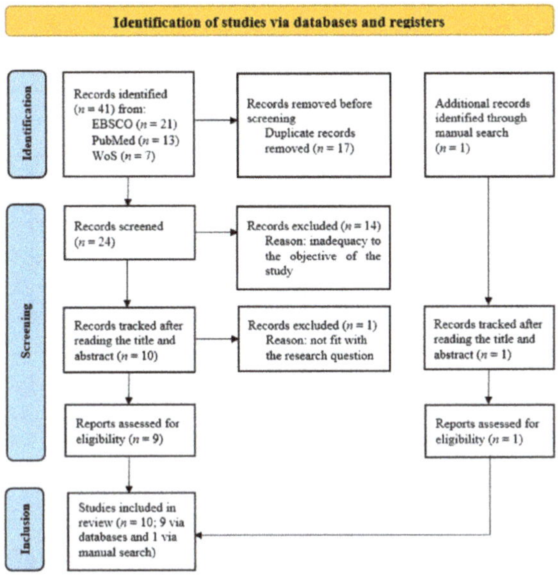

3.4. Quality Analysis of the Articles

The analysis of article quality was conducted using the PRISMA Checklist 2020 (Page, McKenzie et al., 2021). It assessed the articles by examining the consistency and clarity of the title, abstract summary, state of the art, study objectives and/or hypotheses, and method analysis for future research. This checklist was then modified to include 13 items: (1) title; (2) abstract; (3) state of the art; (4) study objectives; (5) study protocol; (6) eligibility criteria; (7) information sources; (8) inclusion/exclusion criteria; (9) information collection; (10) study design; (11) key measures; (12) synthesis of main results; (13) study limitations.

Table 1 rates the articles on a Likert scale from 0 to 2, where 0 corresponds to "not reported/not specified" 1 corresponds to "unclear/reported to some extent" and 2 corresponds to "adequately done". Thus, a score of 0 indicates that the study lacks information on the item in question, a score of 1 reveals that the researchers briefly mentioned the topic without providing detailed explanations, and a score of 2 confirms that the researchers thoroughly discussed the topic, explaining the methodology used in a detailed and understandable manner.

Table 1. Quality analysis of the selected articles.

Articles	1	2	3	4	5	6	7	8	9	10	11	12	13	Total Score
Douglas (2020)	0	2	2	2	2	1	2	2	2	2	2	2	1	22 High
Dursun & Aytaç (2021)	0	2	2	1	2	1	2	2	2	2	2	2	0	20 Average
Gibbs et al. (2017)	0	2	2	1	2	1	2	2	2	2	2	2	2	22 High
Kaminski et al. (2003)	1	2	2	1	2	1	2	2	2	2	2	2	1	22 High
Krope & Ilić (2020)	0	2	2	1	2	2	2	2	2	2	2	1	1	22 High
Krope & Lobnikar (2015)	0	2	2	1	1	1	2	2	2	2	2	1	1	19 Average
Kyriacou et al. (2015)	0	2	2	2	2	2	2	2	2	2	2	2	2	24 High
Oliveira et al. (2014)	0	2	2	2	2	2	2	2	2	2	2	2	1	23 High
Sierra-Arévalo & Nix (2020)	0	2	2	1	2	2	2	2	2	2	2	2	1	22 High
Willits (2014)	0	2	2	2	1	1	2	2	2	2	2	2	2	22 High

Grades: 0—not reported/not specified; 1—unclear/reported to some extent; 2—adequately done.

4. RESULTS

The selected studies underscore the prevalence of aggression directed towards police officers, identifying various contextual and situational factors that may contribute to police victimisation. These factors include challenging working conditions, insufficient protective equipment, and frequent exposure to violence and crime. Such findings highlight the necessity of linking individual police experiences to broader organisational issues, emphasising the systemic nature of police victimisation. As posited by Routine Activities Theory (Cohen & Felson, 1979), victimisation occurs when three key elements converge: a motivated offender, a suitable target (in this case, police officers), and the absence of capable guardianship. Due to the nature of their profession, police officers are frequently placed in environments where these elements coexist, thereby increasing their risk of victimisation (Allen & Jacques, 2014). The inherent exposure to potentially violent encounters further reinforces the need for structural and policy-level interventions to mitigate these risks and enhance officer safety.

Among the included studies (cf. **Table 2**), five conduct an analysis of the prevalence of police victimization and the types of violence in police departments (Dursun & Aytaç, 2021; Kaminski et al., 2003; Krope & Ilić, 2020; Krope & Lobnikar, 2015; Oliveira et al., 2014). Two studies focus on police victimization, correlating it with fatal and non-fatal firearm attacks (Gibbs et al., 2018; Sierra-Avrévalo & Nix, 2020). Another two studies relate police victimization to the organizational structure of police departments, with one also examining the use of Body-Worn Cameras (BWCs) (Douglas, 2020; Willits, 2014). One study compares the prevalence of police victimization between two cities, New York and London (Kyriacou et al., 2015). It is noteworthy that all studies investigated the relationship between police victimization and the areas where it is most prevalent.

One study investigated the working conditions and safety of police detectives, with a particular focus on aspects of violence and police victimization. A sample of 152 police detectives showed that 59.9% of the participants were female, aged between 40 and 59 years (59.8%), with the majority performing administrative duties (63.0%). Among the studied working conditions, the lack of protective equipment stood out (65.0%). Regarding threats and insecurity at work, 39.5% mentioned that their personal safety had been threatened during the practice of their professional activities. In the last 12 months, 30.0% of the participants mentioned at least one episode of aggression or threat committed by superiors or colleagues against another colleague, and 28.0% mentioned having experienced assaults or threats from citizens. The experienced assaults and threats were mainly psychological (38.2%). Considering the nature of the work as a police detective being violent, 83.6% of the participants answered affirmatively, citing exposure to violence and crime (69.5%) and unsuitable working conditions (15.0%) as the main reasons. Out of the 152 participants, 86.2% were diagnosed with one or more illnesses in the last 30 days of the investigation's conclusion, with common mental disorders having a significant presence (37.5%) in the reported diagnoses (Oliveira et al., 2014).

In another study, analysing 363 criminal reports, it was concluded that perpetrators of crimes against police officers tend to be male (92.9%). In 55.5% of the attacks, the ratio was one perpetrator to one police officer, and in 61.1% of the attacks, the officer was alone. Notably, in 29.4% of the incidents, two officers were present, suggesting a potential decrease in assaults on officers with an increase in personnel during an incident. Of the total 531 police victims, 96.8% were male, and 3.2% were female, ranging in age from 21 to 50 years. The analysis revealed that the first five years of a police officer's service are particularly vulnerable to victimization, and an increase in assaults on the same individual corresponds to a decrease in the severity of injuries. On average, this study showed that a police officer is most likely to be assaulted between 11:00 PM and 2:00 AM on a Saturday (21.5%), with the assault occurring in a public place, on the road, street, or in a square (Krope & Lobnikar, 2015).

Although balancing the sample according to sex is difficult to achieve in many countries, it is also important to consider the analysis of victimization depending on whether the perpetrator is male or female. In a study by Rabe-Hemp and Schuck (2007), which compared gender differences in assaults on officers, as well as situational and individual factors of officers related to assaults, it was concluded that female police officers face a greater risk of being attacked in situations of family conflict, especially when the offender is decompensated.

The study of Kyriacou et al. (2015) involves the comparison of police victimization between New York City and London through the analysis of official data from the police departments from 1900 through 1999. The analysis shows that 585 police officers in New York and 160 in London lost their lives while performing their professional duties. The majority of police deaths in New York (88%) resulted from gunshot wounds, while in New York, in the same period of the century, 14 police officers died. These differences presented may likely stem from cultural, socio-economic, and occupational factors. The research demonstrated that during the latter part of the 20th century, there was a decline in police deaths in New York City, which may suggest that some measures implemented by the New York Police Department were successful in significantly reducing the prevalence of intentional and unintentional gunshot that led to the deaths of police officers.

One study examined the relationship between the police organizational structure and assaults on officers, with negative binomial regression models indicating that organizational context and complexity are significant predictors of violence against police officers. In other words, it is necessary to consider the organizational structure for successful policing outcomes. Specifically, with other factors controlled, police departments serving more disadvantaged communities exhibit higher rates of assaults against their officers, while those departments that are functionally and spatially more differentiated show lower rates of police victimization (Willits, 2014).

Another study, which included 339 police officers from various areas of operation, where the majority of participants were male (92.3%) in an age range of 31 to 40 years (71.1%), and 53.7% of them had a service tenure of 11 to 15 years, reveals that in the past year, 39.2% of police officers have been exposed to at least one type of violence in the course of their professional activities. It was observed that among the types of violence to which police officers are exposed, verbal abuse stands out (28.9%), followed by workplace harassment (21.2%) and physical violence (20.9%), respectively. This analysis demonstrates that the levels of depression and anxiety among police officers who have been victims of at least one type of violence were higher compared to officers who have not experienced any type of violence, and the difference between these two groups of violence-exposed and non-exposed police officers is also statistically significant ($p < 0.05$). Thus, violence associated with this profession is a significant risk factor for police officers, and its exposure also adversely affects their mental health (Dursun & Aytaç, 2021)

The study involving the evaluation of the relationship between Body-Worn Cameras (BWCs) and police victimization demonstrates a 20.3% reduction in the expected number of assaults against the police in departments that already deploy BWCs on their patrols, compared to departments that do not use them. Among the investigated policing styles, only the drug arrest rate proved significant ($p < 0.01$), showing a positive association with the overall number of assaults. In other words, on average, an increase of 100 units in the drug arrest rate in a police department would result in an approximately 8.0% increase in assaults against its officers, while keeping the other covariates constant. This study also reveals that for every 1-unit increase in the police department's staffing level, there was a 0.087% increase in the expected count of assaults against the police. On average, assaults against officers would be 9.0% higher for a department with 200 officers compared to one with only 100 officers. However, through a more in-depth analysis of the relationship between police victimization and the staffing level of a department, it was found that departments with a higher staffing level using BWCs tended to experience fewer assaults compared to departments with a higher staffing level but without the use of BWCs (Douglas, 2020).

Another study that determined the trend of assaults on police officers over a longer period in Slovenia shows that the highest number of assaults on police officers was recorded between 2009 and 2013, with a steady decline from 2013 onwards. The number of injured police officers, especially in 2012 and 2011, was quite significant. When determining the reasons for this number, it was found that this was a time of major protests in various Slovenian cities and interventions in Roma camps that involved mass assaults on police officers. However, this study finds that despite a decrease in the number of assault crimes against the police, there is a concerning significant increase in the number of increasingly aggressive and dangerous forms of assaults on police officers (Krope & Ilić, 2020).

In another study, conducted using data from the Gun Violence Archive (GVA) open-source dataset, it is possible to verify that from 2014 to 2019, 249 police officers were fatally shot by suspects, with a total of 1218 being shot or injured by non-fatal gunfire. During this period, the overall number of firearm assaults fluctuated from a minimum of 189 in 2014 to a maximum of 288 in 2016. Over these 6 years, an average of 245 police officers per year were shot by suspects in the line of duty. Of these shot officers, an average of 42 per year (17%) were killed, and only 14% to 21% of firearm assaults against the police each year resulted in fatalities. Analysing longitudinal trends, despite the stable monthly frequency of fatal firearm assaults on police officers from 2014 to 2019, there appears to be a slight upward trend in the monthly frequency of firearm assaults against the police in general. It is observed that the majority of firearm assaults against the police are non-fatal. There is no compelling evidence to suggest that the national rate of firearm assaults against the police has substantially increased in the last 6 years. Additionally, there is substantial variation at the state level in the rates of firearm assault against the police (Sierra-Arévalo & Nix, 2020).

A study that explored violence against police officers who experience assaults in geographical areas smaller than the city level reveals that the most violent cases of aggression against the police follow a pattern among areas with higher rates of concentrated disadvantage, racial homogeneity, and residential mobility. The study's maps demonstrated that areas of aggression against the police, where assailants used firearms and areas of aggression against the police without firearms, appeared to cluster in higher, neighbourhood-level concentrations of concentrated disadvantage. Additionally, the majority of groups involved in injuries and firearm-related assaults were located within and around the city centre (Gibbs et al., 2018).

In a recent study that examined risk factors at the neighbourhood level for severe assaults against the police from a criminal opportunity perspective, it was revealed that the police are more frequently assaulted in areas with a higher propensity for violence, densely populated areas, neighbourhoods with a significant number of university students, and residential mobility, although the latter effect is nearly significant ($p = 0.056$). However, there is a substantial and highly significant relationship between the variable of resource deprivation and assaults against the police. In other words, each unit increase in this variable is associated with a 31% increase in the expected number of assaults. Thus, the risk of assaults occurring is considerably higher in areas characterized by economic problems, family disintegration, and a higher concentration of non-Hispanic black residents. Regarding the variable of arrests, with other controlled variables in the model, each unit increase in this variable is directly associated with a 55% increase in the expected number of assaults against the police. Therefore, neighbourhoods with a higher number of arrests purportedly pose a substantially greater risk to the police in those areas. Thus, the variable of arrests ultimately has a greater effect on the risk of assaults against the police, followed by the variables of violence and resource deprivation. These results demonstrate that increases in the standard deviation of these variables—arrests, violence, and resource deprivation—are associated with increases of 60%, 40%, and 31%, respectively, in the expected number of assaults against the police (Kaminski et al., 2003).

The key findings from the literature available in **Table 2** can be further summarised below in terms of five different important aspects:

Protective Factors & Prevention. Body-Worn Cameras (BWCs) are negatively associated with police victimization and can help prevent violence against officers (Douglas, 2020). Organizational complexity and differentiation (functional and spatial) correlate with lower levels of assaults on police (Willits, 2014).
Risk Factors & Patterns. Violence against police tends to cluster spatially in areas with concentrated disadvantage (Gibbs et al., 2018). Block groups with high arrest density, criminal conditions, and propensity for violence show strong associations with serious attacks on police (Kaminski et al., 2003). Officers with fewer years of service (particularly those with $\leq$5 years) face higher risk of victimization (Krope & Lobnikar, 2015). Organizational context matters: departments serving disadvantaged communities report higher rates of assaults (Willits, 2014).
Temporal & Geographic Variations. Significant variations exist in rates of firearm assaults against police across different states (Sierra-Arévalo & Nix, 2020). Historical trends show higher police occupational death rates in New York compared to London during the 20th century (Kyriacou et al., 2015). In Slovenia, assault rates peaked between 2009–2013, coinciding with large protests and interventions in Roma camps (Krope & Ilić, 2020).
Consequences of Violence. Workplace violence is a significant risk factor for police officers' mental health, associated with depression, anxiety, and stress (Dursun & Aytaç, 2021). Mental health problems are identified as main causes of extended work absence among police officers (Oliveira et al., 2014).
Methodological Insights. Majority of firearm assaults on police are non-fatal, suggesting the importance of tracking both fatal and non-fatal incidents (Sierra-Arévalo & Nix, 2020). Multiple data sources are used across studies: crime reports, census data, surveys, and specialized databases like Gun Violence Archive.

Police victimization is thus influenced by a combination of organizational factors, geographical risks, temporal trends, and mental health consequences, emphasizing the need for comprehensive data tracking and targeted prevention strategies.

Table 2. Summary of studies' characteristics and findings.

Authors/Year	Objectives	Country	Sample	Instruments	Results and Main Findings
Douglas (2020)	Assess the relationship between Body-Worn Cameras (BWCs) and police victimization, focusing on the total number of assaults against police officers, with and without a firearm.	United States of America	$n = 516$ police agencies	Law Enforcement Management and Administrative (LEMAS); Uniform Crime Report (UCR) do Federal Bureau of Investigation (FBI); The Census Bureau; Centers for Disease Control and Prevention (CDC).	Results indicate that the use of Body-Worn Cameras (BWCs) is negatively associated with police victimization. This suggests that BWCs can help prevent the occurrence of general and extreme violence against police across a wide range of law enforcement departments in varying environments.
Dursun & Aytaç (2021)	Reveal the prevalence of violence against police officers and examine the effect of being subjected to violence on their mental health.	Australia	$n = 339$ police officers working in different duty areas	The Workplace Violence Questionnaire; DAS (Depression-Anxiety-Stress).	The survey results show that workplace violence is an important risk factor for police officers. Exposure to violence also negatively affects the mental health of police officers. These results show the importance of implementing policies to prevent violence against police officers.
Gibbs et al. (2018)	Explore violence against police officers who suffer attacks in geographic areas smaller than the city level.	United States of America	-	Census data from the Bureau of the Census.	The maps show that incidents of violence against police tend to cluster spatially, and multivariate analyses suggest that concentrated disadvantage and calls for service, but not other indicators, were significantly related to violence against police.

continued on following page

Table 2. Continued

Authors/Year	Objectives	Country	Sample	Instruments	Results and Main Findings
Kaminski et al. (2003)	Examining risk factors at a block-group-level for serious assaults on police from a criminal opportunity perspective.	United States of America	$n = 603$ aggravated assaults on police	Crime incident reports from the Boston Police Department; Census data from the Bureau of the Census.	Results indicate strong associations between serious attacks on police and block groups characterized by high arrest density, criminal conditions, and propensity for violence.
Krope & Ilić (2020)	Determine the trend of assaults on police officers over a longer period of time.	Slovenia	-	Annual reports of the General Police Directorate concerning assaults on police officers.	The highest number of attacks on police officers was recorded between 2009 and 2013, with a constant drop from 2013 onwards. The number of police officers injured, mainly in 2012 and 2011, was quite significant. When determining the reasons for this number, it was discovered that this was a time of large protests in several Slovenian cities and interventions in Roma camps that involved mass attacks on police officers.
Krope & Lobnikar (2015)	Study the circumstances of attacks on police officers in Slovenia, with emphasis on aggression, place, time, work area and the presence of third parties in the commission of the crime of police aggression.	Slovenia	$n = 363$ criminal charges $n = 531$ police officer victims	Data assembler on criminal charges filed for assaults on police officers.	A total of 531 police victims, of which 514 (96.8%) were men and 17 (3.2%) were women, aged between 21 and 50 years, were recorded. Most of the police officers attacked have 5 years of active service. The greater the number of years of service, the lower the level of risk, that is, the rate of potential victims decreases.

continued on following page

Table 2. Continued

Authors/Year	Objectives	Country	Sample	Instruments	Results and Main Findings
Kyriacou et al. (2015)	To describe the incidences and causes of occupational police deaths in New York City in the United States of America and Greater London in the United Kingdom during the 20th century and assess the relationship between general social violence and violence directed at police officers in these metropolitan areas.	United States of America and United Kingdom	-	Official reference sources; Census of New York and London.	During the 20th century, intentional and unintentional police occupational death rates were significantly higher in New York compared to London. These differences likely arise from various socioeconomic, cultural, and occupational factors. The decline in police deaths in New York during the latter part of the 20th century indicates that at least some measures taken by the New York Police Department were successful in significantly reducing the incidence of intentional and unintentional police deaths.
Oliveira et al. (2014)	Describe the working and safety conditions of civil police officers in the city of Salvador, Bahia, Brazil, focusing on aspects of violence and victimization.	Brazil	$n = 152$ police officers from Salvador, Bahia	Standardized Questionnaire.	High physical and emotional demands are identified in the work of police chiefs, vulnerability in relationships between peers and unsafe conditions in the professional context. Mental health problems are the main causes of absence from work for more than fifteen days.

continued on following page

Table 2. Continued

Authors/Year	Objectives	Country	Sample	Instruments	Results and Main Findings
Sierra-Arévalo & Nix (2020)	Using open-source data from the Gun Violence Archive (GVA), the objective is to analyses reactional and state trends in fatal and non-fatal assaults with firearms by police officers.	United States of America	n = 1467 cases analysed	Gun Violence Archive (GVA).	The results show that (a) the majority of firearm assaults are not fatal, (b) there is no convincing evidence that the national rate of firearm assaults on police has increased substantially over the past 6 years, and (c) there is substantial state-level variation in rates of firearm assaults against police officers.
Willits (2014)	Examine the relationship between police organizational structure and assaults on police in the USA.	United States of America	-	Law of Enforcement Officers Killed and Assaulted (LEOKA); National Incident – Based Reporting System (NIBRS).	Negative binomial regression models suggest that organizational context and organizational complexity are important predictors of violence against police. Specifically, controlling for other factors, police departments that serve more disadvantaged communities report higher rates of assaults on police, while those that are more functionally and spatially differentiated report lower levels of assaults.

5. DISCUSSION

The present scoping review allows to confirm the presence of police victimization, with officers being exposed to physical, verbal, and psychological assaults (Dursun & Aytaç, 2021). In the investigation of factors associated with this significant prevalence of police victimization, the concentrated disadvantage variable, i.e., densely populated areas where there is more violence, more arrests, more resource deprivation, economic problems, and family disruption, demonstrated through statistical analyses to be linked to non-firearm assaults against the police (Gibbs et al., 2017). This conclusion aligns with another study that found resource deprivation, arrests, and areas of violence to be significantly correlated with incidents of aggression against the police. Thus, the risk of serious assaults against the police is substantially higher in areas characterized by criminogenic conditions (Kaminski et al., 2003). However, according to Gibbs et al. (2017), this factor of concentrated disadvantage is not directly related to firearm assaults against the police. Considering the less frequency of assaults involving firearms compared to those resulting in injuries, it was not feasible to draw conclusive findings regarding a more consistent pattern related to this disadvantage variable within these two situations, i.e., injurious assaults or firearm assaults.

Regarding firearm assaults against the police, and according to the study by Sierra-Arévalo and Nix (2020) in the United States of America (USA), the majority of such attacks are not fatal for law enforcement officers in the line of duty. In a comparative study between New York and London, it was revealed that 290 police officers were killed by gunfire in New York, compared to only 14 police officers killed by gunfire in London over the entire study period. This difference is likely attributed to the widespread availability and use of firearms in the USA. Concerning intentional and unintentional mortality rates of police officers, these were substantially higher in New York compared to London for the majority of the 20th century. The majority of police deaths in New York were intentional, while the majority in London were unintentional. This contrast between cities may result from various socio-economic and occupational factors (Kyriacou et al., 2015).

In the face of the potential for more severe victimization, such as death, during the execution of their duties or even during off-duty periods, the adoption of measures and protocols by Security Force institutions is necessary. This is essential both as an institutional tool and procedurally to address police victimization, along with the reflection and awareness of the public safety officer regarding their own security (Silva & Almeida, 2022). According to the authors, in addition to the adoption of measures and protocols by the institutions themselves and the professionals, it is crucial to emphasize that the state bears responsibility for the significant presence of police victimization. For this reason, it has the duty to provide more resources and better working conditions, thereby ensuring the safety of its public safety agents, as well as efficiency in their work (Silva & Almeida, 2022).

Additionally, the deficient provision of working conditions, safety equipment, lack of resources, and the organizational structure of police departments also contribute to instances of victimization among police officers. For instance, in the study by Oliveira et al. (2014), the findings highlight significant issues related to the work environment, exposure to violent situations, and the health of these police officers, prompting further investigations for a better understanding of this professional practice. Measures addressing both the organizational context and the organizational complexity of police departments are significant factors influencing violence against the police. Therefore, the importance of considering organizational structure for achieving positive outcomes in policing is emphasized. This translates into a greater presence of police officers in the community through community policing, which may lead to increased familiarity between the police and neighbourhoods where they operate. Similarly, residents of these neighbourhoods may become more acquainted with the police and less likely to perceive them as a threat (Willits, 2014).

In one study conducted in Slovenia, over the studied timeframe, there was a decrease in assaults against the police. However, while the number of assaults decreased over time, there was a noticeable increase in the number of more dangerous forms of assaults against the police, indicating an elevated risk for officers. In the study conducted by Haoran et al. (2022), the results showed that offenders who engaged in premeditated crimes were more inclined to use a weapon during the commission of their offenses, with a preference for sharp weapons. This raises concerns for police departments, emphasizing the need for measures to manage the situations they face and prevent police victimization (Krope & Ilić, 2020). One such measure that could be implemented is the use of BWCs, which has been associated with lower rates of police victimization. The findings suggest that the use of BWCs may be linked to increased officer safety and underscore the importance and necessity of future evaluations of BWCs to investigate their effects on police victimization outcomes (Douglas, 2020).

Police officers are professionals who must deal with people and violent situations. The violence experienced in the course of their professional activities is a highly relevant risk factor, negatively impacting their mental health. These findings underscore the importance of urgently implementing measures to prevent police victimization (Dursun & Aytaç, 2021). It is crucial to emphasize the need for providing psychosocial support to each assaulted police officer. Public safety institutions should be actively involved in monitoring criminal proceedings against attackers targeting their officers. By doing so, they convey the significance of protecting these police officers (Krope & Lobnikar, 2015).

In a race against time, future studies must explore possible pathways to reduce victimization of police officers in crimes motivated solely by the fact that these agents are performing their duties. To achieve this, it is expected that a risk management approach will be applied to the analysis, identification, assessment, treatment, monitoring, and mitigation of the entire investigation. Thus, the ambition moving forward is a meticulous and in-depth analysis of each factor that may contribute to police victimization and an enhancement in data documentation on the subject. This aims to enable the implementation of more technical measures and protocols to prevent and mitigate the risks associated with this profession (Marty, 2022).

Based on the main findings of this study, several recommendations for practice can be proposed. Cross-cultural research examining law enforcement personnel across different countries is essential for understanding variations in police victimisation. Additionally, comparative analyses of victimisation characteristics among different types of police forces—such as public versus private agencies—and across diverse geographical contexts, including urban and rural settings, are crucial. The implementation of more stringent gun control measures may contribute to reducing firearm-related victimisation of police officers in specific regions. However, the applicability and necessity of such measures may vary considerably depending on cultural and legal contexts.

It is imperative to consider additional factors, including the varying modalities of police organisations and their respective operational environments. Institutions responsible for security forces must develop and implement comprehensive protocols and strategic measures to effectively address police victimisation. The state holds a fundamental responsibility in ensuring that adequate resources are allocated and that working conditions are sufficiently improved to enhance the safety and operational efficiency of public safety personnel. Suboptimal working conditions, a lack of appropriate safety equipment, resource shortages, and deficiencies in organisational structure within police departments are significant contributors to the victimisation of police officers. To mitigate these risks, it is essential to ensure that police departments receive adequate funding and are equipped with the necessary security apparatus, personnel resources, and infrastructural support. Strengthening these areas will contribute to reducing the exposure of police officers to victimisation while fostering a safer and more effective working environment.

The violence encountered in the line of duty has profoundly detrimental effects on the mental health of these individuals, underscoring the critical need for preventive measures to mitigate incidents of victimisation. In addition, the provision of psychosocial support services, including counselling and therapeutic interventions, can play a pivotal role in alleviating the emotional and psychological repercussions experienced by officers subjected to such victimisation. Furthermore, regular assessments of the impact of police victimisation on officers' mental health may be conducted within the police organisation, ensuring that empirical findings inform the development of targeted support and intervention strategies. Such an evidence-based approach can enhance resilience, promote psychological well-being, and foster a supportive organisational culture.

Integrating preventive strategies with comprehensive post-incident support enhances both the well-being and operational effectiveness of public safety officials. For instance, the organisation may implement targeted training programmes that emphasise prevention strategies and self-protection techniques, equipping officers with the necessary skills to mitigate the risk of victimisation incidents. Such initiatives not only contribute to reducing the occurrence of such incidents but also foster resilience and preparedness among personnel, ultimately strengthening their capacity to perform their duties effectively. Law enforcement agencies can take specific measures to reduce victimisation while simultaneously safeguarding the well-being and effectiveness of their personnel. For instance, fostering positive community relations and encouraging citizen engagement can help build trust and create a collaborative environment. Such an approach may contribute to reducing hostility and acts of violence directed towards police officers, ultimately enhancing both public safety and officer resilience.

6. CONCLUSIONS AND FUTURE RESEARCH DIRECTIONS

The literature on violence against police officers during policing duties has developed since the beginning of this century. Information on prevalence, causes, consequences, and potential solutions for this issue is distributed in a scattered manner. This scoping review has aggregated a substantial amount of information, contributing to the direction of research in this area. It focuses on identifying various factors, including structural factors associated with the profession, working conditions, and disadvantaged socioeconomic contexts. Additionally, it explores systemic mechanisms emerging from violence dynamics, leading to the development of stereotypes, increased insecurity, and a negative impact on the trust of both police officers and citizens. The symbolic aspects of violence were also addressed in this scope review, emphasizing the importance of adopting methods that contribute to the prevention of police officer's victimization. This includes addressing both the physical and psychological consequences for the individuals involved, with broader social implications. It can undermine public trust in the agencies responsible for law enforcement, undermine the rule of law, perpetuate a cycle of violence, and hinder the effective functioning of the criminal justice system. The existing literature underscores the multifaceted nature of violence against professionals in police officers, emphasizing the need for comprehensive approaches to address this issue.

Worldwide, the results suggest a notable incidence of police victimization in the police officers, and non-firearm assaults against the police are linked to factors like concentrated disadvantage, resource deprivation, arrests, and areas of violence. Regions characterized by concentrated disadvantage show an elevated likelihood of experiencing severe assaults against the police. The analysis of physical and demographic spaces holds significant relevance in conducting a local security diagnosis. This approach enables the identification of the primary objective and subjective indicators, contributing to the risk of victimization among police officers. The belief is that interventions informed by concrete data can effectively modify both physical and social factors. This, in turn, may improve living conditions, reduce crime rates, and contribute to preventing police victimization.

The availability and use of firearms contribute to the higher number of firearm assaults on police officers in the USA, when compared to countries like the United Kingdom. These differences may be attributed to cultural, socio-economic, and occupational factors.

Emerging research in the field of police officer victimisation has the potential to culminate in a systematic review. Such a review would benefit from an expansion of search criteria to encompass a broader range of sources and studies, thereby facilitating a more comprehensive examination of the factors contributing to police officer victimisation. Particular attention should be directed towards underexplored dimensions of police victimisation, including its impact on mental health outcomes and the role of organisational characteristics in either exacerbating or mitigating the risk of victimisation.

Cross-national collaboration among researchers could further enhance understanding by enabling comparative studies that identify common trends and patterns across diverse contexts. However, it is crucial to acknowledge the potential for publication bias, which is often inherent in scoping reviews, particularly given that the present investigation was confined to studies accessible within identifiable databases, without temporal restrictions. To address these limitations, future research should incorporate self-reporting studies that utilise qualitative methodologies. Such approaches are essential for capturing the nuanced elements that characterise the patterns and dynamics of police officer victimisation while also considering the psychosocial conditions underpinning the profession. Furthermore, an in-depth focus on the physical and mental health implications of victimisation could provide valuable insights for the development of targeted intervention strategies.

The implementation of BWCs has also been associated with reductions in police victimisation rates. This finding underscores the necessity of financial investment in improved working conditions, the provision of safety equipment, and the establishment of effective organisational structures within police departments, all aimed at safeguarding the well-being and operational efficacy of police officers. To cultivate a comprehensive understanding of the determinants and patterns of police victimisation, as well as to devise effective preventive measures, future research should prioritise initiatives focused on mitigating victimisation incidents and enhancing the systematic documentation of relevant data.

ACKNOWLEDGMENT

This work was partially financed by national funds through the Foundation for Science and Technology (FCT) within the framework of the Research Centre for Child Studies (UIDB/00317) of the University of Minho and Research Centre for Justice and Governance (UIDB/05749) of the University of Minho.

REFERENCES

Allen, A., & Jacques, S. (2014). Police officers' theories of crime. *American Journal of Criminal Justice*, *39*(2), 206–227. DOI: 10.1007/s12103-013-9219-1

Allison, P., Mnatsakanova, A., McCanlies, E., Fekedulegn, D., Hartley, T. A., Andrew, M. E., & Violanti, J. M. (2020). Police stress and depressive symptoms: Role of coping and hardiness. *Policing*, *43*(2), 247–261. DOI: 10.1108/PIJPSM-04-2019-0055 PMID: 32714068

Arksey, H., & O'Malley, L. (2005). Scoping studies: Towards a methodological framework. *International Journal of Social Research Methodology*, *8*(1), 19–32. DOI: 10.1080/1364557032000119616

Azevedo, V., Nunes, L., & Sani, A. (2022). Is campus a place of (in)security and crime? Perceptions of higher education students. *European Journal of Investigation in Health, Psychology and Education*, *12*(2), 193–208. DOI: 10.3390/ejihpe12020015 PMID: 35200237

Azevedo, V., Sani, A., Nunes, L., & Paulo, D. (2021). Do you feel safe in the urban space? From perceptions to associated variables. *Anuário de Psicología Juridica*, *31*(1), 75–84. DOI: 10.5093/apj2021a12

Bernardino, R. C., & Bernardino, A. V. S. (2018). Fatores estressores que influenciam na qualidade de vida, gerando danos à saúde do policial militar [Stressors that influence the quality of life, causing damage to the health of military police officers]. *Revista Mosaico*, *09*(2), 02-09. https://doi.org/DOI: 10.21727/rm.v9i2.1304

Cohen, L. W., & Felson, M. (1979). Social change and crime rate trends: A routine activity approach. *American Sociological Review*, *44*(4), 588–608. DOI: 10.2307/2094589

Douglas, S. (2020). The effects of body-worn cameras on violent police victimization. *Policing (Oxford, England)*, *15*(2), 1399–1416. DOI: 10.1093/police/paaa032

Dursun, K., & Aytaç, S. (2021). Workplace violence against police officers and the effects of workplace violence on mental health. *Pakistan Journal of Medical & Health Sciences*, *15*(12), 3491–3493. DOI: 10.53350/pjmhs2115123491

Eliasson, M. N. (2021). Police officers: Invisible victims in the line of duty. In Blasdell, R., Krieger-Sample, L., & Kilburn, M. (Eds.), *Invisible victims and the pursuit of justice: analyzing frequently victimized yet rarely discussed populations* (pp. 41–62). IGI Global Scientific Publishing., DOI: 10.4018/978-1-7998-7348-8.ch003

Ellrich, K., & Baier, D. (2016). Police officers as victims of violence: Findings of a Germany-wide survey. In Baier, D., & Pfeiffer, C. (Eds.), *Representative Studies on Victimisation: Research Findings from Germany (1st ed.). Nomos Verlagsgesellschaft mbH.* (pp. 139-162). https://www.jstor.org/stable/j.ctv941sz3.8

Gibbs, J. C., Lee, J., Moloney, J., & Olson, S. (2018). Exploring the neighbourhood context of serious assaults on police. *Policing and Society*, *28*(8), 898–914. DOI: 10.1080/10439463.2017.1333120

Gondal, A. Q., Ahmad, M., Raza, M. A., Bagum, U., & Husnain, M. (2023). Community policing for violence prevention in Pakistan: A critical review of evidence and future directions. *Research*, *8*(1), 53–75. DOI: 10.1234/re.v8.i1.4

Haoran, X., Verhage, A., & Vandeviver, C. (2022). Violence against on-duty police officers: The role of criminal motivation and premeditation in offenders' choice of weapons. *Policing*, *45*(6), 1023–1036. DOI: 10.1108/PIJPSM-04-2022-0062

Jackson, J. (2009). A psychological perspective on vulnerability in the fear of crime. *Psychology, Crime & Law*, *15*(4), 365–390. DOI: 10.1080/10683160802275797

John-Akinola, Y. O., Ajayi, A. O., & Oluwasanu, M. M. (2020). Experience of stress and coping mechanism among police officers in south western Nigeria. *International Quarterly of Community Health Education*, *41*(1), 7–14. DOI: 10.1177/0272684X19900878 PMID: 31969056

Kaminski, R. J., Jefferis, E., & Gu, J. (2003). Community Correlates of Serious Assaults on Police. *Police Quarterly*, *6*(2), 119–149. DOI: 10.1177/1098611103006002001

Krope, S. F., & Ilić, V. (2020). Assaults on police officers in Slovenia between 2007 and 2017. *Journal of Criminal Justice and Security*, *2*(2), 237–258.

Krope, S. F., & Lobnikar, B. (2015). Assaults on police officers in Slovenia – The profile of perpetrator and assaulted police officers. *Revija za kiminalistiko in kriminologijo*, *4*(4), 300-306. https://www.fvv.um.si/rv/arhiv/2018-2/06_Krope_Ilic_rV _2018-2.pdf

Kyriacou, D. N., Monkkonen, E. H., Peek-Asa, C., Lucke, R. E., Labbett, S., Pearlman, K. S., & Hutson, H. R. (2015). Police deaths in New York and London during the twentieth century. *Injury Prevention : Journal of the International Society for Child and Adolescent Injury Prevention, 12*(21), 219–224. DOI: 10.1136/ip.2005.010827 PMID: 16887942

Marty, O. L. (2022). Os fatores de influência no risco do policial militar ser vítima de crimes em decorrência da sua função [The factors influencing the risk of military police officers becoming victims of crime as a result of their work]. *Brazilian Applied Science Review, 6*(2), 1148–1170. DOI: 10.34115/basrv6n3-026

Meinen, A. R., Girardi, H. U., Martins, J. G. S., Lima, P. L., & Santos, Z. C. (2020). As principais mudanças percebidas pelos consumidores, ocasionadas pela sensação de medo e insegurança desencadeados pela violência urbana [The main changes perceived by consumers, occasioned by the feel of fear and insecurity unleashed by urban violence]. *Faculdade Sant'Ana em Revista, 4*(1), 88-100. https://www.iessa.edu.br/revista/index.php/fsr/article/view/1735

Mello, C. M. A., & Nummer, F. V. (2014). Policial militar: uma profissão de risco [Military police: a risky profession]. *Trabalho da 29ª Reunião Brasileira de Antropologia*. https://www.29rba.abant.org.br/resources/anais/1/1401917965_ARQUIVO_ArtigoRBA2014CesareFernanda.pdf

Minayo, M. C. S., & Adorno, S. (2013). Risco e (in) segurança na missão policial [Risk and (in) security in the police mission]. *Ciencia & Saude Coletiva, 18*(13), 585–593. DOI: 10.1590/S1413-81232013000300002 PMID: 23546183

Minayo, M. C. S., Souza, E. R., & Constantino, P. (2007). Riscos percebidos e vitimização de policiais civis e militares na (in)segurança pública [Perceived risks and victimization of military and civil police in the public (in)security domain]. *Cadernos de Saude Publica, 23*(11), 2767–2779. DOI: 10.1590/S0102-311X2007001100024 PMID: 17952268

Monteiro, V. F., Silva, S. S. C., Ramos, E. M. L. S., & Nascimento, R. G. (2020). Caracterização dos policiais feridos por arma de fogo [Characterization of policemen injured by firearms]. *Research. Social Development, 9*(9), e248997218. DOI: 10.33448/rsd-v9i9.7218

Oliveira, G. M., Araújo, T. M., & Carvalho, F. M. (2014). Insegurança e violência no trabalho dos delegados de polícia civil de Salvador, Brasil [Insecurity and violence in the work of civil police delegates in Salvador, Brazil]. *Revista Psicologia. Diversidade e Saúde, 2*(1), 14–23. DOI: 10.17267/2317-3394rpds.v3i1.372

Page, M. J., McKenzie, J. E., Bossuyt, P. M., Boutron, I., Hoffmann, T. C., Mulrow, C. D., Shamseer, L., Tetzlaff, J. M., Akl, E. A., Brennan, S. E., Chou, R., Glanville, J., Grimshaw, J. M., Hróbjartsson, A., Lalu, M. M., Li, T., Loder, E. W., Mayo-Wilson, E., McDonald, S., & Moher, D. (2021). The PRISMA 2020 statement: An updated guideline for reporting systematic reviews. *BMJ (Clinical Research Ed.)*, *372*(71), n71. Advance online publication. DOI: 10.1136/bmj.n71 PMID: 33782057

Page, M. J., Moher, D., Bossuyt, P. M., Boutron, I., Hoffmann, T. C., Mulrow, C. D., Shamseer, L., Tetzlaff, J. M., Akl, E. A., Brennan, S. E., Chou, R., Glanville, J., Grimshaw, J. M., Hróbjartsson, A., Mayo-Wilson, E., McDonald, S., & McKenzie, J. E. (2021). Prisma 2020 explanation and elaboration: Updated guidance and exemplars for reporting systematic reviews. *BMJ (Clinical Research Ed.)*, *372*(160), n160. Advance online publication. DOI: 10.1136/bmj.n160 PMID: 33781993

Pang, M., & Pavlou, P. A. (2019). On information technology and the safety of police officers. *Decision Support Systems*, *127*, 113143. DOI: 10.1016/j.dss.2019.113143

Peters, M. D., Godfrey, C. M., Khalil, H., McInerney, P., Parker, D., & Soares, C. B. (2015). Guidance for conducting systematic scoping reviews. *International Journal of Evidence-Based Healthcare*, *13*(3), 141–146. DOI: 10.1097/XEB.0000000000000050 PMID: 26134548

Rabe-Hemp, C. E., & Schuck, A. M. (2007). Violence against police officers: Are female officers at greater risk? *Police Quarterly*, *10*(4), 411–428. DOI: 10.1177/1098611107304326

Raigoso-Mayorga, J. A. (2016). Factores de riesgo psicossocial en oficiales de la policía [Psychosocial risk factors among police officers]. *Revista de Educación y Desarrollo*, 36, 33-37. https://www.cucs.udg.mx/revistas/edu_desarrollo/anteriores/36/36_Raigoso.pdf

Richie, B. E. (2022). The effects of violence on communities: The violence matrix as a tool for advancing more just policies. *Daedalus*, *151*(1), 84–96. DOI: 10.1162/daed_a_01890

Sani, A., Azevedo, V., Paulo, D., Magalhães, M., & Nunes, L. (2022). Satisfaction with the police: Perceptions and related variables from an urban community sample. *Social Sciences (Basel, Switzerland)*, *11*(2), 79. DOI: 10.3390/socsci11020079

Sani, A., Nunes, L., Azevedo, V., & Sousa, H. (2020). Campus criminal victimization among higher education students: A diagnosis of local security in Porto. *Journal of Criminal Justice Education*, *31*(2), 250–266. DOI: 10.1080/10511253.2019.1702219

Sani, A., & Nunes, L. M. (2013). *Crime e vitimação: polícia e análise de comunidades urbanas* [Crime and victimization: police and urban community analysis]. Edições Universidade Fernando Pessoa., DOI: 10.14417/ap.609

Saraiva, M., Neves, A. V., Diniz, M., Jota, L., & Ribeiro, P. (2018). A prevenção criminal através do espaço construído (CPTED) em Portugal: Revisão da literatura e redes de conhecimento [Criminal prevention through built space (CPTED) in Portugal: Literature review and knowledge networks]. *Revista Portuguesa de Estudos Regionais, 52*(52), 72–93. DOI: 10.59072/rper.vi52.465

Sierra-Arévalo, M., & Nix, J. (2020). Gun victimization in the line of duty. *Criminology & Public Policy, 19*(19), 1041–1066. DOI: 10.1111/1745-9133.12507 PMID: 38549863

Silva, J. B., & Almeida, A. J. S. (2022). Vitimação policial: Diagnósticos e perspectivas [Vitimação policial: Diagnósticos e perspectivas]. *Revista Brasileira de Segurança Pública, 16*(2), 240–263. DOI: 10.31060/rbsp.2022.v16.n2.1376

Silva, J. B., & Júnior, A. L. S. (2021). Vitimação policial militar no Rio Grande do Norte: Causas, consequências e perspectivas [Military police victimization in Rio Grande do Norte: Causes, consequences and perspectives]. *Vigilantis Semper – Revista Científica de Segurança Pública (RCSP), 1*(1) 77-86.

Silva, M. A., & Bueno, H. P. V. (2017). O suicídio entre policiais militares na polícia militar do Paraná: Esforços para prevenção [Suicide among military police officers in Paraná: Efforts towards prevention]. *Revista de Ciências Policiais APMG, 1*(1), 5-23. https://www.apmg.pr.gov.br/sites/apmg/arquivos_restritos/files/documento/2020-01/01_o_suicidio_entre_policiais_militares_na_policia_militar_do_parana.pdf

Simmler, M., Stempkowski, M., & Markwakder, N. (2021). Punitive attitudes and victimization among police officers in Switzerland: An empirical exploration. *Police Practice and Research, 22*(2), 1191–1208. DOI: 10.1080/15614263.2019.1697264

Singo, C., & Shilubane, H. N. (2022). How do police officers cope with stress? A qualitative study. *Journal of Police and Criminal Psychology, 37*(4), 984–992. DOI: 10.1007/s11896-022-09556-0

Soares, A. M. C. (2014). Dialética da violência: Civilidade e incivilidade [Dialectic of violence: Civility and incivility]. *Prelúdios, 3*(3), 11–35. DOI: 10.9771/revpre.v3i3.14209

Souza, E. R., & Minayo, M. C. S. (2005). Policial, riscos como profissão: Morbimortalidade vinculada ao trabalho [Police officer, risks as a profession: morbidity and mortality linked to work]. *Ciencia & Saude Coletiva, 10*(4), 917–928. DOI: 10.1590/S1413-81232005000400015

Souza, K. L. N. (2017). Os riscos da atividade policial [The risks of police activity]. *Revista CEJ. Brasília*, (73), 63-66. https://www.mpsp.mp.br/portal/page/portal/documentacao_e_divulgacao/doc_biblioteca/bibli_servicos_produtos/bibli_informativo/bibli_inf_2006/Rev-CEJ_n.73.06.pdf

Teles, D. O., Oliveira, R. A., Parnaíba, A. L. O., Rios, M. A., Machado, M. B., Aquino, P. S., Menezes, P. R., Ribeiro, S. G., Soares, P. R. A. L., Biazus Dalcin, C., & Pinheiro, A. K. B. (2024). Assessment of the mental health of police officers: A systematic review of specific instruments. *International Journal of Environmental Research and Public Health, 21*(10), 1300. DOI: 10.3390/ijerph21101300 PMID: 39457273

Tricco, A. C., Lillie, E., Zarin, W., O'Brien, K. K., Colquhoun, H., Levac, D., Moher, D., Peters, M. D., Horsley, T., Weeks, L., Hempel, S., Akl, E. A., Chang, C., McGowan, J., Stewart, L., Hartling, L., Aldcroft, A., Wilson, M. G., Garritty, C., & Strauss, S. (2018). PRISMA extension for scoping reviews (PRISMA-ScR): Checklist and explanation. *Annals of Internal Medicine, 169*(7), 467–473. DOI: 10.7326/M18-0850 PMID: 30178033

Wickramasinghe, N. D., Wijesinghe, P. R., Dharmaratne, S. D., & Agampodi, S. B. (2016). The prevalence and associated factors of depression in policing: A cross sectional study in Sri Lanka. *SpringerPlus, 5*(1776). DOI: 10.1186/s40064-016-3474-9

Willts, D. W. (2014). The organisational structure of police departments and assaults on police officers. *International Journal of Police Science & Management, 16*(2), 140–154. DOI: 10.1350/ijps.2014.16.2.334

Zavala, E. (2020). When the cop is the victim: A test of target congruence theory on intimate partner violence victimization experienced by police officers. *Journal of Interpersonal Violence, 35*(17-18), 3622–3642. DOI: 10.1177/0886260517709801 PMID: 29294764

Zouain, D. M., Neto, V. R., & Zamith, J. L. C. (2006). Policiamento comunitário: Fator estratégico para implantação de política pública de segurança [Community policing: Strategic factor for implementing public security policy]. *Alcance – Univali, 13*(3), 373-388. https://www.redalyc.org/pdf/4777/477748623005.pdf

ADDITIONAL READING

Bent-Goodley, T., St. Vil, C., Cuevas, C. A., & Abbey, A. (2022). Police, violence, and social justice: A call for research and introduction to the special issue. *Psychology of Violence*, *12*(4), 195–200. DOI: 10.1037/vio0000438

Eliasson, M. N. (2023). Police Officers: Invisible victims in the line of duty. In I. Management Association (Ed.), *Research Anthology on Modern Violence and Its Impact on Society* (pp. 783-804). IGI Global Scientific Publishing. https://doi.org/ DOI: 10.4018/978-1-6684-7464-8.ch043

Ellrich, K., & Baier, D. (2017). Post-traumatic stress symptoms in police officers following violent assaults: A study on general and police-specific risk and protective factors. *Journal of Interpersonal Violence*, *32*(3), 331–356. DOI: 10.1177/0886260515586358 PMID: 25990384

Freitas, P., Nunes, L., Gouveia, F., Guerreiro, M., & Sani, A. (2019). Smart cities and security – a preventive approach. A. Muñoz, S. Ouhbi, W. Minker, L. Echabbi, & M. Navarro-Cía. Intelligent Environments 2019. *Workshop Proceedings of the 15th International Conference on Intelligent Environments* (serie 26, pp. 216-221). IOS Press. https://doi.org/. ISBN: 978-1-61499-982-9 (print) lDOI: 10.3233/AISE190045

Management Association. I. (Ed.). (2019). *Police Science: Breakthroughs in Research and Practice*. IGI Global. https://doi.org/DOI: 10.4018/978-1-5225-7672-3

Moreno, A. F., Karanika-Murray, M., Batista, P., Hill, R., Vilalta, S. R., & Oliveira-Silva, P. (2024). Resilience training programs with police forces: A systematic review. *Journal of Police and Criminal Psychology*, *39*(1), 227–252. DOI: 10.1007/s11896-023-09633-y

Nunes, L., & Sani, A. (Eds.). (2021). *Manual de Criminologia e Vitimologia* [Criminology and Victimology Handbook]. Pactor.

Nunes, L., Sani, A., Azevedo, V., Magalhães, M., Maia, R. L., Freitas, P., Estrada, R., Caridade, S., Guerreiro, M.J., Gouveia, F., Oliveira, G., & Dinis, M.A. (2021). *Segurança urbana. Boas práticas para a prevenção criminal* [Urban security. Good practices for crime prevention]. Pactor. ISBN eBook: 978-989-693-139-1 https://doi.org/.DOI: 10.54670/NZUH5479

Sani, A., & Nunes, L. (2020). Tempos modernos: (des)construções da urbe, da segurança e do policiamento. In Estrada, R. (Eds.), *Crime, (In)segurança e mediatização* (pp. 125–139). Edições UFP.

Sani, A. I., Nunes, L., Caridade, S., & Azevedo, V. (2020). Violence in urban community: diagnosis of local security and victimization. In Knudsen, M. L. (Ed.), *Victims of Violence: Support, Challenges and Outcomes* (pp. 291–314). Nova Science Publishers, Inc.

KEY TERMS AND DEFINITIONS

Body-Worn Cameras: Small video recording devices, typically used by law enforcement officers, security personnel, and other professionals to capture audio and video footage of their interactions with the public.

Military Police: Police officers who are part of a militarised security force who are responsible for enforcing laws and regulations, maintaining order and discipline, and conducting investigations related to personnel and activities.

Police Officers: An individual with specific training/preparation to carry out duties that ensure public order through the upholding of the social contract, guaranteeing the observance of rights and duties for all.

Police Victimization: The act or effect of being a victim of violent and/or criminal behaviour committed by others against members of the security forces.

Public Safety: Measures and strategies put in place by government agencies and communities to ensure the protection and well-being of the general public, that may include maintaining law and order, preventing crime, responding to emergencies, enforcing regulations, and promoting overall safety and security in public spaces.

Risk of Victimization: Refers to the likelihood that an individual or group will be targeted or harmed by violence, abuse, criminal behaviour, or other forms of victimization as result of a series of factors, such as demographic characteristics, environmental conditions, or personal behaviours.

Security Forces: Entities that serve citizens to ensure their safety and maintain social order, protecting individuals and property, and enforcing laws or regulations within a specific area. These forces can include military personnel, police officers, private security guards, or any other organization dedicated to ensuring the safety and security of a community or location.

Social Formal Control: Group of entities such as governments, legal systems, and organizations that guarantee compliance with rules, legal sanctions, disciplinary actions, and bureaucratic processes established to regulate behaviours, order and stability in a community.

Violence Against the Police: Any type of physical or verbal aggression or harm directed towards law enforcement officers while they are carrying out their duties, such as assault, threats, harassment, or any form of aggressive behavior that puts the safety and well-being of police officers at risk.

Chapter 7
Developing Strategy for Employee Engagement

Lord Dordunoo
https://orcid.org/0009-0007-7395-2677
Marymount University, USA

ABSTRACT

Employee engagement is now a key driver of organizational success, linked directly to productivity, retention, and performance. As workplaces evolve, organizations recognize the need for clear strategies to enhance engagement by aligning objectives with individual motivations, fostering transparent communication, and cultivating a culture of inclusion, recognition, and trust. This chapter provides a framework integrating research and theory to help leaders build strategies that address current employee needs and future trends. It examines the role of transformational leadership, employee autonomy, and work-life balance in maintaining engagement. Practical applications for organizations of various sizes and industries are discussed, with actionable insights for leaders and HR professionals. Recommendations emphasize the need for ongoing evaluation of strategies in light of remote work trends and increasing workforce diversity. This holistic approach contributes to knowledge on employee engagement and offers a roadmap to improve employee experience and achieve lasting business success.

INTRODUCTION

Employee engagement has gained substantial attention in academic research and organizational practice, primarily due to its direct impact on critical outcomes such as productivity, job satisfaction, and organizational loyalty (Zhenjing et al., 2022). Engaged employees exhibit higher levels of discretionary effort, are more

DOI: 10.4018/979-8-3373-5686-0.ch007

invested in their work, and often demonstrate a stronger emotional connection to their organization, which leads to improved performance and innovation (Bakker & Demerouti, 2007; Schaufeli, 2013). Research consistently shows that companies with higher engagement levels experience lower turnover rates, better customer satisfaction, and enhanced financial performance (Harter, Schmidt, & Hayes, 2002). For example, Gallup (2022) found that organizations with high employee engagement levels outperform their peers by 23% in profitability and experience 18% less turnover (Clifton, 2022).

In today's dynamic and increasingly competitive workplace, fostering employee engagement has become more than just a human resource initiative: it is a critical organizational strategy (Zhenjing et al., 2022). This shift reflects a glowing understanding that employee engagement directly influences overall business success, as engaged employees tend to align their personal goals with those of the organization, driving positive outcomes (Gruman & Saks, 2011). As Saks (2006) emphasizes, employee engagement is not simply about motivation but encompasses deeper elements of commitment, psychological safety, and purpose, all essential for sustaining performance over time. As organizations increasingly prioritize engagement, they position themselves to thrive in a competitive market by leveraging the full potential of their workforce (Osborne & Hammoud, 2017).

Recent studies underscore the need for strategic approaches to engagement, moving beyond transactional incentives to create environments where employees feel valued and supported (Osborne & Hammoud, 2017). For example, transformational leadership, characterized by inspiring and empowering employees, has significantly fostered engagement by creating a culture of trust, collaboration, and continuous development (Tims, Bakker, & Xanthopoulou, 2011). Similarly, Kahn's (1990) foundational work on employee engagement highlights the importance of psychological conditions such as meaningfulness, safety, and availability, which allow employees to bring their full selves to work. Leaders who actively cultivate these conditions tend to see higher levels of engagement, as employees feel personally invested in their work and confident that their contributions are valued (Mazzetti & Schaufeli, 2022).

Given the importance of engagement to organizational success, this chapter aims to define employee engagement comprehensively, examining its multidimensional nature and the factors influencing it. Central to this exploration is identifying key drivers of engagement, including leadership style, organizational culture, job design, and work-life balance (Salanova & Schaufeli, 2008). The chapter also seeks to provide a strategic framework for enhancing engagement in modern organizations, recognizing that engagement must be continuously nurtured in response to evolving employee expectations and organizational challenges. By offering a research-driven approach, this chapter equips leaders and human resource professionals with the tools needed to build and sustain high levels of engagement across diverse workforces.

Problem Statement

Despite the increasing recognition of its critical role in organizational success, many organizations still need to grapple with the complexities of developing effective strategies to foster employee engagement (Santoso, Sulistyaningtyas, & Pratama, 2022). One of the primary challenges lies in the inconsistent levels of engagement across different teams, departments, and even within the same organization (Stein et al., 2021). As Macey and Schneider (2008) argue, employee engagement is a multifaceted construct requiring organizations to address various psychological, social, and environmental factors simultaneously. The absence of a coherent and well-articulated engagement strategy can result in fragmented efforts that fail to achieve sustainable improvements. This fragmentation can manifest in inconsistent leadership practices, poorly designed job roles, and inadequate recognition systems, all contributing to disengagement (Gruman & Saks, 2011).

Inconsistent engagement levels often lead to tangible adverse outcomes for organizations, including high employee turnover, diminished productivity, and deteriorating workplace culture (Harter, 2022). High turnover is a costly consequence of disengagement, as organizations must bear the financial burden of recruiting, onboarding, and training new employees (Harter et al., 2002). Gallup (2022) reported that only 32% of employees are actively engaged, while a staggering 17% are actively disengaged, meaning they are not just unmotivated but also negatively affect the morale of their colleagues (Harter, 2022). This disengagement translates into significant financial losses, with estimates suggesting that disengaged employees cost companies $8.8 trillion annually in lost productivity (Dennison, 2024).

The root of this challenge lies in the difficulty organizations face in aligning employee engagement efforts with broader organizational goals. Engagement is not a one-size-fits-all concept; what motivates and engages one employee may not work for another (Saks, 2006). For example, while some employees may value opportunities for professional development and career growth, others may prioritize work-life balance or recognition from their peers (Shiri et al., 2023). Therefore, developing an engagement strategy requires a nuanced understanding of these diverse needs and how they relate to organizational objectives. Macey and Schneider (2008) emphasize that engagement strategies must be comprehensive and systematic, incorporating clear objectives, measurable outcomes, and continuous feedback loops to be effective. Without such a framework, efforts to improve engagement risk being ad hoc and short-lived, leading to further disengagement and dissatisfaction.

Moreover, leadership plays a crucial role in shaping engagement strategies (Huang et al., 2022). Leaders who fail to model engaged behaviors or do not actively promote a culture of engagement can undermine even the best-intentioned initiatives (Tims et al., 2011). For example, engagement suffers when leaders must communicate organizational goals or provide employees with the resources to succeed (Lewis, 2022). Therefore, organizations must integrate leadership development into their engagement strategies to ensure leaders at all levels understand their role in fostering engagement (Saks & Gruman, 2021). By aligning engagement efforts with employee needs and organizational goals, companies can create a more engaged workforce that drives sustainable success.

Significance of the Study

The study of employee engagement is crucial because it reveals how organizations can leverage their human capital to create a sustainable competitive advantage (Ahmeti, 2023). Human capital, often regarded as one of an organization's most valuable assets, is pivotal in driving innovation, productivity, and long-term success (Bassi & McMurrer, 2007). Engaged employees who are deeply committed to their work and organizational mission consistently demonstrate higher performance, adaptability, and creativity (Bakker & Demerouti, 2007). These employees go beyond merely fulfilling their job responsibilities; they proactively contribute to organizational improvement and often serve as catalysts for innovation. Research by Harter et al., (2002) confirm that highly engaged teams outperform disengaged counterparts across key performance indicators, including profitability, productivity, and customer satisfaction, underscoring the importance of engagement as a strategic lever for organizational success.

Furthermore, engaged employees are more resilient in the face of challenges and are better equipped to handle periods of organizational change or uncertainty (Saks & Gruman, 2021). This resilience is especially critical in today's rapidly changing business landscape, where organizations must continuously adapt to technological advancements, market fluctuations, and shifting workforce demographics (Deep, 2023). Engaged employees tend to have stronger emotional connections to their organizations, which helps them remain motivated during adversity and embrace new growth opportunities (David, 2016). Their positive attitudes and behaviors also contribute to a more supportive and collaborative workplace environment, fostering a culture of trust, openness, and shared purpose (Salanova & Schaufeli, 2008). In this way, engagement enhances individual performance and contributes to a more cohesive and high-performing organizational climate.

The chapter's contribution to the broader conversation on employee engagement lies in its structured approach to developing an engagement strategy that addresses organizational and individual needs. Engagement efforts have often been treated as isolated HR initiatives focused on boosting morale or job satisfaction (De-la-Calle-Durán & Rodríguez-Sánchez, 2021). However, contemporary research suggests that to be effective, engagement strategies must be deeply embedded within the organizational fabric and aligned with its strategic objectives (Saks, 2006). This chapter emphasizes the importance of designing engagement strategies that reflect a holistic understanding of the factors driving engagement at both the individual and organizational levels. Such factors include leadership practices, organizational culture, job design, and opportunities for professional development (Gruman & Saks, 2011). When organizations tailor their engagement efforts to meet these diverse needs, they can cultivate an environment where employees feel motivated to contribute their best efforts, ultimately driving better business outcomes (Thomas, 2009).

A structured engagement strategy can also address the growing need for innovation in today's competitive markets (Pisano, 2015). Engaged employees are more likely to take the initiative, share ideas, and collaborate on creative problem-solving, which is critical for fostering innovation (Bakker & Demerouti, 2007). Moreover, by promoting an environment that encourages open communication and recognizes employee contributions, organizations can tap into the collective intelligence of their workforce, unlocking new solutions and opportunities for growth (Salanova & Schaufeli, 2008). This chapter outlines critical elements of an effective engagement strategy, providing leaders with practical tools to align engagement with broader organizational goals and drive continuous improvement.

Finally, understanding the factors that drive engagement can significantly improve employee retention, performance, and organizational innovation (Stein et al., 2021). Studies show engaged employees are less likely to leave their jobs, reducing turnover and recruitment and training costs (Harter et al., 2002). When employees feel valued and engaged, they are more likely to be productive, creative, and invested in the organization's success (Salanova & Schaufeli, 2008). In this context, engagement serves as both a means and an end: a means to improve organizational performance and innovation and an end in terms of fostering a more resilient and adaptable workforce. This chapter's exploration of engagement strategies offers a comprehensive framework for leaders aiming to harness the full potential of their workforce, ensuring sustained competitive advantage through human capital.

DISCUSSION

Key Drivers of Employee Engagement

Effective employee engagement strategies are multifaceted, addressing various organizational, interpersonal, and individual factors (Toseef et al., 2022). Among these, leadership style, organizational culture, and work-life balance are critical drivers of engagement (Toseef et al., 2022).

Leadership Style

Leadership plays a central role in fostering an environment where employees feel motivated, valued, and committed to their work (Huang et al., 2022). Transformational leadership, which involves inspiring and motivating employees through a compelling vision, individualized support, and intellectual stimulation, has been consistently linked to higher engagement levels (Tims et al., 2011). Transformational leaders focus on task completion and employee development, creating a work environment where individuals feel empowered to contribute beyond their formal job descriptions (Lin et al., 2020). Research has shown that when leaders demonstrate behaviors associated with transformational leadership such as showing genuine concern for employee well-being, articulating a clear and inspiring vision, and encouraging innovation, employees are more likely to be engaged (Mazzetti & Schaufeli, 2022). This is because employees feel a stronger sense of purpose, autonomy, and trust in leadership, which fosters more excellent emotional investment in their work (Lee, Willis, & Tian, 2018).

Organizational Culture

Organizational culture is another significant determinant of employee engagement (Kim & Jung, 2022). A culture that promotes open communication, autonomy, and inclusion provides a foundation for sustained engagement (Stein et al., 2021). Open communication ensures that employees feel heard and valued, essential for fostering a sense of belonging and commitment (Schaufeli, 2013). When employees are given autonomy in their work, they feel trusted and empowered to take ownership of their tasks, leading to increased motivation and engagement (Deci & Ryan, 2000). Inclusion, which involves recognizing and valuing diversity within the workplace, ensures that all employees feel respected and appreciated for their unique contributions, further enhancing engagement (Miller, 2023). Research by Qasim et al. (2022) indicate that employees in inclusive workplaces report higher levels of psychological safety, which is a critical factor in engagement. These cultural factors create a work environment where employees feel secure, valued, and motivated to engage with their tasks fully (Zhenjing et al., 2022).

Work-Life Balance

Work-life balance has also emerged as a crucial driver of employee engagement, particularly in the post-pandemic workforce, where flexible work arrangements have become increasingly common (Chan et al., 2022). The COVID-19 pandemic shifted employee expectations, with many now prioritizing flexibility in their work schedules and locations (Prager, Rhoads, & Martinez, 2022). Studies have shown that flexible work arrangements such as remote work options, flexible hours, or compressed workweeks can significantly enhance employee engagement, improving job satisfaction and reducing stress (Allen, Golden, & Shockley, 2015). Flexibility allows employees to balance their personal and professional lives more effectively, enhancing their overall well-being and productivity (Kossek, Gettings, & Misra, 2021). Organizations that do not offer flexibility risk disengaging employees who seek greater control over their work environments, especially in a job market where flexibility is essential for attracting and retaining talent (Crail & Watts, 2024).

CHALLENGES IN ENGAGEMENT STRATEGY IMPLEMENTATION

While the benefits of employee engagement are well-documented, implementing an effective engagement strategy is fraught with challenges. One of the primary obstacles is the need for more engagement in the broader organizational strategy (Harter, 2024). Too often, engagement is seen as a standalone HR initiative rather than a core component of the organization's overall business strategy (Cohen, 2017). Gruman and Saks (2011) argue that organizations should embed engagement in their strategic vision, clearly linking it to performance metrics, leadership development, and organizational culture. Without this integration, engagement initiatives tend to be reactive, temporary, or superficial, addressing symptoms of disengagement rather than the root causes (Tindall et al., 2020). For example, offering financial incentives to boost engagement may yield short-term results but will likely sustain long-term commitment if deeper issues, such as poor leadership or a toxic culture, are addressed (Macey & Schneider, 2008). Organizations that view engagement as a strategic imperative are more likely to implement comprehensive, sustainable initiatives that align with their long-term goals (Gong et al., 2023).

Engagement Measurement

Another significant challenge in implementing engagement strategies is accurately measuring engagement (Cappelli & Eldor, 2019). Traditional engagement surveys, which often rely on self-reported data, may not capture the full range of employee experiences or the complexity of engagement levels (Huebner & Zacher, 2021). For instance, some employees may show high productivity and appear engaged, but fear or obligation, rather than genuine motivation or commitment, may drive their engagement (Saks & Gruman, 2021). This false engagement can lead to burnout and turnover, even if initial performance metrics suggest high engagement (Saks & Gruman, 2021). Moreover, engagement is not a static concept; it can fluctuate over time based on changes in leadership, work conditions, or personal circumstances (Boccoli, Gastaldi, & Corso, 2022). As a result, relying solely on annual or biannual surveys may not provide a complete or timely picture of engagement, making it difficult to implement targeted interventions when needed (Huebner & Zacher, 2021).

Resistance to Cultural Change

Resistance to cultural change is another barrier to successful engagement strategy implementation (Lawrence, 1969). Both employees and management can resist the changes required to foster a more engaged workforce (Albrecht et al., 2020). For example, introducing a more open and inclusive communication culture may challenge existing organizational power structures or norms, leading to push back from leadership or long-tenured employees (Kahn, 1990). Engagement initiatives often require organizations to significantly restructure work by offering more autonomy or flexible work arrangements, which can encounter resistance from managers accustomed to traditional command-and-control management styles (Schein, 2010). Overcoming this resistance requires a concerted effort to change organizational mindsets and behaviors, which can be long and complex (Hollister et al., 2021). Leadership buy-in is critical to overcoming resistance, as leaders set the tone for how engagement is perceived and prioritized within the organization (Mazzetti & Schaufeli, 2022). Engagement initiatives will likely only succeed with strong leadership support, leading to employee disengagement and dissatisfaction (Alton, 2024).

While the drivers of engagement such as leadership, culture, and work-life balance are well understood, implementing effective engagement strategies requires addressing significant challenges (Osborne & Hammoud, 2017). Organizations must integrate engagement into their overall strategy, develop more nuanced measurement tools, and overcome cultural resistance to change (Bridges, 2024). By tackling these challenges, organizations can create more sustainable engagement initiatives that lead to higher productivity, innovation, and employee satisfaction (Stein et al., 2021).

THEORETICAL APPLICATIONS

Employee engagement is grounded in several foundational organizational behavior theories, each offering a unique lens to understand the factors that drive employee commitment, motivation, and performance (Pincus, 2022). These theoretical frameworks help explain why engagement occurs and the mechanisms through which organizations can foster and sustain high levels of engagement among their workforces (Albrecht, Furlong, & Leiter, 2023). Three of the most influential theories in this area are Kahn's (1990) Engagement Theory, the Job Demands-Resources (JD-R) Model, and the Social Exchange Theory, which highlight the dynamic interplay between employee needs, organizational resources, and perceived support (Albrecht, Connaughton, & Leiter, 2022).

Kahn Engagement Theory

One of the most prominent frameworks for understanding employee engagement is Kahn's (1990) Engagement Theory, which posits that employees are engaged when they experience three psychological conditions: meaningfulness, safety, and availability. Meaningfulness refers to how employees find purpose in their work and feel that their contributions are valued (Martela & Pessi, 2018). When employees perceive that their work aligns with their values and positively impacts others or the organization, they are more likely to be motivated and emotionally invested in their roles (Kahn, 1990). Research by May, Gilson, and Harter (2004) supports this, showing that when employees experience meaningfulness in their work, they report higher levels of engagement and satisfaction. The second condition, safety, involves the feeling that one can express oneself at work without fear of negative consequences (Gallo, 2023). This psychological safety is essential for employees to feel comfortable sharing ideas, taking risks, and engaging fully with their work (Gallo, 2023). Edmondson's (1999) research on psychological safety reinforces this, suggesting that environments where employees feel safe to speak up and innovate are more likely to see higher levels of engagement. Finally, availability refers to the physical and emotional resources employees have at their disposal to meet the demands of their work (Huang et a., 2022). When employees feel they have adequate resources such as time, energy, and support they are more likely to be engaged and perform effectively (Kahn, 1990).

Job Demands-Resources

The Job Demands-Resources (JD-R) Model, developed by Bakker and Demerouti (2007), further elaborates on the factors influencing employee engagement by focusing on the balance between job demands and available resources. According to the JD-R model, job demands are the physical, psychological, or emotional efforts required to perform work tasks, while job resources are the physical, social, and organizational assets that help employees meet these demands (Galanakis & Tsitouri, 2022). The model suggests that high job demands such as excessive workload, time pressure, or emotional strain can lead to stress and burnout if not counterbalanced by sufficient resources (Bakker & Demerouti, 2007). However, when job resources such as supervisor support, autonomy, opportunities for development, and feedback are available, they act as buffers against the adverse effects of job demands, fostering engagement instead of burnout (Bakker, Demerouti, & Euwema, 2005). Job resources help employees cope with job demands and provide intrinsic and extrinsic motivation, enhancing engagement. For example, providing employees autonomy over tasks has increased engagement by fostering a sense of ownership and control (Bakker et al., 2005). In another development, social support from colleagues and supervisors can reduce stress and increase feelings of belonging, boosting engagement (Basit & Nauman, 2023). This model highlights the importance of designing jobs that balance demands and resources to optimize engagement and prevent burnout.

Social Exchange Theory

The third theoretical framework, Social Exchange Theory (SET), provides a relational perspective on employee engagement, emphasizing the reciprocal nature of the employee-employer relationship (Ahmad et al., 2023). According to SET, individuals engage in relationships with a perceived balance of give and take (Ahmad et al., 2023). In the workplace context, when employees perceive that their organization values their contributions and provides support whether through fair compensation, career development opportunities, or a positive work environment they feel obligated to reciprocate through greater engagement and effort (Cropanzano & Mitchell, 2005). Perceived organizational support (POS) is a critical component of this exchange, as it shapes employees' perceptions of how much the organization cares about their well-being and recognizes their contributions (Rhoades & Eisenberger, 2002). Research consistently shows that employees who perceive high levels of organizational support are probable to report higher engagement, as they feel a stronger emotional bond to the organization and are motivated to reciprocate by going above and beyond (Saks, 2006). This sense of reciprocity fosters a virtuous cycle: the more support employees perceive, the more engaged they become, leading to higher performance, and eliciting further organizational support (Weisman et al., 2023). SET thus underscores the importance of building a supportive organizational culture where employees feel valued and appreciated, as this can significantly enhance engagement and commitment.

These theoretical perspectives converge to provide a comprehensive understanding of the drivers of employee engagement. Kahn's Engagement Theory emphasizes that employees must meet certain psychological conditions to be engaged, highlighting the importance of meaningful work, psychological safety, and resource availability (Albrecht et al., 2023). The JD-R Model highlights the importance of balancing job demands with adequate resources, demonstrating that autonomy, social support, and feedback are crucial for maintaining engagement in the face of high demands (Tummers & Bakker, 2021). Social Exchange Theory brings in the relational dimension of engagement, showing how perceived organizational support creates a sense of obligation and commitment, motivating employees to engage more fully with their work (Ahmad et al., 2023). These theories provide a robust framework for understanding and improving employee engagement in modern organizations.

In practice, these theories suggest that organizations should take a holistic approach to engagement, ensuring that employees find their work meaningful and feel supported and empowered to succeed (Ahmad et al., 2023). This includes providing clear communication about the organization's mission and values (to foster meaningfulness), creating a psychologically safe environment (to encourage openness and innovation), balancing job demands with sufficient resources (to prevent burnout), and building relational solid ties through perceived organizational support (to strengthen the employee-organization bond) (Kleynhans et al., 2022). By addressing these multiple dimensions of engagement, organizations can create a work environment where employees feel motivated, valued, and capable of performing at their best.

IMPLICATIONS

Developing an effective employee engagement strategy has profound and far-reaching implications for organizational success (Whittington et al., 2017). It influences critical performance metrics and shapes the organization's long-term trajectory (Zhenjing et al., 2022). Engagement is not merely about improving morale; it is a strategic tool that directly impacts organizational health, innovation, and competitive positioning (Chanana & Sangeeta, 2020). Organizations that successfully implement robust engagement strategies often see a cascade of positive outcomes, from improved employee well-being and job satisfaction to enhanced innovation and a strengthened employer brand that attracts top talent (Whittington et al., 2017).

Employee Well-Being and Job Satisfaction

First, organizations prioritizing employee engagement are more likely to see improvements in employee well-being and job satisfaction, which strongly correlate with reduced absenteeism and lower turnover rates (American Psychological Association (APA), 2023). Engaged employees are typically more satisfied with their jobs because they feel valued, supported, and connected to their work and organizational mission (Harter et al., 2002). Job satisfaction, in turn, is a critical determinant of employee retention (Chen et al., 2023). When employees are satisfied and more likely to remain committed to the organization, turnover rates decrease, reducing the costs associated with recruiting, onboarding, and training new hires (Mazzetti & Schaufeli, 2022). Also, higher engagement levels lead to lower absenteeism, as employees who are invested in their work are more likely to show up consistently and contribute to their teams (Mazzetti & Schaufeli, 2022). In a meta-analysis of engagement across industries, (Harter, 2022) found that organizations with highly engaged employees experienced 37% lower absenteeism and 25-65% lower turnover, depending on the industry. These reductions in absenteeism and turnover decrease direct costs and enhance operational efficiency, as organizations can better maintain continuity in their workforce (Knight, Becan, & Flynn, 2013).

Employee Engagement

Second, employee engagement plays a crucial role in driving organizational innovation (Wan et al., 2022). Engaged employees are more likely to go beyond their basic job requirements, actively contributing creative ideas and solutions to the organization (Salanova & Schaufeli, 2008). Engagement fosters an intrinsic motivation to improve processes, solve problems, and contribute to the organization's success (Pandya, 2024). According to Bakker and Demerouti (2007), engaged employees exhibit higher levels of job crafting, a proactive behavior where individuals shape their roles to better fit their strengths and interests. This process often leads to innovative solutions and improvements that benefit the organization.

Also, engaged employees are more likely to collaborate with colleagues, exchange ideas, and challenge the status quo behaviors essential for fostering a culture of innovation (Carmeli & Spreitzer, 2009). In environments where engagement is high, employees feel psychologically safe to experiment, take risks, and propose new ideas without fear of retribution (Edmondson, 1999). This psychological safety and high organizational commitment create fertile ground for innovation (Gallo, 2023). Companies like Google and 3M, known for their innovative cultures, have long prioritized engagement and autonomy as key drivers of creative performance (Barker Scott & Manning, 2024).

Reputational Benefits

Finally, the reputational benefits of a strong engagement strategy are substantial, particularly when it comes to attracting and retaining top talent (Rodríguez-Sánchez et al., 2020). In an increasingly competitive labor market, where organizations are vying for the best talent, companies known for their strong engagement practices are better positioned to attract high-caliber candidates (Gardner, 2016). Employer branding has become a critical element of recruitment, and organizations prioritizing employee engagement build a reputation as desirable workplace places (Thang & Trang, 2024). Engaged employees reinforce this reputational advantage by generating positive word-of-mouth, as they are more likely to act as brand ambassadors and advocate for the organization within their networks (Bakker & Schaufeli, 2008). Studies have shown that employees who feel engaged and connected to their organization are more likely to recommend their employer to others, enhancing the organization's ability to attract top talent (Harter et al., 2002). This creates a virtuous cycle as organizations attract and retain talented, engaged individuals better positioned to innovate, grow, and further strengthen their competitive advantage.

Diversity, Equity, and Inclusion

Moreover, a strong engagement strategy also contributes to better diversity, equity, and inclusion (DEI) outcomes, which are increasingly important for attracting talent, particularly among younger workers who prioritize DEI in their job search (Dowling et al., 2022). Organizations that foster inclusive engagement strategies, ensuring that all employees regardless of background feel valued and have equal opportunities for development and recognition, are more likely to attract a diverse workforce (Saks, 2006). This diversity, in turn, fuels creativity and innovation, as diverse teams bring a more comprehensive range of perspectives and experiences to problem-solving and decision-making (Page & Vella-Brodrick, 2009). Therefore, by developing and implementing comprehensive engagement strategies, organizations can boost employee satisfaction and innovation and enhance their reputation as leaders in talent development and DEI (Okatta, Ajayi, & Olawale, 2024).

Developing an effective employee engagement strategy has significant implications for organizational success across multiple dimensions (Stein et al., 2021). By fostering engagement, companies can enhance employee well-being and job satisfaction, reduce absenteeism and turnover, and improve retention (Harter, 2017). Engaged employees are more likely to contribute to organizational innovation by proposing creative solutions and engaging in collaborative problem-solving (Wan et al., 2022). Finally, a strong engagement strategy bolsters an organization's employer brand, attracting top talent and reinforcing the cycle of growth and innovation (Adams,

2022). As organizations increasingly recognize the strategic value of engagement, those who implement effective, comprehensive engagement strategies will be better positioned to thrive in the competitive and dynamic business environment of the future (Webb, 2024).

RECOMMENDATION

Organizations aiming to improve employee engagement must adopt a multifaceted strategy addressing leadership, organizational culture, flexibility, and integration into the broader business strategy (Huang et al., 2022). Employee engagement is a complex and dynamic construct that requires a comprehensive approach to foster long-term commitment, satisfaction, and productivity (Osborne & Hammoud, 2017). To achieve this, organizations should implement leadership development programs that emphasize transformational leadership, cultivate a culture of open communication and psychological safety, offer flexible work arrangements, and align engagement initiatives with the overarching business strategy, ensuring sustained engagement through measurable outcomes (Maximo, Stander, & Coxen, 2019). A vast body of research supports these strategies by highlighting their individual and collective impact on engagement levels.

1. Leadership development programs focusing on transformational leadership are crucial for inspiring and motivating employees (Lin et al., 2020). Transformational leaders go beyond transactional management by creating a vision that inspires employees and fosters an emotional connection with their teams (Steinmann, Klug, & Maier, 2018). Research has consistently shown that transformational leadership is strongly associated with higher levels of employee engagement (Tims et al., 2011). This leadership style emphasizes individualized consideration, where leaders show genuine concern for their employees' needs and development, and intellectual stimulation, which encourages creativity and innovation in the workplace (Bass, 1985). When leaders demonstrate these behaviors, employees are more likely to feel valued and motivated, which enhances their engagement (Mazzetti & Schaufeli, 2022). Studies by Mazzetti and Schaufeli (2022) found that transformational leaders directly influence engagement by providing employees with autonomy, empowering them to take initiative, and encouraging them to align their personal goals with the organization's vision. Moreover, leadership development programs focusing on emotional intelligence, empathy, and communication can help leaders create a more supportive and engaging environment for their teams (Goleman, Boyatzis, & McKee, 2004). By investing in transformational leadership development, organizations can

build a cadre of leaders who inspire employees to engage more fully with their work and contribute to the organization's success (Joubert, 2024).

2. Fostering a culture of open communication and psychological safety is essential for improving employee engagement (Gallo, 2023). Kahn's (1990) work on engagement emphasizes the importance of psychological safety, the belief that one can express oneself without fear of negative consequences, as a critical driver of engagement. Employees who feel psychologically safe are likelier to contribute ideas, voice concerns, and take the risks necessary to innovate and grow (Clark, 2020). Edmondson (1999) further underscores the importance of psychological safety, particularly in team environments, where open communication and trust are critical for collaboration and engagement. Organizations can cultivate this culture by encouraging transparency, creating regular feedback channels, and ensuring leaders and managers actively listen to and act on employee input (Detert & Burris, 2007). Research by Carmeli, Brueller, and Dutton (2009) shows that when employees feel their voices are heard and valued, they are more likely to be engaged and committed to the organization. Moreover, fostering psychological safety requires addressing issues of inclusion and equity, as employees from underrepresented groups may face additional barriers to feeling safe and valued in the workplace (Nishii, 2013). Organizations can significantly enhance engagement across diverse teams by creating an inclusive environment where all employees feel their contributions are recognized and respected (Shore, Cleveland, & Sanchez, 2018).
3. Offering flexible work arrangements has become increasingly important considering the changing needs of the modern workforce, particularly with the rise of remote and hybrid work models (Smite et al., 2023). Flexible work options, such as remote work, flexible hours, and compressed workweeks, give employees greater autonomy and control over their work-life balance, enhancing their engagement (Ray & Pana-Cryan, 2021). Research by Allen et al. (2015) demonstrates that flexible work arrangements positively correlate with employee engagement, as they reduce stress, increase job satisfaction, and promote a sense of work-life integration. The rise of remote work, accelerated by the COVID-19 pandemic, has further underscored the importance of flexibility in maintaining engagement (Battisti, Alfiero, & Leonidou, 2022). Employees who can balance their work responsibilities with personal commitments are more likely to experience higher levels of job satisfaction and lower levels of burnout, both of which contribute to sustained engagement (Maslach, Schaufeli, & Leiter, 2001). However, flexibility must be supported by robust digital infrastructure and practical communication tools to ensure that remote employees remain connected to their teams and organizational goals (Gajendran & Joshi, 2012). Moreover, organizations should recognize that employees have different needs

for flexibility based on their circumstances, and a one-size-fits-all approach may not be practical (Ling et al., 2021). Tailoring flexible work policies to meet the workforce's diverse needs can further enhance engagement and productivity (Kossek et al., 2021).

4. Organizations must integrate engagement into their overall business strategy, treating it as a core organizational priority rather than a standalone human resources initiative (Armstrong & Taylor, 2023). This requires establishing clear metrics for measuring and sustaining engagement over time. Saks (2006) argues that many organizations fail to effectively measure engagement, relying on annual surveys that provide a snapshot of employee sentiment but do not capture the complexity of engagement as an ongoing, dynamic process. Instead, organizations should implement continuous feedback mechanisms, such as pulse surveys and real-time analytics, to monitor engagement levels and identify areas for improvement (Johnson, 2018). Furthermore, engagement strategies should be aligned with key business objectives, such as innovation, customer satisfaction, and financial performance, to ensure that engagement efforts contribute to the organization's overall success (Macey & Schneider, 2008). Gruman and Saks (2011) suggest that organizations can help sustain engagement over the long term by integrating it into performance management systems that link engagement to goal setting, development planning, and performance reviews. By making engagement a strategic priority and embedding it into the organization's culture and systems, companies can drive sustained improvements in employee and organizational outcomes (Sroufe, 2017).

Improving employee engagement requires a multifaceted strategy that addresses leadership, organizational culture, flexibility, and strategic integration (Osborne & Hammoud, 2017). Leadership development programs emphasizing transformational leadership can inspire and motivate employees, fostering a deeper connection to the organization (Lin et al., 2020. An open communication and psychological safety culture ensures that employees feel valued and empowered to contribute fully (Hallam, Popovic, & Karimi, 2023). Flexible work arrangements accommodate the workforce's diverse needs, promoting work-life balance and enhancing engagement (Kossek et al., 2021; Ray & Pana-Cryan, 2021). By integrating engagement into the overall business strategy and implementing clear metrics for measuring and sustaining engagement, organizations can ensure that their efforts lead to lasting improvements in employee commitment, satisfaction, and performance (De-la-Calle-Durán & Rodríguez-Sánchez, 2021). Together, these strategies provide a comprehensive framework for driving engagement and achieving long-term organizational success (Canavesi & Minelli, 2021).

FUTURE RESEARCH

While research on employee engagement has significantly expanded over the past few decades, several critical gaps remain in the literature, pointing to areas where future studies can make valuable contributions. As the workplace continues to evolve driven by technological advances, shifts in workforce demographics, and work arrangements, understanding the factors that sustain or hinder engagement in new contexts is essential for scholars and practitioners. These gaps include the effects of remote and hybrid work environments on engagement, the need for longitudinal studies to track engagement over time, and the intersection of diversity and inclusion initiatives with employee engagement. Addressing these gaps can provide more nuanced insights into how organizations can foster and maintain engagement in a rapidly changing world.

1. One of the most pressing areas for future research is the impact of remote and hybrid work environments on employee engagement. The COVID-19 pandemic accelerated the adoption of remote and hybrid work arrangements, fundamentally altering how employees interact with their colleagues, leaders, and the organization (Wyld, 2022). While remote work has offered benefits such as increased flexibility and autonomy, it has also raised questions about how these arrangements influence engagement (Neidlinger, Felfe, & Schubbe, 2022). Virtual communication tools, such as Zoom, Slack, and Microsoft Teams, have become central to maintaining collaboration, but their effects on engagement are still not fully understood (Karl, Peluchette, & Aghakhani, 2022). For example, research suggests that virtual interactions may lack the richness of face-to-face communication, potentially leading to feelings of isolation or disconnection from the organization (Bailey & Kurland, 2002).
2. Furthermore, the effectiveness of virtual communication in fostering team cohesion, trust, and engagement may vary depending on the quality of leadership and the organization's ability to create a supportive digital culture (Gajendran & Joshi, 2012). Studies by Golden and Veiga (2005) indicate that while remote work can enhance job satisfaction and work-life balance for some employees, it can also reduce engagement for others, particularly those who thrive on social interaction and in-person collaboration. Future research should examine how organizations can leverage virtual tools to maintain or enhance engagement in remote and hybrid settings. Key areas of focus include exploring the role of leadership in virtual engagement, the impact of digital communication practices on psychological safety and trust, and the effectiveness of virtual team-building initiatives in promoting a sense of belonging and connectedness.

3. Another critical gap in the engagement literature is the lack of longitudinal research to assess how engagement evolves over time in response to organizational changes and external shocks. Most engagement studies rely on cross-sectional data, capturing engagement levels simultaneously. However, engagement is a dynamic construct that fluctuates based on organizational changes or the broader economic and technological environment (Schaufeli, 2013). For instance, engagement levels may rise or fall during organizational restructuring, mergers, or leadership changes depending on how employees perceive these shifts and their impact on job security, resources, and workplace culture (Alfes et al., 2013). Similarly, external shocks, such as economic downturns or technological disruptions, can influence engagement by altering job demands and resources. Research by Bakker and Demerouti (2007) suggests that during financial uncertainty, employees may experience increased job demands such as heavier workloads or job insecurity that can erode engagement if not counterbalanced by adequate support. Longitudinal studies are needed to track how engagement evolves in response to these changes, identifying the factors that sustain engagement over time and those that lead to disengagement. Such research could provide valuable insights into how organizations can proactively manage engagement during transition periods, ensuring that employees remain motivated and committed despite challenges.
4. In addition, there is a growing need to explore the role of diversity and inclusion (D&I) initiatives in promoting employee engagement, particularly as organizations become more diverse and inclusive. Diversity in the workplace encompasses not only demographic differences such as race, gender, and age but also differences in experiences, perspectives, and cognitive styles (Turi et al., 2022). Research suggests that diverse teams have the potential to be more innovative and productive, but they may also face challenges related to communication, cohesion, and trust (Page & Vella-Brodrick, 2009). Turi et al. (2022) argue that traditional engagement strategies may not be equally effective for all employees, as individuals from different backgrounds may have varying engagement drivers. For instance, employees from underrepresented groups may experience unique barriers to engagement, such as feelings of exclusion or discrimination, which can undermine their ability to contribute to the organization (Roberson, 2006). D&I initiatives, which aim to create a more equitable and inclusive workplace, are critical for addressing these barriers and fostering an environment where all employees feel valued and engaged (United States Department of Defense (defense.gov), 2022). Future research should investigate how specific D&I practices such as mentorship programs, bias training, and equitable access to development opportunities affect engagement across different employee groups. Understanding how diversity dimensions, such as

gender, race, or disability, intersect with engagement can help organizations tailor their strategies to meet the diverse needs of their workforce (United States Equal Employment Opportunity Commission, 2016). Moreover, examining the role of inclusive leadership in fostering engagement across diverse teams could provide insights into how leaders can create an inclusive culture that promotes psychological safety and empowers all employees to engage fully in their work (Nishii, 2013).

While the existing body of research on employee engagement provides valuable insights, several gaps remain critical for future exploration. The impact of remote and hybrid work environments on engagement requires further study, mainly on how virtual communication and collaboration tools influence engagement levels. Longitudinal research is needed to capture the dynamic nature of engagement, tracking how it evolves in response to organizational changes and external shocks such as economic downturns or technological disruption. Furthermore, investigating the role of diversity and inclusion initiatives in promoting engagement is essential, as diverse teams may experience different engagement drivers, and traditional engagement strategies may only be equally effective across some employee groups. Addressing these gaps will enrich the academic understanding of engagement and provide practical guidance for organizations seeking to create more inclusive, adaptive, and engaging workplaces.

CONCLUSION

Developing an effective strategy for employee engagement requires a holistic approach that aligns organizational goals with individual motivations (Osborne & Hammoud, 2017). Such a strategy is more than just a one-size-fits-all solution. Still, it must be dynamic, incorporating multiple layers of organizational life, from leadership and culture to work arrangements and employee well-being (Kossek et al., 2021. A successful engagement strategy leverages leadership practices, cultivates an inclusive and supportive culture, and embraces flexibility to meet the evolving needs of today's diverse workforce. The integration of these components not only boosts individual performance but also enhances organizational resilience and innovation, positioning companies to thrive in an increasingly volatile business environment. In this regard, understanding the complexities of engagement and tailoring strategies to foster it can produce tangible benefits for both employees and the organization.

1. Aligning leadership with employee engagement is critical for ensuring organizational goals resonate with individual motivations. Leadership plays a central role in influencing engagement levels, as it sets the tone for the organizational climate and shapes the employee experience. Research highlights that transformational leadership profoundly impacts engagement by fostering an environment where employees feel inspired, valued, and motivated to achieve their full potential (Tims et al., 2011). Transformational leaders communicate a compelling vision, provide individualized support, and encourage innovation, which are key drivers of engagement. Mazzetti and Schaufeli (2022) found that employees working under transformational leaders exhibit higher levels of engagement because they perceive their work as meaningful and aligned with broader organizational goals. Leaders prioritizing emotional intelligence and fostering open communication can enhance psychological safety, making employees feel more comfortable sharing ideas and taking risks (Goleman et al., 2004). Organizations can create a culture that motivates employees and promotes long-term organizational success by aligning leadership practices with engagement objectives.
2. Cultivating a supportive organizational culture is essential for creating an environment where employees feel connected, valued, and motivated. Organizational culture refers to the shared values, beliefs, and practices that shape how employees interact and work together. A positive, inclusive culture enhances engagement by promoting open communication, trust, and collaboration. Kahn's (1990) research on engagement emphasizes the importance of psychological safety where employees feel safe to express themselves without fear of negative consequences as a core driver of engagement. When employees feel their voices are heard, and their contributions matter, they are more likely to engage with and commit to the organization (Detert & Burris, 2007). Schaufeli (2013) further underscores that organizational culture should prioritize autonomy, mastery, and purpose, contributing to intrinsic motivation and, by extension, employee engagement. Creating a culture promoting these values requires leaders and managers to involve employees in decision-making processes actively, provide opportunities for professional growth, and recognize their contributions. Moreover, a culture that embraces diversity and inclusion is vital in today's workforce, as employees from different backgrounds may have varying engagement drivers (Turi et al., 2022). Companies can enhance engagement across all workforce levels by cultivating an organizational culture that fosters inclusion, trust, and collaboration.
3. Offering flexibility to accommodate the changing needs of the workforce is increasingly seen as a critical component of engagement strategies. Flexibility has become crucial in maintaining employee engagement as work environments evolve, particularly with the rise of remote and hybrid work models. Research

by Allen et al., (2015) show that flexible work arrangements, such as remote work, flexible hours, and compressed workweeks, are associated with higher levels of engagement, job satisfaction, and work-life balance. Employees with greater control over when and where they work can better balance personal and professional demands, reducing stress and preventing burnout (Maslach et al., 2001). The shift to remote work, accelerated by the COVID-19 pandemic, has highlighted the importance of flexibility in maintaining engagement in a rapidly changing work environment. However, it is essential that organizations implement robust communication tools and provide ongoing support to ensure that remote employees remain connected to their teams and the company's mission (Gajendran & Joshi, 2012). Offering flexibility requires a tailored approach, as employees may have different preferences and needs. For example, some employees may prefer remote work for its autonomy, while others may value the social interaction and collaboration of working in a physical office (Golden & Veiga, 2005). By offering a range of flexible work options, organizations can accommodate these diverse preferences and maintain engagement across their workforce.

4. The benefits of a well-developed engagement strategy extend beyond individual performance and have broader implications for organizational resilience and competitiveness. Engaged employees are more productive and likely to contribute innovative solutions, driving organizational success in today's fast-paced and competitive business environment. Research by Salanova and Schaufeli (2008) demonstrates that engaged employees are more proactive, creative, and adaptable to change, making them invaluable assets during organizational transformation or external disruption. Furthermore, engaged employees contribute to a positive organizational climate, improving overall employee morale and reducing turnover (Harter et al., 2002). This strengthens the organization's capacity to respond to challenges, enhancing its resilience. Engaged employees also play a critical role in improving organizational reputation, as companies known for high levels of employee engagement are more attractive to top talent, creating a cycle of sustained growth and innovation (Maylett & Wride, 2017). Therefore, investing in employee engagement improves individual outcomes and is a strategic imperative for ensuring organizational competitiveness and long-term success.
5. Future research and practice should continue exploring how evolving workplace trends affect engagement, ensuring engagement strategies remain relevant and impactful. The workplace is undergoing rapid changes driven by technological advancements, shifting workforce demographics, and evolving employee expectations. For example, the increasing reliance on digital communication tools in remote work environments raises questions about how these tools influence

engagement and how organizations can support employees in maintaining meaningful connections in a virtual world (Bailey & Kurland, 2002). Research should explore how diversity and inclusion initiatives can be integrated into engagement strategies to ensure all employees feel valued and engaged regardless of background. Longitudinal studies are needed to track how engagement evolves over time in response to organizational changes, external shocks, or economic downturns (Schaufeli, 2013). As the nature of work continues to grow, engagement strategies must also adapt, ensuring they remain aligned with organizational goals and the needs of a diverse and dynamic workforce.

In conclusion, developing an effective employee engagement strategy requires a holistic approach that aligns organizational goals with individual motivations. When successfully integrated, leadership, culture, and flexibility are critical components that significantly improve productivity, innovation, and employee satisfaction. Beyond individual performance, the benefits of engagement extend to enhancing organizational resilience and competitiveness, making it a vital component of long-term business strategy. Engagement strategies must evolve alongside workplace trends to remain relevant, requiring continuous research and adaptation. By investing in engagement, organizations can create a more motivated, innovative, and resilient workforce better equipped to meet the challenges of an ever-changing business environment.

REFERENCES

Adams, B. (2022). *Make your employer brand stand out in the talent marketplace*. Havard Business Review.

Ahmad, R., Nawaz, M. R., Ishaq, M. I., Khan, M. M., & Ashraf, H. A. (2023). Social exchange theory: Systematic review and future directions. *Frontiers in Psychology*, *13*, 1015921. DOI: 10.3389/fpsyg.2022.1015921 PMID: 36710813

Ahmeti, F. (2023). Leveraging employee engagement for competitive advantage: Satisfaction and work motivation management. *Journal of Liberty and International Affairs Institute for Research and European Studies*, *9*(2), 216–227. DOI: 10.47305/JLIA2392216a

Albrecht, S. L., Connaughton, S., Foster, K., Furlong, S., & Yeow, C. J. L. (2020). Change engagement, change resources, and change demands: A model for positive employee orientations to organizational change. *Frontiers in Psychology*, *11*, 531944. DOI: 10.3389/fpsyg.2020.531944 PMID: 33240144

Albrecht, S. L., Connaughton, S., & Leiter, M. P. (2022). The influence of change-related organizational and job resources on employee change engagement. *Frontiers in Psychology*, *13*, 910206. DOI: 10.3389/fpsyg.2022.910206 PMID: 35769731

Albrecht, S. L., Furlong, S., & Leiter, M. P. (2023). The psychological conditions for employee engagement in organizational change: Test of a change engagement model. *Frontiers in Psychology*, *14*, 1071924. DOI: 10.3389/fpsyg.2023.1071924 PMID: 36743243

Alfes, K., Shantz, A. D., Truss, C., & Soane, E. C. (2013). The link between perceived human resource management practices, engagement and employee behaviour: A moderated mediation model. *International Journal of Human Resource Management*, *24*(2), 330–351. DOI: 10.1080/09585192.2012.679950

Allen, T. D., Golden, L. S., & Shockley, K. M. (2015). How effective is telecommuting? Assessing the status of our scientific findings. *Psychological Science in the Public Interest*, *16*(2), 40–68. DOI: 10.1177/1529100615593273 PMID: 26403188

Alton, L. (2024). *7 strategies to address employee disengagement*. Retrieved from https://www.shrm.org/topics-tools/news/managing-smart/7-strategies-to-address - employee-disengagement

American Psychological Association (APA). (2023). *2023 work in America survey*. Retrieved from https://www.apa.org/pubs/reports/work-in-america/2023-workplace -health-well- being

Armstrong, M., & Taylor, S. (2023). *Armstrong's handbook of human resource management practice: A guide to the theory and practice of people management* (16th ed.). Kogan Page.

Bailey, D. E., & Kurland, N. B. (2002). A review of telework research: Findings, new directions, and lessons for the study of modern work. *Journal of Organizational Behavior, 23*(4), 383–400. DOI: 10.1002/job.144

Bakker, A. B., & Demerouti, E. (2007). The job demands-resources model: State of the art. *Journal of Managerial Psychology, 22*(3), 309–328. DOI: 10.1108/02683940710733115

Bakker, A. B., Demerouti, E., & Euwema, M. C. (2005). Job resources buffer the impact of job demands on burnout. *Journal of Occupational Health Psychology, 10*(2), 170–180. DOI: 10.1037/1076-8998.10.2.170 PMID: 15826226

Bakker, A. B., & Schaufeli, W. B. (2008). Positive organizational behavior: Engaged employees in flourishing organizations. *Journal of Organizational Behavior, 29*(2), 147–154. DOI: 10.1002/job.515

Barker Scott, B. A., & Manning, M. R. (2024). Designing the collaborative organization: A framework for how collaborative work, relationships, and behaviors generate collaborative capacity. *The Journal of Applied Behavioral Science, 60*(1), 149–193. DOI: 10.1177/00218863221106245

Basit, A. A., & Nauman, S. (2023). How workplace loneliness harms employee well-being: A moderated mediational model. *Frontiers in Psychology, 13*, 1086346. DOI: 10.3389/fpsyg.2022.1086346 PMID: 36726515

Bass, B. M. (1985). *Leadership and performance beyond expectations*. Free Press.

Bassi, L., & McMurrer, D. (2007). Maximizing your return on people. *Harvard Business Review*. PMID: 17348175

Battisti, E., Alfiero, S., & Leonidou, E. (2022). Remote working and digital transformation during the COVID-19 pandemic: Economic-financial impacts and psychological drivers for employees. *Journal of Business Research, 150*, 38–50. DOI: 10.1016/j.jbusres.2022.06.010 PMID: 35706830

Boccoli, G., Gastaldi, L., & Corso, M. (2022). The evolution of employee engagement: Towards a social and contextual construct for balancing individual performance and wellbeing dynamically. *International Journal of Management Reviews, 25*(1), 75–98. DOI: 10.1111/ijmr.12304

Bridges, M. (2024). Integrate KPI's into culture strategies and case studies. Retrieved from https://mark-bridges.medium.com/integrated-kpis-into-culture-strategies-and-case- studies-b3a1fa7b085c

Canavesi, A., & Minelli, E. (2021). Servant leadership and employee engagement: A qualitative study. *Employee Responsibilities and Rights Journal*, *34*(4), 413–435. DOI: 10.1007/s10672-021-09389-9 PMID: 40477632

Cappelli, P., & Eldor, L. (2019). Where measuring engagement goes wrong. *Harvard Business Review*.

Carmeli, A., Brueller, D., & Dutton, J. E. (2009). Learning behaviors in the workplace: The role of high-quality interpersonal relationships and psychological safety. *Systems Research and Behavioral Science*, *26*(1), 81–98. DOI: 10.1002/sres.932

Carmeli, A., & Spreitzer, G. M. (2009). Trust, connectivity, and thriving: Implications for innovative behaviors at work. *The Journal of Creative Behavior*, *43*(3), 169–191. DOI: 10.1002/j.2162-6057.2009.tb01313.x

Chan, X. W., Shang, S., Brough, P., Wilkinson, A., & Lu, C. Q. (2022). Work, life and COVID- 19: A rapid review and practical recommendations for the post-pandemic workplace. *Asia Pacific Journal of Human Resources*. Advance online publication. *10*.1111/1744-7941.12355

Chanana, N., & Sangeeta, . (2020). Employee engagement practices during COVID-19 lockdown. *Journal of Public Affairs*, *21*(4), e2508. DOI: 10.1002/pa.2508 PMID: 33041656

Chen, X., Al Mamun, A., Hussain, W. M. H. W., Jingzu, G., Yang, Q., & Shami, S. S. A. A. (2023). Envisaging the job satisfaction and turnover intention among the young workforce: Evidence from an emerging economy. *PLoS One*, *18*(6), e0287284. DOI: 10.1371/journal.pone.0287284 PMID: 37327240

Clark, T. R. (2020). *The 4 stages of psychological safety: Defining the path to inclusion and innovation*. Berrett-Koehler Publishers.

Clifton, J. (2022). *The world's workplace is broken: Here's how to fix it.* Retrieved from https://www.gallup.com/workplace/393395/world-workplace-broken-fix.aspx

Cohen, E. (2017). *CSR for HR: A necessary partnership for advancing responsible business practices*. Routledge. DOI: 10.4324/9781351278607

Crail, C., & Watts, R. (2024). *15 effective employee retention strategies in 2024*. Retrieved from https://www.forbes.com/advisor/business/employee-retention-strategies/

Cropanzano, R., & Mitchell, M. S. (2005). Social exchange theory: An interdisciplinary review. *Journal of Management, 31*(6), 874–900. DOI: 10.1177/0149206305279602

David, S. (2016). *Emotional agility: Get unstuck, embrace change, and thrive in work and life*. Penguin House.

De-la-Calle-Durán, M. C., & Rodríguez-Sánchez, J. L. (2021). Employee engagement and wellbeing in times of COVID-19: A Proposal of the 5Cs Model. *International Journal of Environmental Research and Public Health, 18*(10), 5470. DOI: 10.3390/ijerph18105470 PMID: 34065338

Deci, E. L., & Ryan, R. M. (2000). The "what" and "why" of goal pursuits: Human needs and the self-determination of behavior. *Psychological Inquiry, 11*(4), 227–268. DOI: 10.1207/S15327965PLI1104_01

Deep, G. (2023). The power of resilience and flexibility in business leadership: Adapting to change. *Magna Scientia Advanced Research and Reviews, 9*(2):086-091.

Dennison, K. (2024). *Gallup says $8.8 trillion is the future cost of low employee engagement*. Retrieved from https://www.forbes.com/sites/karadennison/2024/07/16/gallup-says-88- trillion-is-the-true-cost-of-low-employee-engagement/

Detert, J. R., & Burris, V. (2007). Leadership behavior and employee voice: Is the door open? *Academy of Management Journal, 50*(4), 869–884. DOI: 10.5465/amj.2007.26279183

Dowling, B., Goldstein, D., Park, M., & Price, H. (2022). *Hybrid work: Making it fit with your diversity, equity, and inclusion strategy*. Retrieved from https://www.mckinsey.com/capabilities/people-and-organizational-performance/our- insights/hybrid-work-making-it-fit-with-your-diversity-equity-and-inclusion-strategy

Edmondson, A. (1999). Psychological safety and learning behavior in work teams. *Administrative Science Quarterly, 44*(2), 350–383. DOI: 10.2307/2666999

Gajendran, R. S., & Joshi, A. (2012). Innovation in globally distributed teams: The role of LMX, communication frequency, and innovation climate. *The Journal of Applied Psychology, 97*(6), 1252–1261. DOI: 10.1037/a0028958 PMID: 22708920

Galanakis, M. D., & Tsitouri, E. (2022). Positive psychology in the working environment. Job demands-resources theory, work engagement and burnout: A systematic literature review. *Frontiers in Psychology, 13*, 1022102. DOI: 10.3389/fpsyg.2022.1022102 PMID: 36204770

Gallo, A. (2023). What is psychological safety? *Harvard Business Review*.

Gardner, H. K. (2016). *Smart collaboration: How professionals and their firms succeed by breaking down silos*. Harvard Business Review Press.

Golden, T. D., & Veiga, J. F. (2005). The impact of extent of telecommuting on job satisfaction: Resolving inconsistent findings. *Journal of Management, 31*(2), 301–318. DOI: 10.1177/0149206304271768

Goleman, D., Boyatzis, R., & McKee, A. (2004). *Primal leadership: Learning to lead with emotional intelligence*. Harvard Business Review Press.

Gong, Y., Xiao, J., Tang, X., & Li, J. (2023). How sustainable marketing influences the customer engagement and sustainable purchase intention? The moderating role of corporate social responsibility. *Frontiers in Psychology, 14*, 1128686. DOI: 10.3389/fpsyg.2023.1128686 PMID: 37008858

Gruman, J. A., & Saks, A. M. (2011). Performance management and employee engagement. *Human Resource Management Review, 21*(2), 123–136. DOI: 10.1016/j.hrmr.2010.09.004

Hallam, K. T., Popovic, N., & Karimi, L. (2023). Identifying the key elements of psychologically safe workplaces in healthcare settings. *Brain Sciences, 13*(10), 1450. DOI: 10.3390/brainsci13101450 PMID: 37891818

Harter, J. (2017). *Employee engagement vs. employee satisfaction and organizational culture*. Retrieved from https://www.gallup.com/workplace/236366/right-culture-not-employee- satisfaction.aspx

Harter, J. (2022). *Disengagement persists among U.S. employees*. Retrieved from https://www.gallup.com/workplace/391922/employee-engagement-slump-continues.aspx

Harter, J. (2024). *In new workplace, U.S employee engagement stagnates*. Retrieved from https://www.gallup.com/workplace/608675/new-workplace-employee-engagement- stagnates.aspx

Harter, J. K., Schmidt, F. L., & Hayes, T. L. (2002). Business-unit-level relationship between employee satisfaction, employee engagement, and business outcomes: A meta- analysis. *The Journal of Applied Psychology, 87*(2), 268–279. DOI: 10.1037/0021-9010.87.2.268 PMID: 12002955

Hollister, R., Tecosky, K., Watkins, M., & Wolpert, C. (2021). Why every executive should be focusing on culture change now. *MIT Sloan Management Review*.

Huang, C., Xie, X., Cheung, S. P., & Zhou, Y. (2022). Job demands and resources, positive and negative affect, and psychological distress of social workers in China. *Frontiers in Psychiatry, 12*, 752382. DOI: 10.3389/fpsyt.2021.752382 PMID: 35115957

Huang, Z., Sindakis, S., Aggarwal, S., & Thomas, L. (2022). The role of leadership in collective creativity and innovation: Examining academic research and development environments. *Frontiers in Psychology, 13*, 1060412. DOI: 10.3389/fpsyg.2022.1060412 PMID: 36619078

Huebner, L. A., & Zacher, H. (2021). Following up on employee surveys: A conceptual framework and systematic review. *Frontiers in Psychology, 12*, 801073. DOI: 10.3389/fpsyg.2021.801073 PMID: 34956026

Johnson, S. R. (2018). *Engaging the workplace: Using surveys to spark change*. Association for Talent Development.

Joubert, S. (2024). *Transformational leadership: How to inspire innovation in the workplace*. Retrieved from https://graduate.northeastern.edu/resources/transformational-leadership/

Kahn, W. A. (1990). Psychological conditions of personal engagement and disengagement at work. *Academy of Management Journal, 33*(4), 692–724. DOI: 10.2307/256287

Karl, K. A., Peluchette, J. V., & Aghakhani, N. (2022). Virtual work meetings during the COVID-19 pandemic: The good, bad, and ugly. *Small Group Research, 53*(3), 343–365. DOI: 10.1177/10464964211015286 PMID: 38603094

Kim, J., & Jung, H. S. (2022). The Effect of employee competency and organizational culture on employees' perceived stress for better workplace. *International Journal of Environmental Research and Public Health, 19*(8), 4428. DOI: 10.3390/ijerph19084428 PMID: 35457296

Kleynhans, D. J., Heyns, M. M., Stander, M. W., & de Beer, L. T. (2022). Authentic leadership, trust (in the Leader), and flourishing: Does precariousness matter? *Frontiers in Psychology, 13*, 798759. DOI: 10.3389/fpsyg.2022.798759 PMID: 35432051

Knight, D. K., Becan, J. E., & Flynn, P. M. (2013). The impact of staff turnover on workplace demands and coworker relationships. *Counselor (Deerfield Beach, Fla.), 14*(3), 20–23. PMID: 27540331

Kossek, E. E., Gettings, P., & Misra, K. (2021). The future of flexibility at work. *Harvard Business Review*.

Kossek, E. E., Pichler, S., Bodner, T., & Hammer, L. B. (2011). Workplace social support and work-family conflict: A meta-analysis clarifying the influence of general and work- family-specific supervisor and organizational support. *Personnel Psychology, 64*(2), 289–313. DOI: 10.1111/j.1744-6570.2011.01211.x PMID: 21691415

Lawrence, P. L. (1969). How to deal with resistance to change. *Harvard Business Review.*

Lee, A., Willis, S., & Tian, A. W. (2018). When empowering employees works, and when it doesn't. *Harvard Business Review.*

Lewis, A. (2022). *Good leadership? It all starts with trust.* Harvard Business Publishing. Retrieved from https://www.harvardbusiness.org/good-leadership-it-all-starts-with-trust/

Lin, C. P., Xian, J., Li, B., & Huang, H. (2020). Transformational leadership and employees' thriving at work: The mediating roles of challenge-hindrance stressors. *Frontiers in Psychology, 11*, 1400. DOI: 10.3389/fpsyg.2020.01400 PMID: 32655458

Ling, B., Liu, Z., Chen, D., & Sun, L. (2021). Team flexibility in organizational change context: Antecedents and consequences. *Psychology Research and Behavior Management, 14*, 1805–1821. DOI: 10.2147/PRBM.S332222 PMID: 34754250

Macey, W. H., & Schneider, B. (2008). The meaning of employee engagement. *Industrial and Organizational Psychology: Perspectives on Science and Practice, 1*(1), 3–30. DOI: 10.1111/j.1754-9434.2007.0002.x

Martela, F., & Pessi, A. B. (2018). Significant work is about self-realization and broader purpose: Defining the key dimensions of meaningful work. *Frontiers in Psychology, 9*, 363. DOI: 10.3389/fpsyg.2018.00363 PMID: 29632502

Maslach, C., Schaufeli, W. B., & Leiter, M. P. (2001). Job burnout. *Annual Review of Psychology, 52*(1), 397–422. DOI: 10.1146/annurev.psych.52.1.397 PMID: 11148311

Maximo, N., Stander, M. W., & Coxen, L. (2019). Authentic leadership and work engagement: The indirect effects of psychological safety and trust in supervisors. *SA Journal of Industrial Psychology, 45*(1), 1–11. DOI: 10.4102/sajip.v45i0.1612

May, D. R., Gilson, R. L., & Harter, L. M. (2004). The psychological conditions of meaningfulness, safety and availability and the engagement of the human spirit at work. *Journal of Occupational and Organizational Psychology, 77*(1), 11–37. DOI: 10.1348/096317904322915892

Maylett, T., & Wride, M. (2017). *The employee experience: How to attract talent, retain top performers, and drive results*. John Wiley & Sons.

Mazzetti, G., & Schaufeli, W. B. (2022). The impact of engaging leadership on employee engagement and team effectiveness: A longitudinal, multi-level study on the mediating role of personal- and team resources. *PLoS One*, *17*(6), e0269433. DOI: 10.1371/journal.pone.0269433 PMID: 35767536

Miller, J. (2023). *The power of diversity and inclusion: Driving innovation and success*. Retrieved from https://www.forbes.com/councils/forbesbusinesscouncil/2023/08/16/the-power-of-diversity-and-inclusion-driving-innovation-and-success/

Neidlinger, S. M., Felfe, J., & Schübbe, K. (2022). Should I stay or should I go (to the office)? - Effects of working from home, autonomy, and core self-evaluations on leader health and work-life balance. *International Journal of Environmental Research and Public Health*, *20*(1), 6. DOI: 10.3390/ijerph20010006 PMID: 36612327

Nishii, L. H. (2013). The benefits of climate for inclusion for gender-diverse groups. *Academy of Management Journal*, *56*(6), 1754–1774. DOI: 10.5465/amj.2009.0823

Okatta, C. G., Ajayi, F. A., & Olawale, O. (2024). Enhancing organizational performance through diversity and inclusion initiatives: A meta-analysis. *International Journal of Applied Research in Social Sciences*, *6*(4), 734–758. DOI: 10.51594/ijarss.v6i4.1065

Osborne, S., & Hammoud, M. S. (2017). Effective employee engagement in the workplace. *International Journal of Applied Management and Technology*, *16*(1), 50–67. DOI: 10.5590/IJAMT.2017.16.1.04

Page, K. M., & Vella-Brodrick, D. A. (2009). The 'what', 'why' and 'how' of employee well- being: A new model. *Social Indicators Research*, *90*(3), 441–458. DOI: 10.1007/s11205-008-9270-3

Pandya, J. D. (2024). Intrinsic & Extrinsic motivation & its impact on organizational performance at Rajkot city: A review. *Journal of Management Research and Analysis*, *11*(1), 46–53. DOI: 10.18231/j.jmra.2024.009

Pincus, J. D. (2022). Employee engagement as human motivation: Implications for theory, methods, and practice. *Integrative Psychological & Behavioral Science*, *57*(4), 1223–1255. DOI: 10.1007/s12124-022-09737-w PMID: 36577907

Pisano, G. P. (2015). You need an innovation strategy. *Harvard Business Review*.

Prager, F., Rhoads, M., & Martínez, J. N. (2022). The COVID-19 economic shutdown and the future of flexible workplace practices in the south bay region of Los Angeles County. *Transport Policy*, *125*, 241–255. DOI: 10.1016/j.tranpol.2022.06.004 PMID: 35720050

Qasim, S., Usman, M., Ghani, U., & Khan, K. (2022). Inclusive leadership and employees' helping behaviors: Role of psychological factors. *Frontiers in Psychology, 13*, 888094. DOI: 10.3389/fpsyg.2022.888094 PMID: 35874383

Ray, T. K., & Pana-Cryan, R. (2021). Work flexibility and work-related well- being. *International Journal of Environmental Research and Public Health, 18*(6), 3254. DOI: 10.3390/ijerph18063254 PMID: 33801122

Rhoades, L., & Eisenberger, R. (2002). Perceived organizational support: A review of the literature. *The Journal of Applied Psychology, 87*(4), 698–714. DOI: 10.1037/0021-9010.87.4.698 PMID: 12184574

Roberson, Q. M. (2006). Disentangling the meanings of diversity and inclusion in organizations. *Group & Organization Management, 31*(2), 212–236. DOI: 10.1177/1059601104273064

Rodríguez-Sánchez, J. L., González-Torres, T., Montero-Navarro, A., & Gallego-Losada, R. (2020). Investing time and resources for work-life balance: The effect on talent retention. *International Journal of Environmental Research and Public Health, 17*(6), 1920. DOI: 10.3390/ijerph17061920 PMID: 32187973

Saks, A. M. (2006). Antecedents and consequences of employee engagement. *Journal of Managerial Psychology, 21*(7), 600–619. DOI: 10.1108/02683940610690169

Saks, A. M., & Gruman, J. A. (2021). *Employee engagement. Essentials of job attitudes and other workplace psychological constructs.* Routledge., DOI: 10.4324/9780429325755-12

Salanova, M., & Schaufeli, W. B. (2008). A cross-national study of work engagement as a mediator between job resources and outcomes. *International Journal of Human Resource Management, 19*(1), 116–131. DOI: 10.1080/09585190701763982

Santoso, N. R., Sulistyaningtyas, I. D., & Pratama, B. P. (2022). Transformational leadership during the COVID-19 pandemic: Strengthening employee engagement through internal communication. *The Journal of Communication Inquiry, 01968599221095182*, 01968599221095182. Advance online publication. DOI: 10.1177/01968599221095182

Schaufeli, W. B. (2013). *Employee engagement in theory and practice.* Routledge.

Schein, E. H. (2010). *Organizational culture and leadership* (4th ed.). John Wiley & Sons.

Shiri, R., El-Metwally, A., Sallinen, M., Pöyry, M., Härmä, M., & Toppinen-Tanner, S. (2023). The role of continuing professional training or development in maintaining current employment: A systematic review. *Healthcare (Basel), 11*(21), 2900. DOI: 10.3390/healthcare11212900 PMID: 37958044

Shore, L. M., Cleveland, J. N., & Sanchez, D. (2018). Inclusive workplaces: A review and model. *Human Resource Management Review, 28*(2), 176–189. DOI: 10.1016/j.hrmr.2017.07.003

Smite, D., Moe, N. B., Hildrum, J., Gonzalez-Huerta, J., & Mendez, D. (2023). Work-from- home is here to stay: Call for flexibility in post-pandemic work policies. *Journal of Systems and Software, 195*, 111552. DOI: 10.1016/j.jss.2022.111552 PMID: 36320721

Sroufe, R. (2017). Integration and organizational change towards sustainability. *Journal of Cleaner Production, 162*, 315–329. DOI: 10.1016/j.jclepro.2017.05.180

Stein, D., Hobson, N., Jachimowicz, J. M., & Whillans, A. (2021). How companies can improve employee engagement right now. *Harvard Business Review*.

Steinmann, B., Klug, H. J. P., & Maier, G. W. (2018). The path Is the goal: How transformational leaders enhance followers' job attitudes and proactive behavior. *Frontiers in Psychology, 9*, 2338. DOI: 10.3389/fpsyg.2018.02338 PMID: 30555375

Thang, N. N., & Trang, P. T. (2024). Employer branding, organization's image and reputation, and intention to apply: The moderating role of the availability of organizational information on social media. *Frontiers in Sociology, 9*, 1256733. DOI: 10.3389/fsoc.2024.1256733 PMID: 38650696

Thomas, K. W. (2009). *Intrinsic motivation at work: What really drives employee engagement* (2nd ed.). Berrett-Koehler Publishers.

Tims, M., Bakker, A. B., & Xanthopoulou, D. (2011). Do transformational leaders enhance their followers' daily work engagement? *The Leadership Quarterly, 22*(1), 121–131. DOI: 10.1016/j.leaqua.2010.12.011

Tindall, R., Simmons, M., Allott, K., & Hamilton, B. (2020). Disengagement processes within an early intervention service for first-episode psychosis: A longitudinal, qualitative, multi-perspective study. *Frontiers in Psychiatry, 11*, 565. DOI: 10.3389/fpsyt.2020.00565 PMID: 32595545

Toseef, M., Kiran, A., Zhuo, S., Jahangir, M., Riaz, S., Wei, Z., Ghauri, T. A., Ullah, I., & Ahmad, S. B. (2022). Inspirational leadership and innovative communication in sustainable organizations: A mediating role of mutual trust. *Frontiers in Psychology, 13*, 846128. DOI: 10.3389/fpsyg.2022.846128 PMID: 36003091

Tummers, L. G., & Bakker, A. B. (2021). Leadership and job demands-resources Theory: A systematic review. *Frontiers in Psychology*, *12*, 722080. DOI: 10.3389/fpsyg.2021.722080 PMID: 34659034

Turi, J. A., Khastoori, S., Sorooshian, S., & Campbell, N. (2022). Diversity impact on organizational performance: Moderating and mediating role of diversity beliefs and leadership expertise. *PLoS One*, *17*(7), e0270813. DOI: 10.1371/journal.pone.0270813 PMID: 35877610

United States Department of Defense. (defense.gov), (2022). *Diversity, equity, inclusion, and accessibility strategic plan.* Retrieved from https://media.defense.gov/2022/Sep/30/2003088685/-1/-1/0/DEPARTMENT-OF- DEFENSE-DIVERSITY-EQUITY-INCLUSION-AND-ACCESSIBILITY- STRATEGIC-PLAN.PDF United States Equal Employment Opportunity Commission (2016). *Advancing diversity in law enforcement.* Retrieved from https://www.eeoc.gov/advancing-diversity-law-enforcement

Wan, X., He, R., Zhang, G., & Zhou, J. (2022). Employee engagement and open service innovation: The roles of creative self-efficacy and employee innovative behaviour. *Frontiers in Psychology*, *13*, 921687. DOI: 10.3389/fpsyg.2022.921687 PMID: 36118444

Webb, A. (2024). Bringing true strategic foresight back to business. *Harvard Business Review*.

Weisman, H., Wu, C.-H., Yoshikawa, K., & Lee, H.-J. (2023). Antecedents of organizational identification: A review and agenda for future research. *Journal of Management*, *49*(6), 2030–2061. DOI: 10.1177/01492063221140049

Whittington, J. L., Meskelis, S., Asare, E., & Beldona, S. (2017). *Enhancing employee engagement: An evidence-based approach.* Palgrave Macmillan., DOI: 10.1007/978-3-319-54732-9

Wyld, D. (2022). *The age of remote work: How COVID-19 transformed organizations in real time*. IBM Center for the Business of Government.

Zhenjing, G., Chupradit, S., Ku, K. Y., Nassani, A. A., & Haffar, M. (2022). Impact of employees' workplace environment on employees' performance: A multi-mediation Model. *Frontiers in Public Health*, *10*, 890400. DOI: 10.3389/fpubh.2022.890400 PMID: 35646787

Section 3
Diversity, Equity, and Inclusion and Artificial Intelligence

Chapter 8
A Focus-Group Study on Diversity Equity and Inclusion in Cross-Sector Cyber Intelligent Markets in Relationship to Underrepresented Minorities

Allison Fullwood
https://orcid.org/0009-0000-6484-6396
Capitol Technology University, USA

Eugene J. Lewis
https://orcid.org/0000-0002-2956-0760
Capitol Technology University, USA

ABSTRACT

The cybersecurity industry has historically lacked diversity among its professional workforce. Women and minorities have been underrepresented in STEM for over ten years. Statistics conclude that, in the United States, only 3% of cybersecurity analysts are African American, 11% are women, 5% are Asian, and 2% are Hispanic. A diverse workforce promotes critical thinking with different perspectives, allowing organizations to use unique methods for problem-solving, which positively impacts the success of an organization. This research uses an exploratory research design to define areas within the literature that frequently ignore underrepresented com-

DOI: 10.4018/979-8-3373-5686-0.ch008

munities. The failure to explore this underserved community demonstrates gaps in cross-sector cyber intelligent markets. The cyber intelligent market historically and currently hinders diversity, equity, and inclusion of individuals in underserved communities. The goal of the research is to gain actionable and applicable methodologies and ideals needed to influence change in the cyber intelligence community.

INTRODUCTION

The issue derives from cybersecurity organizations lacking diversity awareness in their hiring, training, and retention efforts (Lyon, 2020). Jordan (2020) postulates that this is even more so given the large cybersecurity workforce gap, paired with the increase and complexity of cyber threats, which has created a greater need than ever to grow, train, and diversify the cybersecurity workforce. As technology has advanced, cyber threats have become more prominent (Guha & Indurkar, 2020) and sophisticated. Burrell (2018) postulates cybersecurity organizations will not have the ability to maximize the use of complex problem-solving necessary to combat the influx of new forms of cyber-attacks without a robust diverse workforce that brings different perspectives and skills to the organization. Meanwhile, many organizations find it difficult to attract, recruit, and retain women and minority cyber professionals (Burrell, 2018) due to a lack of diversity in the pool of applicants. This is often a result of an inadequate number of women and minority STEM college graduates in the pipeline (Burrell, 2018). Therefore, the cybersecurity industry continues to be dominated by white male cyber professionals (ISC2, 2018). Admission to cybersecurity diversity confronts the cybersecurity industry value proposition based on the lack of a diverse cyber workforce.

These arguments by experts and scholars demonstrate an awareness of the need for diversity growth in the cybersecurity industry, but as the cyber landscape continues to change and become more complex, organizations will need to effectively institute functional strategies for the hiring and retention of women and minority cyber professionals. These strategies include (a) filling the cyber workforce gap with women and minorities to enhance diversity (Lyons, 2020), (b) ensuring appropriate education and training (Jordan, 2020) opportunities for women and minority cyber professionals, and (c) integrating diversity, equity, and inclusion into the organizational culture (Lyons, 2020), starting at the top with leadership. According to statistical data from IBM Security (2022), the United States maintained its status as being the costliest country on the average total cost of a data breach. The United States average cost of a data breach was 4.35 million dollars in 2022, which is an increase of 0.9 million, up from 4.24 million in 2021 (IBM, 2022). Out of the organizations studied, eighty-three percent have experienced more than one data breach (IBM,

2022). Oyelami and Azleena (2020) postulate that both the government and private sector are finding themselves victims of cybercrime and cyber-attacks. Overall, cyber-crimes are causing the global economy over $400 billion in losses each year (Oyelami & Azleena, 2020). In the cybersecurity industry of today, cyber intelligent markets need professionals with diverse backgrounds to successfully address organizational problem-solving and decision-making (Burrell, 2018). In this complex cyber landscape, lack of diversity has continuously contributed to an academic discussion of inclusivity within the workplace (Lewis, 2023). In recent years, technological advances have dramatically changed the way organizations address cyber threats and maintain their security posture (Cabaj et al., 2018). According to Lyon (2020), organizations need to attract, recruit, and retain a diverse workforce to improve operations. Leaders must learn operative and managerial strategies (Triplett, 2022) to conserve or salvage a diverse workforce that will position the organization strongly among cyber anomalies often disguised through various ethnic groups. (Lyon, 2020) Manuel and Karloff (2020) classify workforce diversity as "cultivating talent and promoting the full inclusion of excellence across the cyber intelligent market spectrum." Manuel and Karloff's (2020) perspectives put the value of an organization's ability to thrive in a complex cyber industry in ways that positively influence the cybersecurity industry through prominent use of diversity hiring. As cyber entities around the world change, so should its personnel. These influences correlate to the cybersecurity industry as cyber intelligent markets seek to elevate their defenses and overall security to posture against cyber anomalies and threats.

According to Burrell and Nobles (2018), educators and researchers have struggled to understand the underrepresentation of women and minorities in STEM fields. Palmer, Maramba, and Gasman (2013) postulate that there need to be more students pursuing STEM in U.S. institutions. However, large segments of the population, including women and minority individuals, are poorly represented in the STEM academic field (Burrell & Nobles, 2018). Therefore, further inquiry is needed to understand the essential workforce core competencies within the cybersecurity profession. Burrell (2018) explained that experts and scholars postulated about challenges consisting of not having enough diversity among the workforce to provide differentiating expertise, perspectives, viewpoints, and strategies that enhance the breadth of knowledge in the cybersecurity industry. Underprivileged communities are often ignored by most of the populace who have a Science, Technology, Engineering, and Mathematics (STEM) background. Unfortunately, individuals from underprivileged communities are not given an opportunity to experience education in these areas because they are not offered. Many of the institutions in which underserved communities are represented do not have proper funding to formally offer a curriculum conducive to understanding STEM opportunities in Cyber Intelligent Markets. Simms (2023) points out the unique burden majority-black jurisdictions have in maintaining fiscal strength to fund local schools due to legacy, contemporary racist and racialized policies, and racial capitalism. Limited access to quality education and resources within STEM are foundational inequalities (Nkrumah & Scott, 2022). The persistent issue of unequal funding allocation, which disproportionately impacts schools in low-income areas, exacerbates inequities and hinders the ability to provide students with a comprehensive STEM education (Morgan & Amerikaner, 2018). Fewer resources for underserved communities lead to a disproportionate lack of knowledge (Sultan, 2019) in the cyber-intelligent market. The research in Sultan's (2019) article highlights the lack of knowledge of basic cybersecurity concepts in underserved communities. Holland (2018) and Vega (2016) attribute the lack of targeted approach aimed at women and minorities to a trend of wealth disparity. Consequently, the articles create a systemic problem in accepting and prioritizing the need for a diverse workforce. Universally, it is accepted that having a diverse workforce with various backgrounds and perspectives is a key function in the success of the problem-solving and critical thinking needed to mitigate cyber threats and enhance organizational security posture.

The rate of cyber-attacks and associated financial losses invoke the need for a skilled, robust, and diverse workforce (Burrell, 2018) that can help organizations mitigate cyber threats. Crumpler and Lewis (2019) postulate that institutions for education and training in the United States have found it difficult to match the demand for new cyber talent. Quantitative data provided by Cyberseek (2022) determine that there are only enough cybersecurity workers in the United States to fill 68 percent of cybersecurity positions. The underutilization of employees with diverse backgrounds has a significant contribution to the cybersecurity workforce shortage (Lyons, 2020). Lang et. al (2022) point out that widening the diversity of the cybersecurity industry by hiring more women and minorities will greatly aid in closing the cybersecurity workforce gap. Women comprise less than 15 percent of the cybersecurity industry (Jordan, 2022), followed by 9 percent Black, 4 percent Hispanic, 8 percent Asian, 1 percent American Indian or Alaskan Native and Native Hawaiian/Pacific Islander, and 4 percent self-identifying as Other (ISC2, 2018). The statistical data provided by ISC2 (2018), Jordan (2022), and the analysis provided by Lang et al. (2022) introduce a valuable opportunity for cyber intelligent markets to fill their workforce gap with employees that provide different perspectives and approaches necessary to combat increasingly prevalent and complex cyber challenges. Any cyber intelligent market workforce intends to create a robust environment (Jordan, 2022) that allows an organization to understand and identify inadequacies, vulnerabilities, and the necessary modifications to create a strong security posture. In general, the cybersecurity industry should achieve the goal of attracting, recruiting, and retaining women and minority cyber professionals to diversify the cyber workforce (Lyons, 2022). The contents of the information introduce a crucial factor in how the cybersecurity industry attracts and retains a diverse workforce, adequately provides education and training opportunities for women and minorities, and creates a diverse, equitable, and inclusive culture (Lyons, 2022). A disconnect exists within the diversity area of the cyber intelligent markets as there is a shortage of diversity within the cyber intelligent workforce (Burrell, 2018).

PROBLEM STATEMENT

Women and minorities have been underrepresented in STEM and the cybersecurity industry for decades (Lyon, 2020). According to Burrell and Nobles (2018), underrepresentation has been a concern for government, industry, and academia for more than 10 years. Poster (2018) postulates that the future of cybersecurity strongly depends on its ability to attract and retain women and minorities who have long been untapped resources. The Bureau of Labor Statistics (BLS), published by the United States Department of Labor, reported that Black or African American individuals make up only 3% of the information security analysts in the United States (Morgan, 2016a). The BLS also reported that women make up only 11% of the world's information security workforce (Morgan, 2016b). When broadening the scope to the overall computing workforce, women with a minority background are still vastly outnumbered by their counterparts. With 26% of women in the computing workforce, less than 10% of these individuals were women of color (National Association for Woman & Information Technology, 2017). The Central Intelligence Agency (CIA) published its Diversity in Leadership (CDL) study in 2015, which indicated that past intelligence failures have been rooted in the lack of an inclusive cybersecurity workforce. (Lyon, 2020) The study's findings state that building a more diverse leadership and inclusive culture will result in more innovation, better decision-making, and greater mission effectiveness (Overcoming Barriers to Advancement - Diversity in Leadership Study, 2015). Historically, the cybersecurity industry has been primarily represented by white men (Cain, 2021a; Poster, 2018).

Misrepresentation of women and minorities in cybersecurity can be attributed to historical and societal barriers. Inadequate academic availability and preparation (Burrell & Nobles, 2018) experienced by women and minorities contribute to a lack of preparation in STEM fields. Though women and minorities have involved themselves in STEM for centuries (Poster, 2018), discrimination and stereotyping have limited the amount of participation. Cain (2021b) emphasizes that, historically, minority black men have been deemed as being academically inferior in comparison to white students. Similarly, women have battled sexist attitudes deeming that they are biologically ill-suited for the technology industry (Poster, 2018). Despite these claims, there has been no evidence that suggests women and minorities have not been succeeding in the educational field and workplace as their counterparts (Cain, 2021b; Poster, 2018).

The problem statement for this research consist of the following:

(1) What can organizations do to ensure diversity, equity, and inclusion in cyber intelligent markets?
(2) What are some key obstacles that prevent organizations from achieving diversity, equity, and inclusion?

NATURE AND SIGNIFICANCE OF THE STUDY

In today's digital era, research on diversity, equality, and inclusion in the cyber intelligent market workforce is critical (Mujtaba et al., 2023). Cyber Intelligent Markets play a critical role in protecting the global infrastructure of information security (Lis & Mendel, 2019). However, the lack of diversity amongst underserved communities still has become a pressing issue. The underrepresentation of women and individuals from minority groups is substantial, emphasizing the need to have a diverse workforce in cybersecurity (James, 2019). Especially as statistics forecast that the largest demographic over the next 10 years will be women and minority groups (Vespa et al., 2018). Extensive research consistently demonstrates that diverse teams offer a broad range of perspectives and problem-solving approaches. This ultimately leads to stronger defenses against the constantly evolving landscape of cyber threats. By fostering an inclusive environment and promoting equal opportunities, organizations can tap into a wider talent pool, thereby strengthening their cybersecurity posture (Lyon, 2020).

Furthermore, the significance of this study extends beyond the practical advantages of diversity in cybersecurity. It also aligns with the broader societal imperative of promoting equity and inclusivity in the technology industry. The underrepresentation of certain groups in cybersecurity not only limits their economic opportunities but also perpetuates systemic inequalities. Addressing these disparities through academic research serves as an opportunity to further investigate models for promoting diversity and inclusion in all STEM fields. The findings of this research could have a substantial impact on the cybersecurity workforce and the overall technical industry.

FINDING THE GAP IN THE RESEARCH

Detailed demographic statistics include information about an individual's race, gender, socioeconomic background, age, and other characteristics. According to Mountrouidou et al. (2019), the lack of accurate demographic data is a fundamental challenge in assessing cybersecurity diversity, equity, and inclusion. Without this information, effectively determining the cybersecurity workforce composition becomes difficult (Renaud & Coles-Kemp, 2022). To analyze the amount of underrepresentation in the cybersecurity industry, detailed demographic statistics are necessary. Without this information, it is difficult to prove the specific experiences of groups experiencing the most inequities in opportunity, representation, and overall treatment (Munyoka, 2022). Policies and actions to foster diversity and inclusion need specific demographic information to successfully target the appropriate underrepresented minorities. To obtain a clear picture of the cybersecurity workforce, extensive demographic data must be collected. It helps organizations and researchers to correctly analyze the industry's diversity and identify which groups are underrepresented (Arun et al., 2022). However, it is not sufficient to solely collect demographic data. An in-depth study is also required to identify the underlying causes of discrepancies and dissatisfaction in cyber intelligent markets. Without an in-depth study, we cannot determine why underrepresented minorities encounter prejudice or how to effectively address these challenges. To design methods that address these systemic challenges, it is necessary to investigate underlying variables, such as cultural prejudices, recruiting practices, and workplace conditions. A key statistics gap that would hinder progress toward a more equal and inclusive cybersecurity workforce is the absence of accurate demographic data. Burrell & Nobles (2018) highlight the necessity of demographic data.

Diversity, equity, and inclusion research in cybersecurity has shed light on systemic barriers that continue within the cyber intelligent markets. Lyon (2020) highlights certain barriers that hinder the recruitment and retention of underrepresented minorities in the cybersecurity industry. Cultural prejudices, poor representation, discriminatory recruiting practices, and hostile employment settings are factors that make up systemic barriers. Cultural biases can take the form of preconceived conceptions or prejudices, which can have an impact on recruiting choices and workplace relations (Botha-Badenhorst & Veerasamy, 2023). Lack of representation stems from a lack of diversity at all levels of the cybersecurity industry, from entry-level employment to leadership positions. Unconscious prejudice during recruiting can be a huge disadvantage to minority candidates and equates to discriminatory hiring practices (Botha-Badenhorst & Veerasamy, 2023). This results in individuals from underrepresented backgrounds being discouraged. Cultural prejudices can contribute to a sense of a lack of chances and justice, which makes cybersecurity professions less appealing (McGee, 2018). Lack of representation in an organization gives the message that persons from disadvantaged backgrounds have limited opportunities for advancement (Lyon, 2020). Discriminatory hiring practices are barriers in the recruiting process that make it more difficult for minority individuals to gain employment (Baker & Ellis, 2021). Unwelcoming work environments can lead to feelings of isolation and prejudice, discouraging minority individuals from seeking or remaining in cybersecurity employment (McGee, 2018). Recognizing the existence of inequities is not enough. Understanding the reasons and their impact on underrepresented minorities is necessary (McGee, 2018). This research can help to develop methods for effectively breaking down these barriers and creating more inclusive and equitable settings in the cybersecurity industry. Researchers need to explore more in-depth to understand these systemic barriers to make considerable progress toward resolving these problems (Walton et al., 2021).

Understanding the long-term impact of DEI programs within the cybersecurity workforce is critical in the pursuit of a fairer and more inclusive sector. DEI programs are often launched to reduce inequities and create an environment in which people from diverse backgrounds can thrive (Lyon, 2020). It is critical to track the performance of these activities over time to adjust them as needed and achieve long-term change. Organizations and researchers can examine the progress of these initiatives and their influence on the overall diversity landscape in cybersecurity. Despite the adoption of many DEI programs in recent years, there remains a significant research gap in understanding their long-term success. Many studies concentrate on short-term results, such as quick improvements in recruiting practices or diversity measures, but do not provide insight into the long-term sustainability of these changes. This lack of data makes it difficult to form accurate judgments regarding the industry's success in improving the representation of underrepresented groups (Lunn et al., 2021). To bridge the gap, long-term studies that examine the progress of DEI efforts and their impact over long-term periods, such as five, ten, or even twenty years, are needed. These studies should include the full range of DEI initiatives, from recruiting and hiring to professional development and leadership representation. Long-term monitoring of these projects allows researchers to get insights into their success, identify obstacles or limitations, and improve techniques. Understanding the growth of these programs will also assist organizations in ensuring long-term progress toward a more fair and diverse cybersecurity workforce.

THEORETICAL FRAMEWORK EXPLAINED

Process theory is a framework for understanding the complex processes that control interactions between underrepresented minorities and their experiences in the cybersecurity industry (Milli et al., 2021). Process theory helps to understand the phases, relationships, and parts involved in achieving diversity and inclusion. This aims to provide a comprehensive understanding of the different variables influencing underrepresented minorities and their growth in the cyber industry. Expectancy theory can be used to understand the motivating factors of diversity, equality, and inclusion in the cybersecurity profession. The study will use expectancy theory to examine how individuals' expectations and inclusion affect the cybersecurity industry.

Process theory applies to the study of underrepresented populations in the cyber intelligent market because it provides a formal framework for understanding and analyzing the nature of individuals' experiences and growth within the cyber industry. According to Gandini (2019), process theory provides a lens through which these issues may be broken down into recognizable phases and events for underserved populations starting and continuing in cybersecurity jobs. Process theory can help understand the path of individuals from underserved groups in cybersecurity by breaking their experiences into related phases and events. The first stage in which individuals from underrepresented minorities decide to pursue a career in cybersecurity is entry into the field. This stage includes considerations like school options, skill development, and interest in the profession (Corradini, 2020). According to Corradini (2020), the application stage includes the process of looking for work or internships in the cybersecurity industry. This stage involves things like writing resumes, networking, and applying for jobs. Hiring is the stage in which applicants are interviewed and hired (Corradini, 2020). Factors such as hiring prejudice, interview performance, and negotiating abilities are important at this stage. Individuals must also negotiate the working environment after being employed, which includes adjusting to organizational culture, seeking mentoring, and managing workplace issues. Career advancement includes advancement in the cybersecurity profession (Corradini, 2020). It includes actions like pursuing higher education, obtaining certificates, and aiming for promotions. Process theory has been used by researchers to determine the important phases in women's professional path, from educational choices to promotion into leadership roles inside technical organizations (Chowdhury et al., 2019). Educational growth, skill development, recruitment into technical organizations, promotions, and obtaining leadership roles are all part of this process. Chowdhury et al. (2019) have determined how women grow within certain occupations by using process theory for their research.

Expectancy theory is a useful framework for understanding the motives and expectations of individuals and organizations in the cyber field. Expectancy theory applied in the context of underrepresented populations in cybersecurity helps to understand why underrepresented minorities choose to pursue careers in the field or why organizations choose to prioritize diversity and inclusion efforts. Underrepresented minorities might base their decision to enter and remain in the cybersecurity industry on things such as career advancement, work satisfaction, and cultural inclusion. These individuals may have different expectations for professional advancement, mentorship opportunities, and job stability (Kuhl, 2021). Expectancy theory helps to understand these expectations. Expectancy theory helps to determine if individuals feel their cybersecurity efforts will result in career advancement, work satisfaction, or solving challenges (Kuhl, 2021). Individuals are more likely to be inspired to pursue professions in the sector if they see a strong relationship between their activities and positive results. Expectancy theory can also be applied to organizations as well. Organizations in the cybersecurity industry that prioritize diversity, equity, and inclusion often invest in building a safe and welcoming workplace (Eccles & Wigfield, 2020). The organizations feel that these investments will enhance improved problem-solving, innovation, and organizational success. Expectancy theory helps organizations examine the expected results of diversity and inclusion activities. Organizations expect that encouraging a diverse workforce will result in higher performance, bigger talent pools, and more innovative solutions for cybersecurity challenges (Eccles & Wigfield, 2020). The original information about expectancy theory by Victor Vroom (1964) provides a framework for understanding motivation and expectations in different organizational situations. Vroom's expectation theory postulates that people make choices and decisions based on their views of the relationship between their efforts, performance, and desired results (Lloyd & Mertens, 2018).

METHODOLOGY

This paper uses a qualitative research design that integrates both exploratory research and focus group methodology to explore concepts of underserved communities, which is the primary objective of this study. The complex problem consists of a lack of representation of women and minorities in cyber intelligent markets. The researcher uses qualitative research methodology to provide data related to individuals' lived experiences. The researcher will use a semi-structured constructivist focus group study. The researcher will conduct a survey study of nine individuals. The two questions from the problem statement will be asked to each participant along with three additional questions related to the study. Once the researcher receives all the contents from each participant. An assessment of the themes and patterns will be provided in the study's findings and results.

The exploratory research methodology utilizing the focus group application has a strong correlation to process theory (Mohr, 1982), which tends to see the world in terms of situations, events, people, and the processes that connect them (Maxwell, 1996). According to Maxwell (1996) using process theory explanation is largely based on how situations and events influence others. The practice of this phenomenon is rooted in understanding the underserved community perspectives of individuals through their lived experiences (Lyon, 2020). In using the focus group qualitative research method, researchers can explore a certain phenomenon in-depth and determine the context, implementation, and results of various decisions (Cronin, 2014; Lyon, 2020). Lazar et al. (2017) postulates that focus groups provide an effective method for gathering a range of opinions in groups that are not large. Krueger (1994) suggests small groups of five to seven participants for a more in-depth conversation, while Robinson (2002) suggests a slightly larger group of eight to twelve participants. When choosing the participants for these focus groups, the participants should have an interest in the topic and be willing to participate constructively (Lazar et al., 2017). Individuals in homogenous groups have a common background and experiences that can help promote discussion and give viewpoints that represent shared context (Lazar et al, 2017). The focus group chosen in this study are individuals in a homogenous group that have similar career backgrounds.

The utilization of a focus group methodology in exploratory research aligns closely with process theory, emphasizing the interconnectedness of situations, events, and people (Maxwell, 1996). Through understanding how these elements influence one another, researchers gain insights into underserved community perspectives and lived experiences. By employing focus groups, researchers can delve deeply into phenomena, examining context, implementation, and outcomes. Recommendations from scholars like Lazar et al. (2017), Krueger (1994), and Robinson (2002) suggest smaller group sizes for productive discussions, while Lazar et al. (2017) emphasizes the importance of selecting participants who are genuinely interested and willing to contribute constructively. Homogeneous groups, as suggested by Lazar et al. (2017), can foster discussion by providing shared context and experiences. Ultimately, these approaches facilitate a comprehensive understanding of complex phenomena and promote inclusive dialogue within research endeavors.

FINDINGS AND RESULTS

The findings and results from the focus group consist of qualitative exploratory research design responses. See Table 1. Survey Participants. Participants, Position Title, and Years of Service.

Table 1. Survey Participants, Position Title, and Years of Service

PARTICIPANTS	POSITION TITLE	YEARS of SERVICE
Participant 1	Chief Operations Officer	15 Years
Participant 2	Project Director	2 Years
Participant 3	Chief of Critical and Emerging Technologies	20 Years
Participant 4	Computer Network Defense Analyst	5 Years
Participant 5	Forensic Analyst	14 Years
Participant 6	Computer Network Defense Analyst	15 Years
Participant 7	Cryptologic Intel Analyst	4 Years
Participant 8	Network Systems Engineer	6 Years
Participant 9	Computer Network Defense Analyst	15 Years

QUESTION 1:

(1) What can organizations do to ensure diversity, equity, and inclusion in cyber intelligent markets? See Table 2. DEI in Cyber-Intelligent Markets

Table 2. DEI in Cyber-Intelligent Markets

Common Theme	Participant Similarities	Summary of Similarities
Recognition of the Need for DEI	Participants 1, 2, 4-6	The participants emphasize the importance of recognizing the need for diversity, equity, and inclusion in various contexts. This recognition is seen as a fundamental step towards creating an inclusive environment. (5 out of 9 participants)
Organizational Self-Reflection and Strategy	Participants 1, 3-5	The participants stress the need for organizations to look inward, assess their current state, and formulate strategies to address imbalances. This includes examining demographics, cultural biases, recruitment practices, and the effectiveness of implemented strategies. (4 out of 9 participants)
Stepping Out of Comfort Zones	Participants 1, 3, 4, 6	The idea of organizations stepping out of their comfort zones is a recurring theme. It involves going beyond traditional recruitment practices, reaching out to underrepresented groups, and actively engaging with communities to promote diversity and inclusion. (4 out of 9 participants)

continued on following page

Table 2. Continued

Common Theme	Participant Similarities	Summary of Similarities
Education and Outreach Programs	Participants 1, 2, 4, 6	The importance of educational initiatives and outreach programs is highlighted. This includes reaching out to schools and colleges with diverse student populations, providing education and certifications, and conducting workshops to generate interest in fields like cyber intelligence. (4 out of 9 participants)
Zero-Tolerance Policy for Discrimination	Participants 1, 5, 7	There is a consistent call for organizations to adopt a zero-tolerance policy for any form of discrimination within the workplace. The focus is on creating a safe and welcoming environment where individuals are evaluated based on their job performance rather than discriminatory factors. (3 out of 9 participants)
Community Engagement and Partnerships	Participants 1, 3, 6	Several participants emphasize the need for organizations to engage with communities, including underprivileged communities and minority colleges. This involves setting up internships, recruitment events, and workshops to connect with potential talent. (3 out of 9 participants)
Continuous Improvement and Monitoring	Participants 1, 3, 6, 7	The idea of continuous improvement is present in the emphasis on regularly assessing the effectiveness of diversity strategies. This involves monitoring progress, addressing biases, and actively working towards creating a more inclusive environment. (4 out of 9 participants)
Inclusivity in the Hiring Process	Participants 1, 2, 4, 8	The participants stress the importance of ensuring inclusivity in the recruiting and hiring processes. This includes eliminating biases, creating diverse teams, and fostering an environment where individuals from various backgrounds can envision themselves as part of the organization. (4 out of 9 participants)
Empowering and Supporting Underrepresented	Participants 1, 3, 9	There is a shared theme of empowering and supporting underrepresented groups, whether it is through resource groups, outreach programs, educational support, or creating opportunities for skill development. (3 out of 9 participants)
Addressing Biases and Stereotypes	Participants 1, 3, 7	The need to address biases and stereotypes is highlighted in multiple paragraphs, emphasizing the importance of education and training to overcome these challenges within the workforce. (3 out of 9 participants)

QUESTION 2:

(2) What are some key obstacles that prevent organizations from achieving diversity, equity, and inclusion? See Table 3. Obstacles Preventing DEI in Cyber-Intelligent Markets

Table 3. Obstacles Preventing DEI in Cyber-Intelligent Markets

Common Theme	Participant Similarities	Summary of Similarities
Culture and Bias	Participants 1-5, 8	Participants consistently highlight the impact of organizational culture and biases as obstacles to diversity, equity, and inclusion. This includes biases in recruitment, management, and decision-making processes. (6 out of 9 participants)
Recruitment Challenges	Participant 1-9	Multiple participants mention recruitment as a key obstacle. This includes issues related to internal recruitment practices, socioeconomic background considerations, and biases in the hiring process. (9 out of 9 participants)
Leadership Accountability	Participants 3, 5, 7	The role of leadership is emphasized, with mentions of the leaders driving organizational culture and the lack of accountability at the leadership level. Participants discuss the need for leaders to take responsibility for fostering diversity, equity, and inclusion. (3 out of 9 participants)
Systemic Failures	Participants 2, 3, 5, 7	Several participants touch upon systemic failures, including the impact of systemic biases, racism, and the structural obstacles that hinder the progress of underrepresented individuals within organizations. (4 out of 9 participants)
Resistance to Change	Participants 2-9	Resistance to change is a recurring theme, where organizations or their leadership may resist acknowledging the need for a more diverse and inclusive environment. This resistance can manifest as intentional exclusion or a lack of willingness to update policies. (8 out of 9 participants)
Inclusivity and Unconscious Bias	Participants 2-5, 8, 9	The need for inclusivity for all individuals is highlighted, and the presence of unconscious biases is acknowledged. This includes biases related to gender, race, age, and other factors that can limit opportunities for certain groups. (6 out of 9 participants)
Resource Constraints	Participants 2-6, 9	Resource-related challenges, including the lack of funds, time, proper training, and the right people to support diversity, equity, and inclusion programs, are mentioned in several paragraphs. (6 out of 9 participants)
Discriminatory Practices	Participants 3-5, 7-9	Discriminatory practices, both overt and subtle, are identified as obstacles, ranging from discriminatory leadership to the perpetuation of stereotypes that limit the hiring and promotion of certain individuals. (6 out of 9 participants)
Communication and Program Implementation	Participants 5-8	The importance of effective communication and the implementation of meaningful diversity, equity, and inclusion programs are stressed. Mere claims without action or substantial programs are seen as hindrances. (4 out of 9 participants)
Unwillingness to Address Issues	Participant 5	Some participants discuss a general unwillingness within organizations to address DEI issues, whether due to discomfort, perceived lack of benefits, or adherence to outdated practices. (16 out of 9 participants)

THEMES AND PATTERNS

The responses shared by the participants provide valuable information on the patterns and themes that characterize discussions surrounding diversity, equity, and inclusion in organizational contexts. An important recurring theme that emerges from the responses is the recognition of the critical need to acknowledge the urgency of implementing DEI initiatives. Participants assert that the absence of such recognition may result in the lack of authenticity and commitment in fostering inclusivity, possibly leading to ineffective outcomes. Once this acknowledgement is established, organizations should engage in introspection to evaluate cultural biases, recruitment practices, incentives, and other factors that may contribute to imbalanced demographics. This introspective process serves as the foundation for developing strategies to rectify inequalities and promote diversity within the organization.

The participants also highlight the importance of stepping out of one's comfort zone to facilitate meaningful progress in DEI initiatives. They advocate for comprehensive approaches that go beyond superficial measures, stressing the need to address systemic barriers and biases that hinder inclusivity. This multi-faceted perspective emphasizes the complexity of fostering DEI within organizational contexts and underscores the importance of proactive, strategic planning. To create an environment that is welcoming and safe for all employees, free from discrimination and biases, participants advocate for engaging with underrepresented communities and adopting inclusive recruitment practices to attract diverse talent pools and create pathways for marginalized groups to enter the workforce. They also stress the importance of instituting zero-tolerance policies towards discrimination and basing employee evaluations solely on job performance. Moreover, participants highlight the need for organizations to invest in education and training to promote awareness and understanding of DEI principles among employees. This emphasis on cultivating a culture of inclusivity aligns with broader efforts to foster supportive workplace environments that value diversity and individuality.

However, participants also identify various obstacles and challenges that hinder progress towards DEI within organizational contexts. These include discriminatory leadership, unconscious biases, and resistance to change, among others. To overcome these obstacles effectively, participants emphasize the importance of addressing systemic failures and investing resources in DEI initiatives. Overall, the responses highlight the importance of acknowledging the need for DEI initiatives, engaging in introspection, stepping out of one's comfort zone, adopting comprehensive approaches, and investing in education and training to promote DEI principles. The discourse also underscores the obstacles and challenges that impede progress towards DEI and emphasizes the need to address systemic failures and invest resources in DEI initiatives to overcome these challenges effectively.

FUTURE RESEARCH ON THE SUBJECT

It would be valuable to broaden the scope of inquiry to advance research on Diversity, Equity, and Inclusion in the cybersecurity industry regarding underrepresented minorities. This could involve expanding the sample size, encompassing diverse demographics, and considering varied geographic regions (Lazar et al., 2017; Robson, 2002). Utilizing larger sample sizes in quantitative studies enables more rigorous statistical analyses and promotes the generalizability of results (Lazar et al., 2017). By including a wider range of participants, researchers can more effectively portray the diversity of experiences, perspectives, and obstacles encountered by underrepresented minorities in the field, leading to more thorough and nuanced insights (Lazar et al. 2017; Krueger, 1994). Expanding the scope of demographic analysis beyond conventional categories like race and gender can offer a more comprehensive perspective on DEI concerns in the realm of cybersecurity (Krueger, 1994). For example, researchers may delve into intersectional identities, encompassing a range of factors such as ethnicity, socioeconomic status, sexual orientation, disability status, and veteran status, to investigate how diverse aspects interact to shape individuals' experiences and prospects in the field (Lazar et al, 2017). Adopting an intersectional lens enables researchers to uncover distinct obstacles and opportunities for fostering inclusion that may be disregarded in analyses that concentrate exclusively on single dimensions of identity (Can, 2021b). Broadening the research area to include rural regions, smaller cities, and developing areas can provide insights into the worldwide scope of DEI challenges in cybersecurity (Kuhl, 2021; Gandini, 2019). By analyzing the impact of cultural, economic, and infrastructural factors on the accessibility of education, employment opportunities, and support networks for underrepresented minorities in various contexts, researchers can formulate targeted interventions and policy suggestions that cater to the unique needs of individuals globally.

The integration of diversity, equity, and inclusion (DEI) principles within cross-sector intelligent markets is not merely a moral imperative but a strategic necessity for fostering innovation, resilience, and competitiveness (Cabaj et al., 2018; Chowdhury, Adam, & Skinner, 2019). For underrepresented minorities, this integration can serve as a powerful lever for systemic change, offering both challenges and opportunities (Walton et al., 2021). As intelligent markets continue to evolve, leveraging artificial intelligence, big data, and other technological advancements, they hold the potential to dismantle traditional barriers to entry and create more equitable pathways for participation and success (McGee, 2018; Habibzadeh et al., 2019). However, this potential can only be realized through deliberate and sustained efforts to embed DEI at every level of market operation and decision-making (Sultan, 2019). This involves not only ensuring diverse representation but also fostering an inclusive culture that values and leverages diverse perspectives, and implementing equitable practices that recognize and address systemic biases and inequalities (Morgan, 2016). Ultimately, the success of cross-sector intelligent markets in enhancing the participation and prosperity of underrepresented minorities will depend on their ability to translate DEI principles from theory into practice, creating ecosystems that are not only more innovative and dynamic but also more just and inclusive.

In addition to advancing theoretical knowledge and methodological approaches in academia, expanded research efforts can have practical implications for academic research and industry practice (Gandini, 2019). By creating more representative and inclusive datasets, researchers can generate evidence-based insights and recommendations that resonate with a wider audience of stakeholders, such as policymakers, educators, employers, and advocacy groups (Lazar et al., 2017). Moreover, prioritizing the voices and perspectives of underrepresented minorities in research can contribute to empowering marginalized communities, promoting social justice, and fostering greater equity and inclusion within the cybersecurity industry and beyond (Kuhl, 2021; McGee, 2018). Incorporating variables like sample size, demographics, and geographic representation into research methodology and data gathering can significantly improve the quality, validity, and significance of studies exploring DEI in the realm of cybersecurity (Lazar et al. 2017; Krueger, 1994). By adopting a comprehensive and inclusive approach to investigation, the academic community can play a vital role in advancing our understanding, shaping policies and practices, and fostering positive societal progress towards building a cybersecurity workforce that is diverse, equitable, and inclusive.

CONCLUSION

The effects of diversity, equity, and inclusion (DEI) initiatives on underrepresented minority communities within the United States are multifaceted and significant (Corradini, 2020). Firstly, such efforts foster a sense of belonging and representation among marginalized groups, leading to increased engagement and productivity (Cabaj et al., 2018; Chowdhury, Adam, & Skinner, 2019). When individuals from underrepresented backgrounds feel valued and respected, they are more likely to contribute their unique perspectives and talents to their communities and workplaces (Triplett, 2022). Additionally, DEI initiatives can mitigate systemic barriers that hinder the advancement of minority groups, such as discriminatory hiring practices and unequal access to opportunities (McGee, 2018; Habibzadeh et al., 2019). By promoting fairness and equal treatment, these initiatives help level the playing field and create pathways for upward mobility (Mountrouidou et al., 2019). Furthermore, DEI efforts contribute to broader societal benefits by challenging stereotypes, promoting cross-cultural understanding, and fostering social cohesion (Renaud &. Coles-Kemp, 2022). However, it's crucial to recognize that the impact of DEI initiatives may vary depending on factors such as implementation strategies, organizational culture, and the broader socio-political context (Sultan, 2019; Mountrouidou et al., 2019). Continuous evaluation and refinement of these initiatives are essential to ensuring that they effectively address the needs and aspirations of underrepresented minority communities in the United States.

In conclusion, this research revealed the complex picture of diversity, equality, and inclusion in the cyber-intelligent market. The research revealed the need for DEI policies as well as the presence of structural hurdles, particularly for those from marginalized backgrounds. The survey emphasized the importance of professional growth opportunities and equitable hiring and recruitment practices. Finally, this study adds to the expanding body of information about DEI in the cybersecurity industry. It fills data gaps, gives vital insights, and lays the foundation for future research and policy development. The findings highlight the significance of enacting more inclusive DEI policies to establish a diverse and equitable cybersecurity workforce. Adherence to these findings promotes a cybersecurity industry that maximizes the potential of diverse workforce while being competitive and innovative in the face of changing challenges. Lastly, this study recommends a more inclusive and diverse cybersecurity workforce, which aligns with the larger objective of establishing a fair and equitable society.

REFERENCES

Arun, A., Rao, A., Hecht, A., Garcia, A., Oluyomi, A., Lapidus, A., & Miggins, J. (2022). Fifth annual summer research summit on Health Equity organized by the Center of excellence in Health Equity, Training and research, Baylor College of medicine, Houston, Texas 77030, USA on May 17, 2022. [IJMA]. *International Journal of MCH and AIDS*, *11*(2s). Advance online publication. DOI: 10.21106/ijma.596

Baker, D., & Ellis, L. (2021). Future directions in digital information: scenarios and themes. In *Future Directions in Digital Information* (pp. 1–15). Chandos Publishing. DOI: 10.1016/B978-0-12-822144-0.00001-X

Botha-Badenhorst, D., & Veerasamy, N. (2023). Examining barriers to entry: Disparate gender representation in cybersecurity within Sub-Saharan Africa. *International Conference on Gender Research*, 6(1), 47–55. DOI: 10.34190/icgr.6.1.1148

Burrell, D. (2018). An Exploration of the Cybersecurity Workforce Shortage. *International Journal of Hyperconnectivity and the Internet of Things*, *2*(1), 29–41. DOI: 10.4018/IJHIoT.2018010103

Burrell, D. N., & Nobles, C. (2018). Recommendations to develop and hire more highly qualified women and minorities cybersecurity professionals. In International Conference on Cyber Warfare and Security (pp. 75-81). Academic Conferences International Limited.

Cabaj, K., Kotulski, Z., Księżopolski, B., & Mazurczyk, W. (2018). Cybersecurity: Trends, issues, and challenges. *EURASIP Journal on Information Security*, *2018*(1), 1–3. DOI: 10.1186/s13635-018-0080-0

Cain, C. C. (2021a). Beyond the IT Artifact - Studying the Underrepresentation of Black Men and Women in IT. *Journal of Global Information Technology Management*, *24*(3), 157–163. DOI: 10.1080/1097198X.2021.1954315

Cain, Curtis C. (2021b) "Establishing a research agenda for broadening participation of Black men in computing, informatics, and engineering," Technology in Society, Elsevier, vol. 67(C).

Chowdhury, N. H., Adam, M. T., & Skinner, G. (2019). The impact of time pressure on cybersecurity behaviour: A systematic literature review. *Behaviour & Information Technology*, *38*(12), 1290–1308. DOI: 10.1080/0144929X.2019.1583769

Corradini, I. (2020). *Building a cybersecurity culture in organizations: How to bridge the gap between people and digital technology* (Vol. 284). Springer Nature. DOI: 10.1007/978-3-030-43999-6

Cronin, C. (2014). Using case study research as a rigorous form of inquiry. *Nurse Researcher, 21*(5), 19–27. DOI: 10.7748/nr.21.5.19.e1240 PMID: 24877907

Eccles, J. S., & Wigfield, A. (2020). From expectancy-value theory to situated expectancy-value theory: A developmental, social cognitive, and sociocultural perspective on motivation. *Contemporary Educational Psychology, 61*, 101859. DOI: 10.1016/j.cedpsych.2020.101859

Eisenhardt, K. M. (1989). Agency Theory: An Assessment and Review. *Academy of Management Review, 14*(1), 57–74. DOI: 10.2307/258191

Gandini, A. (2019). Labour process theory and the gig economy. *Human Relations, 72*(6), 1039–1056. DOI: 10.1177/0018726718790002

Guha, E. A., & Indurkar, S. K. (2020). A Study on Cyber Crime and Data Breach Management. *Research Journal of Engineering and Technology, 11*(2), 113–117. DOI: 10.5958/2321-581X.2020.00020.3

Habibzadeh, H., Nussbaum, B. H., Anjomshoa, F., Kantarci, B., & Soyata, T. (2019). A survey on cybersecurity, data privacy, and policy issues in cyber-physical system deployments in smart cities. *Sustainable Cities and Society, 50*, 101660. DOI: 10.1016/j.scs.2019.101660

Hagan, J., McCarthy, B., Herda, D., & Cann Chandrasekher, A. (2018). Dual-process theory of racial isolation, legal cynicism, and reported crime. *Proceedings of the National Academy of Sciences of the United States of America, 115*(28), 7190–7199. DOI: 10.1073/pnas.1722210115 PMID: 29987006

Holland, J. (2016). The Average Black Family Would Need 228 Years to Build the Wealth of a White Family Today. The Nation Journal. Retrieved from https://www.thenation.com/article/the-average-black-family-wouldneed-228-years-to-build-the-wealth-of-a-white-family-today

ISC2 (2018). Innovation Through Inclusion: The Multicultural Cybersecurity Workforce. https://www.isc2.org/-/media/Files/Research/Innovation-Through-Inclusion-Report.ashx

James, S. L. (2019). The underrepresentation of females in the United States cybersecurity workforce: a multiple-case study (Doctoral dissertation, Capella University).

Jordan, C. A. (2022). Exploring the Cybersecurity Skills Gap: A Qualitative Study of Recruitment and Retention from a Human Resource Management Perspective (Order No. 29320493). Available from ProQuest Central; ProQuest Dissertations & Theses Global; ProQuest One Academic; Publicly Available Content Database. (2706752761). https://www.proquest.com/dissertations-theses/exploring-cybersecurity-skills-gap-qualitative/docview/2706752761/se-2

Krueger, R. A. (1994). *Focus Groups: A practical guide for applied research.* Sage Publications.

Kuhl, J. (2021). The expectancy-value approach within the theory of social motivation: Elaborations, extensions, critique. Expectations and actions, 125-160.

Lazar, J., Feng, J. H., & Hochheiser, H. (2017). *Research methods in human-computer interaction.* Elsevier Science & Technology.

Lis, P., & Mendel, J. (2019). Cyberattacks on critical infrastructure: An economic perspective. *Economics and Business Review, 5*(2), 24–47. DOI: 10.18559/ebr.2019.2.2

Lloyd, R., & Mertens, D. (2018). Expecting more out of expectancy theory: History urges inclusion of the social context. *International Management Review, 14*(1), 28–43.

Lunn, S., Zahedi, L., Ross, M., & Ohland, M. (2021). Exploration of intersectionality and computer science demographics: Understanding the historical context of shifts in participation. [TOCE]. *ACM Transactions on Computing Education, 21*(2), 1–30. DOI: 10.1145/3445985

Lyon, V. (2020). *Exploring Strategies for Recruiting and Retaining Diverse Cybersecurity Professionals* (Order No. 27738274). Available from ProQuest Dissertations & Theses Global; ProQuest One Academic. (2378121182). https://www.proquest.com/dissertations-theses/exploring-strategies-recruiting-retaining-diverse/docview/2378121182/se-2

Manuel, B. A., & Karloff, D. B. (2020). Recruit and retain a diverse workforce. *Nature Reviews. Chemistry, 4*(9), 435–437. DOI: 10.1038/s41570-020-0214-z PMID: 37127964

Maxwell, J. A. (1996). *Qualitative research design: An interactive approach.* Sage Publications, Inc.

McGee, K. (2018). The influence of gender, and race/ethnicity on advancement in information technology (IT). *Information and Organization, 28*(1), 1–36. DOI: 10.1016/j.infoandorg.2017.12.001

McSweeney, K. (2018). Motivating cybersecurity compliance in critical infrastructure industries: A grounded theory study (Doctoral dissertation, Capella University).

Milli, S., Lieder, F., & Griffiths, T. L. (2021). A rational reinterpretation of dual-process theories. *Cognition*, *217*, 104881. DOI: 10.1016/j.cognition.2021.104881 PMID: 34536658

Mohr, L. (1982). *Explaining organizational behavior*. Jossey-Bass.

Morgan, I., & Amerikaner, A. (2018). Funding Gaps 2018: An Analysis of School Funding Equity Across the U.S. Education Trust. Retrieved from https://edtrust.org/resource/funding-gaps-2018/

Morgan, S. (2016a, April, 7). African Americans Underrepresented In The Cybersecurity Field. Forbes. https://www.forbes.com/sites/stevemorgan/2016/04/07/african-americans-underrepresented-in-the-cybersecurity-field/?sh=58ea932115c9

Morgan, S. (2016b, March 28). Calling All Women: The Cybersecurity Field Needs You And There's A Million Jobs Waiting. Forbes. https://www.forbes.com/sites/stevemorgan/2016/03/28/calling-all-women-the-cybersecurity-field-. needs-you/#2470db2a381c

Mountrouidou, X., Vosen, D., Kari, C., Azhar, M. Q., Bhatia, S., Gagne, G., Maguire, J., Tudor, L., & Yuen, T. T. (2019). Securing the human: A Review of Literature on Broadening Diversity in Cybersecurity Education. *Proceedings of the Working Group Reports on Innovation and Technology in Computer Science Education*, 157–176. DOI: 10.1145/3344429.3372507

Mujtaba, B. G., Vardarlıer, P., & Gündoğmuş, E. (2023). Digital Diversity and Inclusion: Agenda for Data Privacy and Protection. In *Multidimensional and Strategic Outlook in Digital Business Transformation: Human Resource and Management Recommendations for Performance Improvement* (pp. 39–52). Springer International Publishing. DOI: 10.1007/978-3-031-23432-3_4

Munyoka, W. (2022). Inclusive Digital Innovation in South Africa: Perspectives from disadvantaged and marginalized communities. *Sustainability (Basel)*, *14*(9), 5372. DOI: 10.3390/su14095372

National Association for Woman & Information Technology. (2017). https://www.ncwit.org/

Nkrumah, T., & Scott, K. A. (2022). Mentoring in STEM higher education: A synthesis of the literature to (re)present the excluded women of color. *International Journal of STEM Education*, *9*(1), 50. Advance online publication. DOI: 10.1186/s40594-022-00367-7 PMID: 35919282

Overcoming Barriers to Advancement - CIA Diversity in Leadership Study. (2015). Retrieved from https://www.cia.gov/library/reports/dls-report.pd

Pekrun, R. (2019). Expectancy–value theory of anxiety: Overview and implications. *Anxiety*, ●●●, 23–42.

Poster, W. R. (2018). Cybersecurity needs women. *Nature*, *555*(7698), 577–580. DOI: 10.1038/d41586-018-03327-w

Renaud, K., & Coles-Kemp, L. (2022) Accessible and inclusive cyber security: a nuanced and complex challenge. SN Computer Science, 3 (5). p. 346. 346. ISSN 2661-8907 ()DOI: 10.1007/s42979-022-01239-1

Robson, C. (2002). *Real World Research: A Resource for Social Scientists and Practitioner-Researchers / C.* Robson.

Security, I. B. M. (2022). Cost of a Data Breach Report 2022. https://www.ibm.com/downloads/cas/3R8N1DZJ

Simms, A. (2023). Fiscal Fragility in Black Middle-Class Suburbia and Consequences for K–12 Schools and Other Public Services. *The Russell Sage Foundation Journal of the Social Sciences : RSF*, *9*(2), 204–225. DOI: 10.7758/RSF.2023.9.2.09

Stamenkov, G. (2023). Recommendations for improving research quality: Relationships among constructs, verbs in hypotheses, theoretical perspectives, and triangulation. *Quality & Quantity*, *57*(3), 2923–2946. DOI: 10.1007/s11135-022-01461-2

Sultan, A. H. M. A. D. (2019). Improving Cybersecurity Awareness in Underserved Populations. Center for Long Term Cybersecurity, UC Berkely. https://cltc. berkeley. edu/wpcontent/uploads/2019/04/CLTC_Underserved_Populations. Pdf.

Tornede, A., Wever, M., Werner, S., Mohr, F., & Hüllermeier, E. (2020, September). Run2Survive: A decision-theoretic approach to algorithm selection based on survival analysis. In *Asian Conference on Machine Learning* (pp. 737-752). PMLR.

Triplett, W. J. (2022). Addressing Human Factors in Cybersecurity Leadership. *Journal of Cybersecurity and Privacy*, *2*(3), 573–586. DOI: 10.3390/jcp2030029

Upadhyay, A. (2023). Rising board gender diversity and incentives of female directors. *Journal of Corporate Finance*, *80*, 102386. DOI: 10.1016/j.jcorpfin.2023.102386

Vega, T. (2016). Blacks Will Take Hundreds of Years to Catch Up to White Wealth. https://money.cnn.com/2016/08/09/news/economy/blacks-white-wealth-gap/index.html

Vespa, J., Armstrong, D. M., & Medina, L. (2018). *Demographic turning points for the United States: Population projections for 2020 to 2060*. US Department of Commerce, Economics and Statistics Administration, US Census Bureau.

Walton, S., Wheeler, P. R., Zhang, Y., & Zhao, X. (2021). An integrative review and analysis of cybersecurity research: Current state and future directions. *Journal of Information Systems*, *35*(1), 155–186. DOI: 10.2308/ISYS-19-033

Chapter 9
Towards a Framework to Support Employees With Hidden Disabilities

Jim W. Tebrake
https://orcid.org/0000-0003-3699-6041
Marymount University, Canada

ABSTRACT

It is estimated that 16% of American workers have a hidden disability. The large share of employees with a hidden disability highlights the need to provide support to these employees. Supporting employees with hidden disabilities can lead to positive reputational and economic benefits for organizations. This study outlines a framework that organizations can use to establish and administer an effective performance support program for employees with hidden disabilities. Through a review of the literature on organizational obligations to disabled employees and factors that constitute effective employee performance support programs, the research shows that an effective program is framed around: 1. Understanding the legal, moral, and ethical obligations 2. Defining the scope of the program appropriate for the organization 3. Engaging all employees through training, outreach, and change management 4. Developing a confidential disclosure process 5. Investing in appropriate accommodations and applying where required 6. Monitoring and evaluating the effectiveness of the program

DOI: 10.4018/979-8-3373-5686-0.ch009

INTRODUCTION

I cannot think straight; I have no energy; I feel tense when I am around others; I want to be left alone. These are common refrains in many workplaces. In the past, employers often associated these types of statements with disgruntled or discouraged workers. However, over the last few decades, the understanding of these statements has improved considerably (Kulkarni, 2021; Sadat & Najie, 2023). What is behind most of these statements is not bitter or disenfranchised employees but rather hidden disabilities such as depression, anxiety, or attention-deficit/hyperactivity disorder (Centers for Disease Control and Prevention [CDC], 2023; Gulyamova et al., 2023).

Approximately 27% of Americans have a disability (Centers for Disease Control and Prevention, 2023). In many cases these disabilities are hidden from the employer (Kulkarni, 2021; Solovieva et al., 2011). If left unaccommodated, hidden disabilities can reduce productivity, increase absenteeism, and negatively impact firm performance (Gulyamova et al., 2023; Sadat & Najie, 2023). Research has shown that implementing workplace accommodations can boost individual productivity by up to 53% and generate substantial organizational benefits (Bebon, 2023; Solovieva et al., 2011). Recognizing the economic and societal value of supporting employees with disabilities, many governments have mandated reasonable accommodation through legislation, such as the Americans with Disabilities Act (ADA) of 1990 (Americans with Disabilities Act of 1990; Jones, 2008).

By providing appropriate accommodations, organizations reduce their legal risk and enhance their internal culture and external reputation (Hartnett et al., 2011; Americans with Disabilities Act of 1990). Programs such as flexible work arrangements, stress management initiatives, and task prioritization supports have been found to significantly enhance the performance of individuals with conditions like ADHD, anxiety, and depression (Mellifont et al., 2016; Chajduga & Ingaldi, 2021). These initiatives often lead to reduced absenteeism, improved employee presence, and greater staff retention (Sadat & Najie, 2023). Empowering employees to effectively manage their disabilities fosters higher job satisfaction, improved quality of life, and stronger team collaboration, as diverse teams draw upon a broader range of perspectives and talents (Hartnett et al., 2011; Solovieva et al., 2011).

Given that hidden disabilities impact a significant portion of the workforce, there is a pressing need to design support programs that allow these employees to thrive (Tamini et al., 2023). Supporting employees with hidden disabilities is crucial for fostering a culture of inclusion and ensuring organizations meet both their ethical obligations and legal requirements (Jones, 2008; Americans with Disabilities Act of 1990). A study reviewing 50 organizational support programs found that 26 had a favorable and substantial impact on organizational performance (Tamini et al., 2023). These programs contribute to improved firm performance, employee retention, and productivity and promote broader cultural acceptance and equity within the workplace (Sadat & Najie, 2023).

However, despite growing awareness of hidden disabilities and the documented benefits of accommodations, there remains a significant gap in the academic and practical literature: few frameworks exist to guide organizations on how to systematically develop and implement support programs for employees with hidden disabilities (Gulyamova et al., 2023; Kulkarni, 2021). This paper addresses that gap by proposing an actionable, research-informed framework for organizations seeking to better support employees with hidden disabilities.

PROBLEM STATEMENT

Despite legislative advances such as the Americans with Disabilities Act (ADA), many organizations continue to struggle with systematically supporting employees with hidden disabilities. Approximately 27% of Americans have a disability, and roughly 80% of these are considered hidden disabilities (Centers for Disease Control and Prevention [CDC], 2023). Hidden disabilities—such as depression, anxiety, ADHD, chronic fatigue syndrome, and PTSD—can significantly impact workplace productivity, job satisfaction, employee retention, and organizational performance if unaddressed (Gulyamova et al., 2023). Employers who implement accommodations report direct benefits, including retaining qualified staff and increasing worker productivity, with most estimated benefits exceeding $1,000 per employee annually (Solovieva et al., 2011). Yet, despite these clear advantages, organizations often encounter difficulties implementing performance support programs. Sometimes, employees with hidden disabilities may not want help or even be aware that they need help (Kulkarni, 2021). In other cases, the programs may result in employee resentment (Calhoun & McCarthy, 2022). Currently, there is a lack of organizational frameworks to guide the structured implementation of support programs for employees with hidden disabilities (Kulkarni, 2021). This paper addresses this critical gap by proposing a research-informed, actionable framework for developing and sustaining effective organizational support systems.

SIGNIFICANT OF THE PROJECT

The significance of supporting employees with hidden disabilities extends beyond legal compliance. Inclusive workplaces that accommodate diverse needs are more innovative, resilient, and competitive (Sadat & Najie, 2023; Gulyamova et al., 2023). Organizations that fail to implement effective support programs face consequences such as higher legal liability risks, increased employee attrition, diminished workforce morale, and reputational damage (Jones, 2008; Americans with Disabilities Act of 1990). From an academic perspective, this study contributes to the fields of organizational behavior, human resource management, and disability inclusion by providing a systematic, theory-driven approach to framework development. From a practical perspective, it offers organizations a blueprint to foster employee well-being, improve retention rates, enhance productivity, and build more inclusive and sustainable workplaces.

METHODOLOGY

This study follows a conceptual framework development approach supported by a structured literature review. Peer-reviewed journal articles, organizational case studies, policy documents, and government reports published primarily between 2000 and 2024 were examined. Databases consulted included JSTOR, EBSCOhost, Google Scholar, and ProQuest. Inclusion criteria focused on literature that addressed disability accommodation practices, organizational leadership and change management, hidden disability typologies and challenges, legal frameworks, including the ADA, and organizational behavior and employee performance. Sources were selected based on their relevance to hidden disability support, organizational implementation strategies, and empirical findings related to accommodation outcomes. A thematic analysis method was used to identify key concepts and best practices that inform the proposed framework.

PROJECT THEORIES

The proposed framework draws on four foundational theories. First, compliance with the Americans with Disabilities Act (ADA) remains a critical legal foundation requiring organizations to provide reasonable accommodations to qualified employees unless doing so would impose an undue hardship (Americans with Disabilities Act, 1990; Scroggins, 2007). Second, Kotter's Eight-Step Change Management Model offers a structured method for embedding disability inclusion into organizational culture, beginning with urgency creation and culminating in cultural reinforcement (Rajan & Ganesan, 2017; Baloh et al., 2017). Third, Authentic Leadership Theory highlights the importance of transparency, ethical behavior, and compassion in fostering trust and employee engagement, particularly among vulnerable populations (Shirey, 2006; Ribeiro et al., 2018; Cao et al., 2020). Finally, Psychological Safety and Stigma Reduction frameworks stress the need to create a culture where individuals feel secure disclosing hidden disabilities without fear of judgment or retaliation (Toth & Dewa, 2014; Stuart, 2006).

LITERATURE REVIEW

Understanding Legal Obligations

Organizations have a legal obligation to accommodate employees with disabilities. The Americans with Disabilities Act (ADA) prohibits discrimination against individuals with disabilities in various areas of public life, including employment (Koruda, 2016). The ADA defines a *disability* as a physical or mental impairment that substantially limits one or more major life activities (Americans With Disabilities Act, 1990; Koruda, 2016). The ADA ensures that individuals with conditions like ADHD, Dyslexia, Anxiety disorders, Depression, Chronic Fatigue Syndrome, and PTSD are protected from discrimination and are entitled to reasonable accommodations in the workplace (Americans With Disabilities Act, 1990). Under the ADA, employers are prohibited from discriminating against qualified individuals with disabilities in all aspects of employment, including hiring, firing, promotions, job assignments, training, benefits, and other terms and conditions of employment (Americans With Disabilities Act, 1990).

Several studies have been conducted to determine the effectiveness of the American Disabilities Act. Research in the early 2000s found that the ADA led to a 7.2 percentage point decrease in employment rates for men with disabilities but no change in wages for these men (DeLeire, 2000). Others concluded that the ADA could enhance access to health care, personal assistance, employment, and the electoral process for individuals with disabilities (Batavia & Schriner, 2001). Within the first decade of the ADA, there was progress in its implementation and enforcement, but challenges remained in ensuring compliance across various sectors (Percy, 2001). The ADA has improved workplace accommodations for adults with learning disabilities, although more research is needed to fully understand their experiences and needs (Gerber & Price, 2003). While most of the studies that evaluated the effectiveness of the ADA are dated and further research is required, there is strong evidence, at least within the first decade of the ADA, that its implementation leads to better outcomes for employees.

Defining Reasonableness

The ADA requires employers to provide reasonable accommodations to qualified employees with disabilities unless doing so would impose undue hardship on the organization (Americans with Disabilities Act, 1990; Tran, 2021; Scroggins, 2007). The definition of reasonableness is critical for organizations and needs to be established and communicated before designing and implementing a program. Organizations should periodically re-examine their definition of 'reasonableness' since they have observed that it changes over time (Crampton & Hodge, 2003). Reasonable accommodations can vary depending on the individual's needs and the nature of the job. However, they may include modifications to workspaces, flexible work arrangements, assistive technologies, modified work schedules, or adjustments to job duties (McDowell & Fossey, 2015). The ADA only provides for reasonable accommodation for known disabilities (Americans With Disabilities Act, 1990). Managers' adherence to the ADA is influenced by the information available to them, organizational policies, attitudes, and the suite of accommodations in place within the organization (Oermann, 2016). Once established, the 'reasonableness' definition should be codified in organizational policies, procedures, and guidelines.

Types of Hidden Disabilities

A *hidden disability* is an impairment that is not visible or apparent to others that impacts an individual's ability to function (Rock et al., 2013). In most cases, there are no physical markers or apparent signs of a hidden disability (Couzens et al., 2015). A *hidden disability* is a neurological handicap that interferes with a person's ability to receive, process, store, and retrieve information, often not easily recognized, accepted, or considered serious (Rafeeque, 2018). Hidden disabilities tend to impact an individual's cognitive function, sensory perceptions, mood, sense of worth, and emotions (Syma, 2019). Examples of hidden disabilities include ADHD (Attention-Deficit/Hyperactivity Disorder), Dyslexia, Autism Spectrum Disorder, Anxiety disorders, Depression, Chronic Fatigue Syndrome, and Post-traumatic Stress Disorder (PTSD) (*Hidden/Invisible Disabilities*) (Moore, 2009). Hidden disabilities can impact all aspects of an individual's life, including their educational pursuits, relationships, work, and social interactions. Students with hidden disabilities like ADHD face obstacles in higher education, such as psychological distress, poor social skills, and cognitive deficits (Wolf, 2001). The challenges associated with hidden disabilities vary from person to person, often requiring tailored accommodations and support to help individuals navigate their environments effectively. For example, youth with disabilities face challenges in disclosing their condition and requesting workplace accommodations (Lindsay et al., 2019). The hidden disabilities that have the most significant impact on employee performance include ADHD (Attention-Deficit/Hyperactivity Disorder), Dyslexia, Autism Spectrum Disorder, Anxiety disorders, Depression, Chronic Fatigue Syndrome, and post-traumatic stress disorder (PTSD) (Moore, 2009).

Individuals with Attention-Deficit/Hyperactivity Disorder (ADHD) have difficulty managing time, prioritizing tasks, and meeting deadlines (Fisher & Baum, 2010). This can impact an individual's ability to complete a task since their attention is quickly drawn from one task to another. Inconsistencies in work quality, missed deadlines, and difficulties in time management could result, potentially impacting their overall performance and productivity (Attention-deficit/hyperactivity disorder, 2020).

Dyslexia is another common hidden disability found in the workplace. *Dyslexia* is a language-processing disability that impacts learning (*Definition of Dyslexia,* 2018). Individuals with dyslexia have difficulty understanding written text, spelling, and decoding words (*Definition of Dyslexia,* 2018). Generally, reading and absorbing text takes them longer than individuals without dyslexia. In the workplace, a person with dyslexia might experience challenges in reading emails, reports, or other written materials quickly and accurately (*Definition of Dyslexia,* 2018). This can lead to slower comprehension, increased time spent on written communication tasks, and potential misunderstandings due to misinterpretation of written information, affecting their efficiency and ability to perform specific job tasks (*Definition of Dyslexia,* 2018).

Autism Spectrum Disorder (ASD) is a complex developmental condition that affects behavior and social interaction (Signs and symptoms of autism spectrum disorders, 2022). Individuals with ASD may have difficulty expressing themselves, adapting to change, or observing social cues (Signs and symptoms of autism spectrum disorders, 2022). In a work setting, they may struggle with team collaboration, understanding social dynamics, or coping with sensory-challenging workspaces (noise, lighting) (Signs and symptoms of autism spectrum disorders, 2022). All of which could lead to anxiety or discomfort and impact an individual's performance and productivity. People with ASD may also have different ways of learning, moving, or paying attention (Signs and symptoms of autism spectrum disorders, 2022).

Anxiety disorder can manifest itself in excessive worry, fear, or nervousness (Anxiety Disorders, 2024). In the workplace, anxiety disorders can cause employees to avoid specific tasks or situations (Anxiety Disorders, 2024). For example, someone with social anxiety may find networking events or public speaking engagements extremely distressing, affecting their ability to build professional relationships or present ideas confidently (Anxiety Disorders, 2024). These symptoms can decrease job satisfaction, reduce productivity, and increase absenteeism if not adequately managed or supported (Anxiety Disorders, 2024). Anxiety disorder is persistent, in which individuals with AD can experience AD for months (Anxiety Disorders,

2024). Common symptoms include feeling restless, having difficulty concentrating, being irritable, or having unexplained pains (e.g., stomachaches, headaches) (Anxiety Disorders, 2024).

Depression is a mood disorder that can cause persistent feelings of sadness, hopelessness, and loss of interest in activities (Depressive disorder (depression), 2024). Employees experiencing depression lack energy and motivation and can find it difficult to concentrate (Depressive disorder (depression), 2024). They may struggle to initiate tasks, meet deadlines, and maintain consistent attendance (Depressive disorder (depression), 2024). Depression can impact interpersonal relationships, affecting teamwork and overall job performance (Depressive disorder (depression), 2024). Depression is different from regular mood changes and feelings about everyday life (Depressive disorder (depression), 2024).

Chronic Fatigue Syndrome (CFS), also known as Myalgic Encephalomyelitis (ME), is a complex and debilitating condition characterized by extreme exhaustion and cognitive difficulties (e.g., brain fog) (Myalgic encephalomyelitis/chronic fatigue syndrome (ME/CFS), 2023). Employees with CFS may struggle with the energy levels required to perform tasks efficiently throughout the day (Myalgic encephalomyelitis/chronic fatigue syndrome (ME/CFS), 2023). They may need accommodations such as flexible work schedules or rest breaks to manage their symptoms and maintain productivity (Myalgic encephalomyelitis/chronic fatigue syndrome (ME/CFS), 2023). According to an Institute of Medicine (IOM) report, an estimated 836,000 to 2.5 million Americans suffer from ME/CFS; however, most of them have not been diagnosed (Valdez et al., 2019; Myalgic encephalomyelitis/chronic fatigue syndrome (ME/CFS), 2023).

Post-traumatic Stress Disorder (PTSD) is a mental health condition that can develop after experiencing or witnessing a traumatic event (What is post-traumatic stress disorder (PTSD)?, 2022). Symptoms of PTSD may include intrusive memories, flashbacks, hypervigilance, and avoidance of triggers associated with the trauma (What is post-traumatic stress disorder (PTSD)?, 2022). In a work environment, employees with PTSD may face challenges related to concentration, emotional regulation, and coping with stressors that remind them of the traumatic experience (What is post-traumatic stress disorder (PTSD)?, 2022). For instance, loud noises in the office or certain types of work-related situations may trigger anxiety or distress, impacting their ability to focus, interact with colleagues, and feel safe in their work environment (What is post-traumatic stress disorder (PTSD)?, 2022). Providing a supportive and understanding workplace culture along with appropriate accommodations can help individuals with PTSD manage their symptoms and thrive professionally (What is post-traumatic stress disorder (PTSD)?, 2022).

Importance of Training, Awareness, and Change Management in Program Development

Implementing a new performance support program constitutes a significant change for an organization (Becker et al., 2011). Organizations considering introducing a new program should ensure the program's rollout includes a change management function, focusing on the new policies, procedures, operations, budgets, and culture being introduced into the organization. Organizations should identify a suitable change management model to organize the change management function. Kotter's is one change management model that could be used to introduce the new program (Baloh et al., 2017). Kotter's model starts with creating a sense of urgency in the organization (Rajan & Ganesan, 2017). In the context of a performance support program for employees with hidden disabilities, the sense of urgency could be the narrative that accommodating employees with hidden disabilities could result in increased productivity and a more inclusive workplace. Once a sense of urgency is created, leaders could engage key stakeholders (create a guiding coalition), including employees with hidden disabilities, in the planning and decision-making process. The form of inclusive management helps build buy-in and commitment to change (Rajan & Ganesan, 2017). Kotter's model also emphasizes the need to develop a vision, purpose, goals, and timeline for the program (developing a strategy) (Rajan & Ganesan, 2017). Cultural change could be facilitated by reinforcing the importance of the program through ongoing communication (communicating the purpose), celebrating successes, and integrating inclusive practices into daily operations. Awareness and training are two ways to deal with this change (Medina & Gamero, 2017). Employers should provide training and raise awareness among managers, supervisors, and employees about disability rights and reasonable accommodations (empowering action). Training can help resentment, promote understanding, and ensure compliance with ADA requirements (Kulkarni & Lengnick-Hall, 2011). Finally, the change management function should celebrate wins and use the momentum from the early wins to drive additional organizational and cultural change (Farkas, 2013).

Stigma and Employee Resentment

While the benefits of implementing an employee performance support program for individuals with hidden disabilities are clear, it's important to acknowledge the potential challenges. In many organizations, there is a stigma associated with hidden disabilities (Ali et al., 2016). The stigma surrounding disabilities may lead to reluctance among employees to disclose their needs or seek accommodations. Less disclosure can lead to lower accommodation and a decrease in the program's overall effectiveness (Balsiger, 2022). Overcoming stigma and misconception often involves cultural change, where employees are encouraged to discuss their challenges without fear of judgment or discrimination (While & Clark, 2010). Investment may also be required as organizations upgrade training, adapt workplaces, or purchase adaptive technologies (Medina & Gamero, 2017).

The stigma associated with disability can lead to social exclusion and mental distress, highlighting the need for policies that promote social participation and mental health interventions (Trani et al., 2016). Individuals with intellectual disabilities are aware of stigma and employ various adaptation strategies, such as aligning with minority groups and distancing from stigmatizing services (Jahoda & Marková, 2004). Perceptions of disability and stigmatization evolve with age and are influenced by cultural socialization practices emphasizing empathy and sensitivity to maintain relationships (Kayama, 2017). Organizational change and investment in core training are essential to reduce inequality and improve care for people with intellectual disabilities (While & Clark, 2010). Employers who hire youth with disabilities use strategies like diversity training, open communication, and mentoring to create an inclusive workplace culture (Lindsay et al., 2019). While training is essential, education policies alone often fail to create a culture of care and inclusion due to ableism and a lack of enforcement, necessitating a more substantial moral commitment to the rights of individuals with disabilities (Hale, 2015). Organizations must invest in training, promote organizational change, and implement inclusive strategies to overcome stigma and create a culture of trust.

Employee support programs can also result in resentment among employees (Calhoun & McCarthy, 2022). An examination of flexible work arrangements in public accounting firms found that it led to resentment among employees and higher turnover intentions among employees (Calhoun & McCarthy, 2022). Most workplaces have healthy internal competition among employees (Jones et al., 2017). Employees compete for promotions, recognition, and bonus compensation. If a particular group of employees is treated differently, resentment may arise. The 'unsupported' employees may attribute a colleague's promotion to the 'special attention' he or she receives from the program (Tzafrir & Hareli, 2009). Ironically, the program intended to create a level playing field within the organization may be perceived as doing the opposite, disadvantaging employees who do not require support. The issue of 'employee resentment' needs to be acknowledged by management if the performance support program will have the desired impact on the organization (Dasborough, 2006; Schrader et al., 2014). Program acceptance increases if senior leadership within the organization is supportive in displays positive attitudes towards individuals with disabilities (Lindsay et al., 2019b) They note that it is essential that the organization identifies a senior sponsor for the performance support program, an individual who has personal experience with hidden disabilities (Luu, 2019).

Employee Disclosure

To implement a performance support program for employees with hidden disabilities, organizations must implement a process that enables employees to confidentially report if they have a disability. Confidentiality determines employee willingness to report disabilities, request accommodations, and comfort in disclosing information to co-workers (Syma, 2019). The US Centers for Disease Control and Prevention (2023) reports that 1 in 4 (27%) of Americans have some form of disability. Disability prevalence estimates have a likely downward bias as some individuals do not identify due to stigma or because they have never been diagnosed with a disability (Feder, 2022). Students who participated in a disability awareness program reported increased self-awareness of their disabilities and improved their understanding of the need for accommodations (Wynants & Dennis, 2017). Self-awareness influences individuals to act more consistently with societal norms and to provide accurate reports about themselves (Wicklund, 1979). This suggests that increased self-awareness of one's disability could lead to more open disclosure and accommodation requests.

Self-awareness of disabilities is crucial for career development; it helps individuals overcome challenges and match their abilities and skills with suitable occupations and work environments (Orok, 2019). Self-awareness of one's condition affects beliefs, attitudes, and actions and can shape attitudes and behaviors about seeking accommodations or support (Fabrigar et al., 2006). An 11-week skills-based group study conducted with high school students demonstrated that understanding one's skill level increased awareness of learning disabilities (Lau et al., 2015). Organizations that engage in efforts to raise self-awareness through education, exposure, and personal narratives can help their employees better understand their disabilities, and increased self-awareness can lead to increased disclosure, ultimately enabling them to receive appropriate accommodations and support.

Stigma and discrimination contribute to employment inequity for people with mental disabilities, affecting their ability to participate in the workforce and contributing to social isolation (Stuart, 2006). People with psychiatric disabilities face complex decisions about disclosing their disability to employers, with disability identity and appropriate job matching being critical factors in their decision-making process (Dalgin & Gilbride, 2003). Individuals with functional neurological disorders experience stigma, leading to lower levels of disclosure of disabilities (Foley et al., 2022). First, Organizations must create a culture and trust that helps eliminate stigma and create an environment where individuals feel comfortable disclosing their hidden disabilities (Toth & Dewa, 2014). Disclosure of hidden disabilities within an organization will be aided by an organizational culture that embraces diversity and removes the stigma of special treatment or 'milking the system' that is often associated with employees with disabilities.

Organizations interested in establishing performance support programs for employees should develop a business process and secure system where employees can voluntarily and confidentially disclose their disability status. The information collected should align with the initial scoping exercise conducted by the organization, which examined the type of disabilities that could impact performance within their specific organization. Linking the system to an earlier exercise of organizational research into the issue will build credibility. The process should clearly outline and explain the purpose, confidentiality measures, and how the information will be used to provide reasonable accommodations (Syma, 2019). The process should allow for anonymous feedback, where employees can share concerns, suggestions, or experiences related to disability inclusion without fear of retaliation (Gonzales, 2024). An effective organizational performance support program for employees with hidden disabilities must include a process (supported by a system) that allows individuals to confidentially report their hidden disabilities as well as seek support if they suspect they have a hidden disability.

Accommodations

Organizations are legally obligated to provide reasonable accommodations for employees with disabilities (hidden or otherwise) (Americans With Disabilities Act, 1990). The term 'reasonable' is left open to interpretation. Several studies have defined the term reasonable; most conclude that the definition differs from firm to firm and depends on the industry, organizational culture, organizational leadership, and the legal framework under which the firm operates (Roig, 2016). Most rational firms will implement programs for which the marginal benefit exceeds the marginal cost (Telwatte et al., 2017). For example, if the program costs $500 for the last employee to participate and the benefit from increased productivity is $550 for that person, it would be rational for the firm to introduce the program. While performance support programs for employees with hidden disabilities can improve productivity and firm performance, organizations must ensure that the marginal benefits outweigh the costs and that the program is designed and implemented to foster inclusion and diversity rather than employee resentment. While the term reasonable accommodation is fluid from organization to organization, most organizations should implement a suite of accommodations that do not place undue hardship on the organization. The suite of accommodations should be developed based on the disclosed disabilities provided by employees and the organization's analysis of how the disabilities can impact an individual's ability to perform the tasks required by the organization. While not exhaustive, most organizations implement the following accommodations: flexible work arrangements, assistive technologies, and job restructuring or reassignments (McDowell & Fossey, 2015).

Flexible work arrangements, a key accommodation implemented by many organizations, have significantly increased, especially given the positive remote working experience during the COVID-19 pandemic (Yeves et al., 2022). These arrangements are associated with better physical health, reduced absenteeism, and fewer somatic symptoms, which can be particularly beneficial for employees with disabilities (Shifrin & Michel, 2021). Employees with disabilities experience higher job satisfaction in decentralized organizational contexts, suggesting that flexibility in work structure can positively impact their work experience (Baumgärtner et al., 2015). Informal flexible work arrangements, such as those negotiated between the employee and their line manager, can positively affect employee performance through increased job satisfaction and organizational commitment (Menezes & Kelliher, 2017). There is a lack of research on how flexible work arrangements can be leveraged to support return-to-work (RTW) and prevent disability, indicating a need for future studies to explore these arrangements as a potential mechanism for disability prevention (Ekberg et al., 2016). Flexible employment schemes have been shown to increase job satisfaction and reduce absenteeism among disabled workers and carers of disabled people, highlighting the role of flexible work arrangements in improving job quality (Giovanis & Ozdamar, 2019). The provision of flexible work arrangements improves work-life balance, which is crucial for the overall well-being of employees, including those with disabilities (Shanker, 2022).

Assistive technologies play a crucial role in supporting employees with hidden disabilities, often requiring significant investment (Stumbo et al., 2009). These technologies are identified as foundational support that equates individuals with disabilities to their non-disabled peers in terms of education, employment, and independent living (Stumbo et al., 2009). Studies suggest that assistive technology provides numerous benefits for individuals with disabilities, including increased independence, skill acquisition, improved educational and employment outcomes, enhanced participation in daily activities, and overall improved quality of life (Morash-Macneil et al., 2018). Assistive technologies help individuals with disabilities by improving their independence and participation in various life activities, with studies showing increases in independence, employment, and life satisfaction (O'Day & Corcoran, 1994). In educational settings, assistive technologies support students with disabilities by enabling them to better engage with the curriculum and participate in learning activities, often leading to improved educational outcomes (Lewis & Lewis, 1998). For individuals with learning disabilities, assistive technologies can leverage their strengths and help compensate for their challenges, thus facilitating academic success and daily living (Lewis & Lewis, 1998). Aging individuals with disabilities benefit from assistive technologies as they sustain health, quality of life, and independence despite the challenges of aging with a disability (Agree, 2014). Despite the benefits, there are obstacles to AT use, such as abandonment due to lack of training, support, or relevance to the user's needs and preferences (Alper & Raharinirina, 2006). Access to assistive technologies is a significant issue, with many individuals facing challenges such as lack of funds, information, and support systems and difficulties in obtaining necessary equipment (O'Day & Corcoran, 1994).

The recruitment of and job reassignments for individuals with disabilities offer a range of benefits that extend beyond compliance with legal mandates. Restructuring jobs and providing social support and coping skills training can enhance productivity (Yandrick, 1997). Disability management programs effectively and efficiently ensure job retention and reintegration for individuals with disabilities (Lefever et al., 2018). Lindsay et al. (2018) systematically reviewed the benefits of hiring individuals with a disability. They reviewed 39 articles and found several benefits from hiring individuals with disabilities, including profitability, competitive advantage, ability awareness, and inclusive work culture. In addition to organizational benefits, the authors also identified several benefits for employees, including an expanded social network, higher incomes, a sense of community, and greater self-confidence (Lindsay et al., 2018).

Monitoring and Evaluation

An organization requires data to design, evaluate, and monitor the effectiveness of an employee performance support program. A critical data element is the prevalence of hidden disabilities across the workforce. This data will provide insight into the scale of the issue and help prioritize support strategies based on the most prevalent disabilities. This information should be collected through an anonymous survey administered to all employees. The collection of prevalence information should be done in two stages. First, organizations should ask employees if a licensed professional has diagnosed them with a hidden disability. For employees who have not been diagnosed, the organization should ask employees to self-assess if they have a hidden disability. If the employee suspects they have a disability, the organization would need to either accept their self-diagnosis or provide an avenue for the employee to be evaluated to confirm (or not) if they have a disability. It is recommended that firms only accept diagnosed disabilities. Only accepting diagnosed disabilities should limit the program's cost in the long term and help address resentment since individuals who are not accommodated are more accepting of accommodation when a licensed professional has diagnosed them. If an employee self-diagnoses a hidden disability, the organization should provide an avenue for the employee to be assessed by a licensed professional. It would be prudent for the organization to absorb some of the cost of the assessment but not the entire cost.

The organization should also develop performance indicators for all employees. This would include metrics such as job performance ratings, attendance records, feedback from supervisors and colleagues, and any existing accommodations or support they have utilized. Understanding how hidden disabilities impact job performance and well-being is crucial for designing targeted interventions and support mechanisms tailored to the specific needs of these employees (Sibisanu et al., 2024). These metrics are also needed to understand the point at which the program's benefits (in terms of improved performance) outweigh the program's cost. Given the personal nature of the collected information, appropriate privacy-preserving techniques should be adopted (Hornberger, 2021). Only individuals with a need to know should have access to the data. Employee surveys, existing HR data, and employee records can provide insights into program performance metrics, accommodation requests, and program participation. Organizations can also establish focus groups with all employees and relevant stakeholders, including disability support specialists, to discuss the program's state and how it may need to evolve. The qualitative data from the focus groups will enrich the data collected from the employee survey and human resource administrative systems.

While these metrics are helpful, there is a darker side to these types of real-time employee data analytics (Siahaan et al., 2023). Active monitoring of employee performance can create a sense of mistrust between the employee and the organization (Siahaan et al., 2023). It would be important that any monitoring program is co-developed between staff and management to ensure that both parties are comfortable with the data being collected and analyzed and how it is being used (Siahaan et al., 2023).

Importance of Leadership

Authentic leadership is the 'glue' that holds together a healthy workplace and is essential for preventing burnout and absenteeism (Shirey, 2006). Authentic leadership is increasingly recognized as pivotal in enhancing employee performance and commitment, including programs aimed at supporting employees with hidden disabilities (Shirey, 2006). Authentic leadership is characterized by genuineness, trustworthiness, and compassion, essential to ensure employee acceptance of a performance support program for employees with hidden disabilities (Ribeiro et al., 2018). Authentic leadership positively influences employees' affective commitment and individual performance (Ribeiro et al., 2018). Authentic leadership can enhance the relationship between organizational programs and employees' perceptions of these programs (Cao et al., 2020). Transformational leadership, which shares similarities with authentic leadership, improves the self-concept of employees with disabilities, thereby reducing stigma and increasing the likelihood that employees will be willing to disclose their disabilities (Kensbock & Boehm, 2016).

RECOMMENDATIONS

Key Findings

The research reviewed highlights that performance support programs for employees with hidden disabilities provide significant benefits when implemented effectively. Studies show that such programs contribute to improved productivity, better staff retention, better work-life balance, and a more inclusive workplace culture (Solovieva et al., 2011; Gulyamova et al., 2023). However, the success of these programs depends heavily on thoughtful implementation, strong organizational culture, and careful management of potential stigma and resentment among employees. Organizations that fail to plan carefully risk undermining the very inclusion and productivity goals they seek to advance. The literature strongly suggests that positive outcomes arise when organizations adhere to a systematic framework that addresses legal compliance, cultural change, disclosure processes, accommodations, and monitoring and evaluation.

Based on the literature review and analysis, six actionable recommendations are proposed to guide organizations seeking to implement performance support programs for employees with hidden disabilities:

1. Understand Legal, Moral, and Ethical Obligations: Organizations must first ensure they fully understand their legal responsibilities under the Americans with Disabilities Act (1990) and similar legislation. In addition to legal compliance, organizations should embrace the moral and ethical imperatives to support all employees and promote workplace fairness and dignity.
2. Conduct Organizational Analysis of Hidden Disabilities: Organizations need to examine the types of hidden disabilities that may exist within their workforces. This involves analyzing how different disabilities may impact the tasks carried out by employees and understanding the support needed for these individuals to succeed (Moore, 2009; Lindsay et al., 2019).
3. Develop Confidential Disclosure Processes: A confidential and trustworthy system for employees to disclose hidden disabilities must be established (Syma, 2019). Disclosure is a foundational step for tailoring accommodations appropriately, and confidentiality plays a crucial role in encouraging participation.

4. Invest in and Implement Reasonable Accommodations: Organizations must identify and provide reasonable accommodations tailored to employee needs. These accommodations might include flexible work arrangements, assistive technologies, job restructuring, or modified duties (McDowell & Fossey, 2015; Stumbo et al., 2009).
5. Monitor and Evaluate Program Outcomes: Program performance must be continuously assessed through both quantitative and qualitative measures. Metrics should include participation rates, accommodation effectiveness, employee satisfaction, and organizational outcomes such as productivity and retention (Hornberger, 2021). Transparent reporting of findings to employees is important to sustain program credibility and momentum.

The recommended steps should be carried out sequentially. Organizations must first understand their legal obligations before determining accommodation strategies. Cultural change must precede requests for disclosure, as employees are unlikely to self-identify in unsupportive environments. Similarly, accommodation strategies should be developed only after needs are properly identified through disclosure processes. Continuous monitoring and evaluation are essential for ensuring long-term success and adapting programs as organizational needs evolve. Underpinning the implementation (and conducted concurrently with each step above) is promoting and supporting the necessary cultural change through awareness and training. Successful implementation depends on creating an inclusive organizational culture. Organizations must invest in structured change management strategies that involve leadership engagement, disability awareness training, and proactive communication (Kulkarni & Lengnick-Hall, 2011; Baloh et al., 2017). Without a culture of acceptance, employees may be reluctant to disclose disabilities or participate fully in the program.

The proposed framework for implementing a performance support program is summarized in Table 1:

Table 1. Performance support for employees with hidden disabilities framework.

Understanding the legal, ethical, and moral obligations. (Step 1)				**Monitoring and evaluating program outcomes. (Step 5 – ongoing)**
	Conducting an organizational analysis of hidden disabilities. (Step 2)	Developing a confidential disclosure process (Step 3)	Investing in and implementing reasonable accommodations (Step 4)	

Training, Awareness, and Change Management

It is recommended that organizations implement performance support programs for employees with hidden disabilities. While the programs can be expensive to implement, the research shows that if done effectively, there will generally be a positive return on the investment and a host of intangible benefits to the organizations. Research also shows that if not managed effectively, the introduction of the program can lead to resentment among employees and between employees and management. The proposed framework outlined in this study is an effective model that organizations can use to develop and deliver a program that generates positive benefits for employees and the organization.

FUTURE RESEARCH

While the current research provides strong evidence supporting the implementation of performance support programs for employees with hidden disabilities, several areas warrant further exploration. First, empirical studies are needed to assess the specific return on investment (ROI) for different types of accommodations, assistive technologies, and flexible work arrangements across various industries. Quantitative cost-benefit analyses would provide stronger business cases for investment in support programs.

Second, longitudinal research should examine the long-term impacts of organizational culture change efforts on disability disclosure rates, employee satisfaction, and productivity outcomes. Particularly important is the need to study how sustained stigma-reduction efforts affect hidden disability inclusion over time.

Third, comparative analyses of different framework models across industries, countries, and organizational sizes would enrich the literature by identifying context-specific best practices. Cross-sector studies could also shed light on how leadership styles, organizational structures, and policy environments influence the success or failure of hidden disability support programs.

Finally, additional research should explore the experiences of employees who choose not to disclose their hidden disabilities. Understanding these barriers could inform more effective communication strategies and cultural initiatives to create truly inclusive workplaces.

Future inquiry into these areas will help refine the proposed framework and ensure that organizations continue to evolve their practices in ways that meaningfully support employees with hidden disabilities.

REFERENCES

Agree, E. M. (2014). The potential for technology to enhance independence for those aging with a disability. *Disability and Health Journal*, *7*(1), S33–S39. DOI: 10.1016/j.dhjo.2013.09.004 PMID: 24456682

Ali, A., King, M., Strydom, A., & Hassiotis, A. (2016). Self-reported stigma and its association with socio-demographic factors and physical disability in people with intellectual disabilities: Results from a cross-sectional study in England. *Social Psychiatry and Psychiatric Epidemiology*, *51*(3), 465–474. DOI: 10.1007/s00127-015-1133-z PMID: 26498927

Alper, S., & Raharinirina, S. (2006). Assistive Technology for Individuals with Disabilities: A Review and Synthesis of the Literature. *Journal of Special Education Technology*, *21*(2), 47–64. DOI: 10.1177/016264340602100204

Americans With Disabilities Act of 1990, 42 U.S.C. § 12101 et seq. (1990)

Baloh, J., Zhu, X., & Ward, M. (2017). Implementing team huddles in small rural hospitals: How does the Kotter model of change apply? *Journal of Nursing Management*, *26*(5), 571–578. DOI: 10.1111/jonm.12584 PMID: 29250892

Balsiger, A. J. (2022). *Differing Perspectives of Stigma from Students with Disabilities and Access Workers*. Practicing Anthropology., DOI: 10.17730/0888-4552.44.4.26

Batavia, A., & Schriner, K. (2001). The Americans With Disabilities Act as Engine of Social Change: Models of Disability and the Potential of a Civil Rights Approach. *Policy Studies Journal: the Journal of the Policy Studies Organization*, *29*(4), 690–702. DOI: 10.1111/j.1541-0072.2001.tb02122.x

Baumgärtner, M. K., Dwertmann, D. J. G., Boehm, S., & Bruch, H. (2015). Job Satisfaction of Employees with Disabilities: The Role of Perceived Structural Flexibility. *Human Resource Management*, *54*(2), 323–343. DOI: 10.1002/hrm.21673

Becker, K., Antuar, N., & Everett, C. (2011). Implementing an employee performance management system in a nonprofit organization. *Nonprofit Management & Leadership*, *21*(3), 255–271. DOI: 10.1002/nml.20024

Bérubé, N. (2023). *Hidden challenges: An invisible disabilities learning activity*. Organization Management Journal., DOI: 10.1108/OMJ-03-2023-1810

Calhoun, M., & McCarthy, V. (2022). An unintended consequence of flexible work arrangements (FWAs) – the role of peer resentment on turnover intentions in public accounting. *Journal of Accounting & Organizational Change*. https://doi.org/DOI: 10.1108/JAOC-03-2022-0045

Cao, M., Zhao, Y., & Zhao, S. (2020). Facilitating the Implementation Process of High-Performance Work Systems: The Role of Authentic Leadership. *Frontiers in Psychology*, *11*, 550711. Advance online publication. DOI: 10.3389/fpsyg.2020.550711 PMID: 33071868

Chajduga, T., & Ingaldi, M. (2021). Hiring Disabled People to Avoid Staff Turnover and Enhance the Sustainability of Production. *Sustainability (Basel)*, *13*(19), 10577. Advance online publication. DOI: 10.3390/su131910577

Chiu, M., Yang, X., Wong, H., & Li, J. (2015). The mediating effect of affective stigma between face concern and general mental health—The case of Chinese caregivers of children with intellectual disability. *Research in Developmental Disabilities*, *36C*, 437–446. DOI: 10.1016/j.ridd.2014.10.024 PMID: 25462504

Cook, L., Foley, J., & Semeah, L. M. (2016). An exploratory study of inclusive worksite wellness: Considering employees with disabilities. *Disability and Health Journal*, *9*(1), 100–107. DOI: 10.1016/j.dhjo.2015.08.011 PMID: 26471518

Corbière, M., Zaniboni, S., Dewa, C., Villotti, P., Lecomte, T., Sultan-Taïeb, H., Hupé, J., & Fraccaroli, F. (2019). Work productivity of people with a psychiatric disability working in social firms. *Work (Reading, Mass.)*, *62*(1), 151–160. DOI: 10.3233/WOR-182850 PMID: 30689598

Couzens, D., Poed, S., Kataoka, M., Brandon, A., Hartley, J., & Keen, D. (2015). Support for Students with Hidden Disabilities in Universities: A Case Study. *International Journal of Disability Development and Education*, *62*(1), 24–41. DOI: 10.1080/1034912X.2014.984592

Crampton, S., & Hodge, J. W. (2003). The ADA and Disability Accommodations. *Public Personnel Management*, *32*(1), 143–154. DOI: 10.1177/009102600303200108

Dalgin, R. S., & Gilbride, D. (2003). Perspectives of people with psychiatric disabilities on employment disclosure. *Psychiatric Rehabilitation Journal*, *26*(3), 306–310. DOI: 10.2975/26.2003.306.310 PMID: 12653452

Dasborough, M. (2006). Cognitive asymmetry in employee emotional reactions to leadership behaviors. *The Leadership Quarterly*, *17*(2), 163–178. DOI: 10.1016/j.leaqua.2005.12.004

DeLeire, T. (2000). The Wage and Employment Effects of the Americans with Disabilities Act. *The Journal of Human Resources*, *35*(4), 693–715. DOI: 10.2307/146368

Ekberg, K., Pransky, G., Besen, E., Fassier, J., Feuerstein, M., Munir, F., & Blanck, P. (2016). New Business Structures Creating Organizational Opportunities and Challenges for Work Disability Prevention. *Journal of Occupational Rehabilitation*, *26*(4), 480–489. DOI: 10.1007/s10926-016-9671-0 PMID: 27704343

Evans, H. (2019). 'Trial by fire': Forms of impairment disclosure and implications for disability identity. *Disability & Society*, *34*(5), 726–746. DOI: 10.1080/09687599.2019.1580187

Farkas, M. (2013). Building and Sustaining a Culture of Assessment: Best Practices for Change Leadership. *RSR. Reference Services Review*, *41*(1), 13–31. DOI: 10.1108/00907321311300857

Foley, C., Kirkby, A., & Eccles, F. J. R. (2022). A meta-ethnographic synthesis of the experiences of stigma amongst people with functional neurological disorder. *Disability and Rehabilitation*, *46*(1), 1–12. DOI: 10.1080/09638288.2022.2155714 PMID: 36519449

Froehlich-Grobe, K., Lee, J., & Washburn, R. (2013). Disparities in obesity and related conditions among Americans with disabilities. *American Journal of Preventive Medicine*, *45*(1), 83–90. DOI: 10.1016/j.amepre.2013.02.021 PMID: 23790992

Gerber, P., & Price, L. A. (2003). Persons with Learning Disabilities in the Workplace: What We Know So Far in the Americans with Disabilities Act Era. *Learning Disabilities Research & Practice*, *18*(2), 132–136. DOI: 10.1111/1540-5826.00069

Giovanis, E., & Ozdamar, O. (2019). Accommodating Employees with Disabilities: The Role of Flexible Employment Schemes in Europe. *Labor: Public Policy & Regulation EJournal*. https://doi.org/DOI: 10.2139/ssrn.3441925

Gulyamova, S. T., Aziz, S. F. A., Omar, N., & Mohd, R. (2023). Workplace-Related Socioeconomic Issues Associated with Job Performance and Productivity among Employees with Various Impairments: A Systematic Literature Review. *Social Sciences (Basel, Switzerland)*, *12*(5), 275. Advance online publication. DOI: 10.3390/socsci12050275

Hale, C. (2015). Urban special education policy and the lived experience of stigma in a high school science classroom. *Cultural Studies of Science Education*, *10*(4), 1071–1088. DOI: 10.1007/s11422-013-9548-x

Hartnett, H., Stuart, H., Thurman, H., Loy, B., & Batiste, L. (2011a). Employers' perceptions of the benefits of workplace accommodations: Reasons to hire, retain and promote people with disabilities. *Journal of Vocational Rehabilitation*, *34*(1), 17–23. DOI: 10.3233/JVR-2010-0530

Hokke, S., Bennetts, S., Crawford, S., Leach, L., Hackworth, N., Strazdins, L., Nguyen, C., Nicholson, J., & Cooklin, A. (2020). Does flexible work 'work' in Australia? A survey of employed mothers' and fathers' work, family and health. *Community Work & Family*, *24*(4), 488–506. DOI: 10.1080/13668803.2019.1704397

Hornberger, R. C. (2021). Creating a sense of digital privacy in the private sector. *Information Security Journal: A Global Perspective, 30*, 30–56. https://doi.org/DOI: 10.1080/19393555.2020.1797948

Jahoda, A., & Marková, I. (2004). Coping with social stigma: People with intellectual disabilities moving from institutions and family home. *Journal of Intellectual Disability Research*, *48*(Pt 8), 719–729. DOI: 10.1111/j.1365-2788.2003.00561.x PMID: 15494061

Jones, J. L., Davis, W., & Thomas, C. H. (2017). Is Competition Engaging? Examining the Interactive Effects of Goal Orientation and Competitive Work Environment on Engagement. *Human Resource Management*, *56*(3), 389–405. DOI: 10.1002/hrm.21773

Jones, M. K. (2008). Disability and the labour market: A review of the empirical evidence. *Journal of Economic Studies (Glasgow, Scotland)*, *35*(5), 405–424. DOI: 10.1108/01443580810903554

Kayama, M. (2017). Development of children's understandings of physical disabilities and stigmatization in a Japanese cultural context: Reflections of children in second through sixth grades. *Children and Youth Services Review*, *83*, 190–200. DOI: 10.1016/j.childyouth.2017.10.039

Kensbock, J. M., & Boehm, S. (2016). The role of transformational leadership in the mental health and job performance of employees with disabilities. *International Journal of Human Resource Management*, *27*(14), 1580–1609. DOI: 10.1080/09585192.2015.1079231

Koruda, E. (2016). More carrot, less stick: Workplace wellness programs & the discriminatory impact of financial and health-based incentives. *BCJL & Soc. Just.*, *36*, 131.

Kulkarni, M. (2021). Hiding but hoping to be found: Workplace disclosure dilemmas of individuals with hidden disabilities. *Equality, Diversity and Inclusion*. Advance online publication. DOI: 10.1108/EDI-06-2020-0146

Kulkarni, M., & Lengnick-Hall, M. (2011). Socialization of people with disabilities in the workplace. *Human Resource Management*, *50*(4), 521–540. DOI: 10.1002/hrm.20436

Lau, W.-F. K., Ortega, K., & Sharkey, J. (2015). Disability Awareness Training with a Group of Adolescents with Learning Disabilities. *Contemporary School Psychology*, *19*(3), 145–156. DOI: 10.1007/s40688-014-0030-x

Lefever, M., Decuman, S., Perl, F., Braeckman, L., & van de Velde, D. (2018). The efficacy and efficiency of Disability Management in job-retention and job-reintegration. A systematic review. *Work (Reading, Mass.)*, *59*(4), 501–534. DOI: 10.3233/WOR-182709 PMID: 29733052

Lewis, R. B., & Lewis, R. B. (1998). Assistive Technology and Learning Disabilities: Today's Realities and Tomorrow's Promises. *Journal of Learning Disabilities*, *31*(1), 16–26. DOI: 10.1177/002221949803100103 PMID: 9455174

Lindsay, S., Cagliostro, E., Albarico, M., Mortaji, N., & Karon, L. (2018). A Systematic Review of the Benefits of Hiring People with Disabilities. *Journal of Occupational Rehabilitation*, *28*(4), 634–655. DOI: 10.1007/s10926-018-9756-z PMID: 29392591

Lindsay, S., Cagliostro, E., & Carafa, G. (2018). A systematic review of workplace disclosure and accommodation requests among youth and young adults with disabilities. *Disability and Rehabilitation*, *40*(25), 2971–2986. DOI: 10.1080/09638288.2017.1363824 PMID: 28797182

Lindsay, S., Cagliostro, E., Leck, J., Shen, W., & Stinson, J. (2019a). Disability disclosure and workplace accommodations among youth with disabilities. *Disability and Rehabilitation*, *41*(16), 1914–1924. DOI: 10.1080/09638288.2018.1451926 PMID: 29558221

Lindsay, S., Cagliostro, E., Leck, J., Shen, W., & Stinson, J. (2019b). Employers' perspectives of including young people with disabilities in the workforce, disability disclosure and providing accommodations. *Journal of Vocational Rehabilitation*, *50*(2), 141–156. Advance online publication. DOI: 10.3233/JVR-180996

Luu, T. (2019). Relationship between benevolent leadership and the well-being of employees with disabilities. *Journal of Business Research*, *99*, 282–294. Advance online publication. DOI: 10.1016/j.jbusres.2019.03.004

Madaus, J. W., Gerber, P., & Price, L. A. (2008). Adults with Learning Disabilities in the Workforce: Lessons for Secondary Transition Programs. *Learning Disabilities Research & Practice*, *23*(3), 148–153. DOI: 10.1111/j.1540-5826.2008.00272.x

Madyaningrum, M., Sonn, C., & Fisher, A. T. (2021). Disability organizations as empowering settings: Challenging stigmatization, promoting emancipation. *American Journal of Community Psychology*. Advance online publication. DOI: 10.1002/ajcp.12560 PMID: 34633668

Maness, D. L., & Khan, M. (2015). Disability evaluations: More than completing a form. *American Family Physician*, *91*(2), 102–109. PMID: 25591211

McDowell, C., & Fossey, E. (2015). Workplace Accommodations for People with Mental Illness: A Scoping Review. *Journal of Occupational Rehabilitation*, *25*(1), 197–206. DOI: 10.1007/s10926-014-9512-y PMID: 24841728

Medina, F., & Gamero, N. (2017). *The Socialization Process: Helping Organizations Integrate People with Disabilities into the Workplace*. 139–152. https://doi.org/DOI: 10.1007/978-3-319-66393-7_9

Mellifont, D., Smith-Merry, J., & Scanlan, J. (2016). Pitching a Yerkes–Dodson curve ball?: A study exploring enhanced workplace performance for individuals with anxiety disorders. *Journal of Workplace Behavioral Health*, *31*(2), 71–86. DOI: 10.1080/15555240.2015.1119654

Menezes, L. D., & Kelliher, C. (2017). Flexible Working, Individual Performance, and Employee Attitudes: Comparing Formal and Informal Arrangements. *Human Resource Management*, *56*(6), 1051–1070. DOI: 10.1002/hrm.21822

Morash-Macneil, V., Johnson, F., & Ryan, J. B. (2018). A Systematic Review of Assistive Technology for Individuals With Intellectual Disability in the Workplace. *Journal of Special Education Technology*, *33*(1), 15–26. DOI: 10.1177/0162643417729166

Muster, C. (2023). Disability Justice and the Americans with Disabilities Act: An Opportunity for Social Work. *Social Work*, *68*(4), 331–339. Advance online publication. DOI: 10.1093/sw/swad034 PMID: 37537135

Nevala, N., Pehkonen, I., Koskela, I., Ruusuvuori, J., & Anttila, H. (2015). Workplace Accommodation Among Persons with Disabilities: A Systematic Review of Its Effectiveness and Barriers or Facilitators. *Journal of Occupational Rehabilitation*, *25*(2), 432–448. DOI: 10.1007/s10926-014-9548-z

O'Day, B. L., & Corcoran, P. J. (1994). Assistive technology: Problems and policy alternatives. *Archives of Physical Medicine and Rehabilitation*, *75*(10), 1165–1169. DOI: 10.1016/0003-9993(94)90096-5 PMID: 7944926

Oermann, M. (2016). Supporting Students With Disabilities. *Nurse Educator, 41 1*, 28. https://doi.org/DOI: 10.1097/NNE.0000000000000231

Orok, A. B. M. (2019). Conceptualizing Self-Awareness as a Correlate for Career Development of Students with Disabilities. *Journal of Education and Training Studies*, *5*, 33–40.

Percy, S. (2001). CHALLENGES AND DILEMMAS IN IMPLEMENTING THE AMERICANS WITH DISABILITIES ACT: LESSONS FROM THE FIRST DECADE. *Policy Studies Journal: the Journal of the Policy Studies Organization*, *29*(4), 633–640. DOI: 10.1111/j.1541-0072.2001.tb02115.x

Rafeeque, A. (2015). A STUDY ON LEARNING DISABILITIES AMONG STUDENTS IN WAYANAD DISTRICT OF KERALA STATE. *Sai Om Journal of Commerce & Management: A Peer Reviewed. National Journal*, *2*, 6–16.

Rajan, R., & Ganesan, R. (2017). A critical analysis of John P. Kotter's change management framework. *Asian Journal of Research in Business Economics and Management*, *7*(7), 181–203. DOI: 10.5958/2249-7307.2017.00106.2

Ribeiro, N., Duarte, A., & Filipe, R. (2018). How authentic leadership promotes individual performance. *International Journal of Productivity and Performance Management*, *67*(9), 1585–1607. Advance online publication. DOI: 10.1108/IJPPM-11-2017-0318

Ribeiro, N., Gomes, D., & Kurian, S. (2018). Authentic leadership and performance: The mediating role of employees' affective commitment. *Social Responsibility Journal*, *14*(1), 213–225. DOI: 10.1108/SRJ-06-2017-0111

Rock, P., Roiser, J., Riedel, W., & Blackwell, A. (2013). Cognitive impairment in depression: A systematic review and meta-analysis. *Psychological Medicine*, *44*(10), 2029–2040. DOI: 10.1017/S0033291713002535 PMID: 24168753

Roessler, R. (2002). Improving Job Tenure Outcomes for People with Disabilities. *Rehabilitation Counseling Bulletin*, *45*(4), 207–212. DOI: 10.1177/00343552020450040301

Roig, R. A. (2016). *Reasonableness in the Concept of Reasonable Accommodation*. 42–59. https://doi.org/DOI: 10.17561/tahrj.v0i6.2929

Sadat, L. A., & Najie, I. 'Ainun. (2023). The Effect of Appropriate Placement of Employees with Disabilities on Work Productivity. *Syntax Literate ; Jurnal Ilmiah Indonesia*. https://doi.org/DOI: 10.36418/syntax-literate.v8i11.14046

Santuzzi, A. M., Waltz, P. R., Finkelstein, L. M., & Rupp, D. (2014). Invisible Disabilities: Unique Challenges for Employees and Organizations. *Industrial and Organizational Psychology: Perspectives on Science and Practice*, *7*(2), 204–219. DOI: 10.1111/iops.12134

Scheid, T. (2005). Stigma as a barrier to employment: Mental disability and the Americans with Disabilities Act. *International Journal of Law and Psychiatry*, *28*(6), 670–690. DOI: 10.1016/j.ijlp.2005.04.003 PMID: 16112732

Scroggins, W. A. (2007). The Measurement and Structural Modeling of the Reasonableness of Workplace Accommodations. *Employee Responsibilities and Rights Journal*, *19*(4), 279–291. DOI: 10.1007/s10672-007-9056-9

Shanker, A. (2022). *Flexible work arrangements and its impact on Work-Life Balance.* MANAGEMENT INSIGHT - THE JOURNAL OF INCISIVE ANALYSERS., DOI: 10.21844/mijia.18.2.3

Shifrin, N. V., & Michel, J. S. (2021). Flexible work arrangements and employee health: A meta-analytic review. *Work and Stress*, *36*(1), 60–85. DOI: 10.1080/02678373.2021.1936287

Shirey, M. R. (2006). Authentic leaders creating healthy work environments for nursing practice. *American Journal of Critical Care : An Official Publication. American Journal of Critical Care*, *15*(3), 256–267. DOI: 10.4037/ajcc2006.15.3.256

Siahaan, J., Mardhiyah, A., & Siregar, O. (2023). THE INFLUENCE OF PERCEIVED ORGANIZATIONAL SUPPORT AND WORK ENGAGEMENT ON EMPLOYEE PERFORMANCE. *MARGINAL JOURNAL OF MANAGEMENT ACCOUNTING GENERAL FINANCE AND INTERNATIONAL ECONOMIC ISSUES*. https://doi.org/DOI: 10.55047/marginal.v3i1.875

Solovieva, T., Dowler, D. L., & Walls, R. (2011). Employer benefits from making workplace accommodations. *Disability and Health Journal*, *4*(1), 39–45. DOI: 10.1016/j.dhjo.2010.03.001 PMID: 21168806

Stuart, H. (2006). Mental illness and employment discrimination. *Current Opinion in Psychiatry*, *19*(5), 522–526. DOI: 10.1097/01.yco.0000238482.27270.5d PMID: 16874128

Stumbo, N. J., Martin, J. K., & Hedrick, B. N. (2009). Assistive technology: Impact on education, employment, and independence of individuals with physical disabilities. *Journal of Vocational Rehabilitation*, *30*(2), 99–110. DOI: 10.3233/JVR-2009-0456

Syma, C. (2019). Invisible disabilities: Perceptions and barriers to reasonable accommodations in the workplace. *Library Management*, *40*(1/2), 113–120. Advance online publication. DOI: 10.1108/LM-10-2017-0101

Tamimi, M., Badrut, M., & Sopiah, T. (2023). The effect of organizational support on employee performance: A systematic literature review. *Asian Journal of Economics and Business Management*. https://doi.org/DOI: 10.53402/ajebm.v2i2.337

Taylor, J. P., Avellone, L., Cimera, R., Brooke, V., Lambert, A., & Iwanaga, K. (2021). Cost-benefit analyses of employment services for individuals with intellectual and developmental disabilities: A scoping review. *Journal of Vocational Rehabilitation*, *54*(2), 193–206. DOI: 10.3233/JVR-201130

Telwatte, A., Anglim, J., Wynton, S. A., & Moulding, R. (2017). Workplace Accommodations for Employees With Disabilities: A Multilevel Model of Employer Decision-Making. *Rehabilitation Psychology*, *62*(1), 7–19. DOI: 10.1037/rep0000120 PMID: 28045280

Tezzoni, L. I., McCarthy, E., Davis, R. B., Harris-David, L., & O'Day, B. (2001). Use of Screening and Preventive Services Among Women With Disabilities. *American Journal of Medical Quality*, *16*(4), 135–144. DOI: 10.1177/106286060101600405 PMID: 11477958

Thakker, D. A., & Solomon, P. (2004). Factors Influencing Managers' Adherence to the Americans with Disabilities Act. *Administration and Policy in Mental Health*, *26*(3), 213–219. DOI: 10.1023/A:1021366731025 PMID: 10339836

Toth, K., & Dewa, C. (2014). Employee Decision-Making About Disclosure of a Mental Disorder at Work. *Journal of Occupational Rehabilitation*, *24*(4), 732–746. DOI: 10.1007/s10926-014-9504-y PMID: 24604575

Tran, T. (2021). Diagnosis of Attention Deficit/Hyperactivity Disorder in Children and Adolescents: A Helpful Guide. *Pediatric Nursing*, *47*(4), 202. DOI: 10.62116/PNJ.2021.47.4.202

Trani, J., Ballard, E., & Peña, J. B. (2016). Stigma of persons with disabilities in Afghanistan: Examining the pathways from stereotyping to mental distress. *Social Science & Medicine*, *153*, 258–265. DOI: 10.1016/j.socscimed.2016.02.024 PMID: 26928587

Tzafrir, S., & Hareli, S. (2009). Employees' emotional reactions to promotion decisions. *Career Development International*, *14*(4), 351–371. DOI: 10.1108/13620430910979844

Valdez, A. R., Hancock, E. E., Adebayo, S., Kiernicki, D. J., Proskauer, D., Attewell, J., Bateman, L., DeMaria, A.Jr, Lapp, C., Rowe, P., & Proskauer, C. (2019). Estimating Prevalence, Demographics, and Costs of ME/CFS Using Large Scale Medical Claims Data and Machine Learning. *Frontiers in Pediatrics, 6*, 412. Advance online publication. DOI: 10.3389/fped.2018.00412 PMID: 30671425

von Schrader, S., Malzer, V., & Bruyere, S. (2014). Perspectives on Disability Disclosure: The Importance of Employer Practices and Workplace Climate. *Employee Responsibilities and Rights Journal, 26*(4), 237–255. DOI: 10.1007/s10672-013-9227-9

While, A., & Clark, L. (2010). Overcoming ignorance and stigma relating to intellectual disability in healthcare: A potential solution. *Journal of Nursing Management, 18*(2), 166–172. DOI: 10.1111/j.1365-2834.2009.01039.x PMID: 20465744

Wolf, L. (2001). College Students with ADHD and Other Hidden Disabilities. *Annals of the New York Academy of Sciences, 931*(1), 385–395. Advance online publication. DOI: 10.1111/j.1749-6632.2001.tb05792.x PMID: 11462755

Wynants, S. A., & Dennis, J. M. (2017). Embracing Diversity and Accessibility: A Mixed Methods Study of the Impact of an Online Disability Awareness Program. *Journal of Postsecondary Education and Disability, 30*(1), 33–48.

Yandrick, R. M. (1997). High-demand, low-control jobs reduce productivity and increase workplace disability costs. *Behavioral Healthcare Tomorrow, 6*(3), 40–44. PMID: 10169460

Yeves, J., Bargsted, M., & Torres-Ochoa, C. (2022). Work schedule flexibility and teleworking were not good together during COVID-19 when testing their effects on work overload and mental health. *Frontiers in Psychology, 13*, 998977. Advance online publication. DOI: 10.3389/fpsyg.2022.998977 PMID: 36389483

Zablotsky, B., Black, L. I., & Blumberg, S. (2017). Estimated Prevalence of Children With Diagnosed Developmental Disabilities in the United States, 2014-2016. *NCHS Data Brief, 291*, 1–8. PMID: 29235982

Chapter 10
Building Trust in Artificial Intelligence

Lord Dordunoo
https://orcid.org/0009-0007-7395-2677
Marymount University, USA

ABSTRACT

This paper examines transformative developments in Artificial Intelligence (AI), focusing on advancements in technical methodologies, ethical challenges, and real-world applications. Innovations like transformer models and generative AI have revolutionized fields such as natural language processing (NLP) and image synthesis, leading to breakthroughs in language translation and content creation. These advancements benefit sectors like healthcare, where AI aids diagnostics and personalized treatment, and finance, where it enhances fraud detection and risk assessment. However, ethical concerns, particularly around bias, accountability, and transparency, have emerged. Biases from unrepresentative datasets risk discriminatory outcomes, highlighting critical ethical issues. The paper argues that embedding fairness and accountability in AI is essential for trust and responsible use. It emphasizes the need for comprehensive ethical frameworks to guide socially sound AI development and recommends transparent, fair practices and further research into regulatory strategies.

INTRODUCTION

AI has undergone rapid advancements in recent years, fundamentally reshaping industries through innovations such as transformer models and deep reinforcement learning, two transformative methodologies in machine learning (Brown et al., 2020; Vaswani et al., 2017). Transformer models, first introduced by Vaswani et al.

DOI: 10.4018/979-8-3373-5686-0.ch010

(2017), revolutionized natural language processing (NLP) by enabling AI to process entire sentences and paragraphs contextually, improving capabilities in tasks like translation and sentiment analysis (Vaswani et al., 2017). These models underpin highly sophisticated NLP applications, including automated customer service and language generation. At the same time, deep reinforcement learning has advanced the development of systems that can optimize complex decision-making processes, particularly useful in sectors like robotics and autonomous driving (Brown et al., 2020; Radford et al., 2019). Together, these technologies empower practical applications across various industries, allowing for innovations ranging from predictive analytics in finance to automated multimedia content creation (Radford et al., 2019).

Despite these advancements, the evolution of AI also brings intricate ethical, practical, and societal challenges (Mitchell et al., 2019). One prominent issue is algorithmic bias, which arises when models trained on unrepresentative or biased data produce unfair or discriminatory outcomes (Barocas, Hardt, & Narayanan, 2019; Buolamwini & Gebru, 2018). Such biases, especially in applications in sensitive domains like hiring or criminal justice, can perpetuate inequality and undermine trust in AI systems (Mehrabi et al., 2021; Obermeyer et al., 2019). Another significant challenge is the threat to privacy, as AI systems often rely on vast amounts of personal data, raising concerns about data security and user consent (Mitchell et al., 2019; Raji et al., 2020). These ethical dilemmas have catalyzed societal debates around the responsible development of AI, intensifying calls for greater transparency and accountability in AI deployment (Binns, 2018; Dignum, 2017).

As AI becomes increasingly embedded in everyday life, from personal devices to public infrastructure establishing clear ethical guidelines has become essential for ensuring the technology's sustainable growth and public acceptance (Floridi et al., 2018; Whittlestone et al., 2019). These guidelines are a matter of technical rigor and societal responsibility to protect user rights and align AI's evolution with broader human values (Friedman & Hendry, 2019). This paper builds upon these considerations, exploring frameworks and practices that can help guide AI's ethical development as it expands its reach and impact (Bryson, 2019; Amodei et al., 2016).

Problem Statement

One of the most pressing challenges in AI today is addressing inherent biases within machine learning models, which have the potential to lead to discriminatory outcomes that disproportionately affect marginalized groups (Buolamwini & Gebru, 2018). These biases often originate in training data lacking adequate diversity or reflect historical and systemic prejudices, causing models to produce results that inadvertently replicate and reinforce existing social inequities (Barocas et al., 2019). For instance, if a model is trained predominantly on data from one demographic, its predictions may systematically disadvantage individuals from underrepresented groups, underscoring the critical need for inclusive and representative datasets (Mehrabi et al., 2021). The opacity of many complex AI systems, particularly deep learning models, poses significant challenges in interpreting and explaining their decisions, thereby undermining transparency and accountability in automated decision-making processes (Guidotti et al., 2018).

This opacity not only makes it difficult for users to understand how specific outcomes are generated but also limits the ability of developers and regulators to identify and rectify biases, further complicating the ethical deployment of AI systems (Rudin, 2019). As a result, addressing these issues is essential to ensure AI systems operate in ways that are both ethical and equitable, fostering public confidence in AI's growing role across high-stakes domains, from healthcare to criminal justice (Dignum, 2017; Mehrabi et al., 2021). Ultimately, creating more transparent, accountable, and unbiased AI systems is foundational to developing technology that aligns with societal values and promotes trust in its widespread adoption (Binns, 2018).

Significance of the Study

Addressing bias and enhancing explainability in AI is essential for ensuring both technical integrity and positive societal impact, as unmitigated biases or a lack of transparency can lead to outcomes that erode public trust and violate ethical standards (Veale & Binns, 2017). AI's influence in high-stakes domains such as healthcare and criminal justice underscores the importance of addressing these issues, as biased or opaque algorithms can lead to severe consequences, including discriminatory practices or incorrect diagnoses that may disproportionately affect marginalized groups (Obermeyer et al., 2019). For example, studies have shown that algorithms trained on unrepresentative data can produce biased recommendations in patient care or flawed assessments in recidivism predictions, highlighting the need for greater transparency and fairness (Mehrabi et al., 2021).

Implementing transparent AI systems and accountability measures can help mitigate these risks, ensuring that AI technologies serve diverse populations equitably and reliably (Dignum, 2017). Transparent algorithms enable users and developers to understand how decisions are made, facilitating identifying and correcting any biases, while accountability frameworks allow for responsible oversight and compliance with ethical standards (Raji et al., 2020). Therefore, this study focuses on how ethical AI contributes to foundational work on creating fair, transparent AI systems, reinforcing technically sound practices that align with societal values and regulatory expectations (Floridi, 2020). By promoting these practices, this research advocates for AI systems that respect individual rights and operate responsibly, paving the way for AI to support rather than undermine social justice and ethical standards (Binns, 2018).

METHODOLOGY

This study compiles recent scholarly literature on ethical AI practices, concentrating on advancements in key areas such as AI fairness, accountability, and transparency, all essential for developing responsible AI systems (Veale & Binns, 2017). By conducting a systematic review of relevant research, the study aims to identify and synthesize existing methods for bias mitigation in machine learning (ML), regulatory frameworks governing AI, and interpretability techniques designed to make AI models more understandable and accessible to users (Rudin, 2019). Bias mitigation strategies are essential, as they address how machine learning models can inadvertently produce discriminatory outcomes, often due to historical biases embedded within training data (Mehrabi et al., 2021).

In addition, the study explores emerging regulatory frameworks that seek to address ethical concerns surrounding AI by enforcing standards for transparency and accountability, thus reducing the potential for harm in high-stakes applications such as healthcare, finance, and criminal justice (Raji et al., 2020). The analysis emphasizes frameworks like Fairness, Accountability, and Transparency in Machine Learning (FATML), which serve as a foundational approach to ethical AI (Singhal et al., 2024). The FATML framework outlines standards that promote equitable and responsible AI development by emphasizing the need to understand, evaluate, and disclose algorithmic processes and their potential impacts on different demographic groups (Barocas et al., 2019).

Through a critical evaluation of these sources, this study contributes to an in-depth understanding of the ethical dimensions of modern AI, shedding light on how principles like fairness and transparency can be operationalized in real-world applications (Dignum, 2017). Such an approach highlights current best practices and identifies gaps where further research and policy intervention are needed to address ongoing ethical challenges in AI development (Floridi, 2020). Ultimately, this research underscores the importance of fostering AI systems that align with societal values and legal standards, promoting technology that respects individual rights and serves the public good (Binns, 2018).

LITERATURE REVIEW

Fairness, Accountability, and Transparency in Machine Learning (FATML)

The FATML framework prioritizes fairness, accountability, and transparency as fundamental principles for developing ethical AI systems, aiming to create technology that aligns with societal expectations of justice and responsibility (Binns, 2018; Floridi et al., 2018). Fairness within this framework focuses on reducing algorithmic bias, ensuring that AI systems treat all users equitably without regard to race, gender, age, or other demographic characteristics that could introduce discriminatory outcomes (Anderson & Visweswaran, 2024; Holstein et al., 2019). To achieve fairness, developers must rigorously assess datasets and model outputs to detect and mitigate any embedded biases, thereby promoting fair treatment across diverse user groups (Mehrabi et al., 2021). This emphasis on fairness is critical, as studies have shown that biased algorithms can exacerbate existing societal inequalities, underscoring the need for deliberate strategies to counteract these risks (Mehrabi et al., 2021; Raji et al., 2020).

Accountability, as defined within FATML, highlights the obligation of AI developers, companies, and relevant stakeholders to acknowledge and address any adverse effects that AI systems may produce, holding them responsible for the outcomes of their technologies (Binns, 2018; Raji et al., 2020). Accountability mechanisms include establishing clear lines of responsibility, ensuring that developers are answerable for AI-related decisions, and implementing regular audits to monitor compliance with ethical standards (Holstein et al., 2019; Raji et al., 2020). This aspect of FATML is essential because it provides a framework for attributing responsibility when AI systems cause harm, encouraging organizations to prioritize ethical practices in their AI development processes (Raji et al., 2020). Moreover, accountability fosters trust among users, who can feel more confident in AI applications, knowing that systems are in place to address potential harms (Binns, 2018; Whittlestone et al., 2019).

Transparency is a core component of the FATML principles, as it ensures that the processes driving AI decisions are understandable and accessible to both users and stakeholders, thus enabling informed scrutiny (Binns, 2018; Doshi-Velez & Kim, 2017). Transparent AI systems provide detailed information about how algorithms make decisions, enabling users to understand the factors influencing outcomes and to seek redress if they believe the system has acted unfairly (Doshi-Velez & Kim, 2017; Holstein et al., 2019). Transparency enhances user trust and facilitates accountability by giving stakeholders visibility into AI processes, which can help detect and correct unethical practices (Raji et al., 2020; Whittlestone et al., 2019). This visibility is especially important in sensitive healthcare or criminal justice applications, where opaque decision-making can have significant consequences (Anderson & Visweswaran, 2024; Floridi et al., 2018).

The FATML principles of fairness, accountability, and transparency establish a comprehensive ethical foundation for AI development, guiding developers toward systems that are trustworthy, responsible, and aligned with social values (Binns, 2018; Raji et al., 2020). Embedding these principles into AI design, developers can create practical and just technology, fostering trust and promoting equitable outcomes across diverse applications (Holstein et al., 2019; Mehrabi et al., 2021). As AI continues to evolve, adhering to FATML principles will be crucial to ensuring that these technologies serve society responsibly and ethically (Anderson & Visweswaran, 2024; Whittlestone et al., 2019).

Equity Theory

Equity theory, established in social psychology, posits that fairness is perceived when there is a balance between the inputs individuals contribute such as effort, time, or resources and the outcomes they receive in return, like recognition, rewards, or benefits (Adams, 1965; Dordunoo, 2025; Mehrabi et al., 2021). This theory has been instrumental in understanding human perceptions of fairness and is increasingly relevant in AI, where algorithmic decisions can profoundly affect diverse groups (Holstein et al., 2019; Mehrabi et al., 2021). In applying equity theory to AI, it becomes essential to design algorithms that equitably distribute benefits and potential harms, ensuring that no demographic is disproportionately advantaged or disadvantaged by AI outcomes (Holstein et al., 2019; Raji et al., 2020). This focus on balanced outcomes is critical, as studies show that biased algorithms can perpetuate or even exacerbate existing societal inequalities, further emphasizing the importance of fairness-oriented frameworks like equity theory in algorithmic design (Anderson & Visweswaran, 2024; Raji et al., 2020).

Equity theory highlights the significance of equitable treatment across different user demographics, which is fundamental in mitigating bias and preventing discrimination within AI systems (Binns, 2018; Mehrabi et al., 2021). For instance, in applications like hiring or loan approval algorithms, equitable treatment would require that AI systems make decisions that do not unfairly privilege or disadvantage individuals based on race, gender, or socioeconomic status (Binns, 2018; Holstein et al., 2019). By centering equity, AI developers can aim to address disparities within data or decision processes, thus fostering fairer outcomes that resonate with a broader range of users (Raji et al., 2020; Whittlestone et al., 2019). Ensuring such equity is essential because if users perceive AI systems as mistreating certain groups, it can erode public trust and hinder the technology's adoption (Jobin, Ienca, & Vayena, 2019; Whittlestone et al., 2019).

Applying equity theory in AI development involves creating algorithms that account for the diverse needs of user groups, which helps to produce outcomes that are fairer and more inclusive (Floridi et al., 2018; Mehrabi et al., 2021). Equity-centered AI models may incorporate bias audits and fairness testing to identify and address potential disparities in how algorithms impact various demographics (Holstein et al., 2019; Mehrabi et al., 2021). For example, re-weighting data to ensure representation of minority groups or adjusting models to account for historical inequities can lead to fairer algorithmic results (Doshi-Velez & Kim, 2017; Raji et al., 2020). These strategies, grounded in equity theory, help ensure that AI systems effectively serve all user demographics, promoting outcomes reflecting society's values of fairness and justice (Binns, 2018; Floridi et al., 2018).

Equity theory also provides a structured framework for evaluating AI systems' effectiveness in achieving balanced, fair interactions across different user groups, which is increasingly recognized as an essential component of ethical AI (Holstein et al., 2019; Mehrabi et al., 2021). This framework allows developers to systematically assess whether AI decisions maintain equity by comparing the input-to-outcome ratios across various demographics and adjusting algorithms when disparities are found (Binns, 2018; Holstein et al., 2019). Such evaluations are crucial for fairness and help build public confidence in AI by demonstrating a commitment to ethical standards (Anderson & Visweswaran, 2024; Jobin et al., 2019). By embedding equity principles into AI development, the technology can better align with societal expectations of fairness, fostering an ethical and trustworthy AI ecosystem (Floridi et al., 2018; Whittlestone et al., 2019).

Value-Sensitive Design (VSD)

The Value-Sensitive Design (VSD) framework is a critical approach in technology development that advocates for the incorporation of ethical and societal values directly into the design process, making it particularly relevant to the ethical challenges in artificial intelligence (Friedman & Hendry, 2019; Umbrello & van de Poel, 2020). VSD aims to bridge the gap between technological innovation and societal needs by ensuring that technology development considers the values of the communities it impacts (Friedman & Hendry, 2019). This framework emphasizes the importance of designing AI systems that respect fundamental ethical principles, such as fairness, accountability, and transparency, which are central to building responsible AI (Raji et al., 2020; Umbrello & van de Poel, 2020). As AI becomes increasingly integrated into various aspects of daily life, VSD provides a structured methodology for aligning technological development with the broader social values it serves (Friedman & Hendry, 2019).

A core tenet of VSD is to create design processes that prioritize inclusivity, considering users' diverse perspectives and values to ensure that technology aligns with broader social and moral considerations (Friedman & Hendry, 2019; Umbrello & van de Poel, 2020). This inclusivity means developing systems that actively respect the cultural, social, and ethical values of different user groups, thereby reducing the risk of alienation or discrimination (Holstein et al., 2019. Amalgamation of a wide range of values into AI design, developers can create systems that better serve and represent society, addressing potential biases or imbalances that might emerge if only limited perspectives are considered (Holstein et al., 2019; Jobin et al., 2019). Inclusivity within VSD also requires that AI systems are adaptable to different contexts and sensitive to the specific needs of marginalized or vulnerable groups, ensuring that technology serves all users equitably (Raji et al., 2020).

The VSD framework encourages developers to embed core values, such as fairness, privacy, and autonomy, within AI systems from the early design stages, thereby proactively addressing ethical concerns (Friedman & Hendry, 2019; Umbrello & van de Poel, 2020). For example, by prioritizing privacy, developers can build AI models that protect user data, aligning with societal expectations for data security and confidentiality (Doshi-Velez & Kim, 2017; Jobin et al., 2019). Furthermore, incorporating fairness as a design principle can help reduce biases in AI outcomes, ensuring that decisions made by these systems do not disproportionately impact specific groups based on race, gender, or other demographic factors (Mehrabi et al., 2021; Raji et al., 2020). By embedding such values from the outset, VSD guides developers in creating ethically responsible AI and establishes a foundation for AI systems that reflect and uphold societal standards (Friedman & Hendry, 2019; Umbrello & van de Poel, 2020).

Ultimately, the VSD approach promotes the development of AI systems that resonate with core human values, fostering public trust and acceptance through ethically aligned design (Friedman & Hendry, 2019; Umbrello & van de Poel, 2020). When users recognize that AI technologies are built with ethical considerations, they are more likely to trust and engage with these systems, enhancing technology's positive impact on society (Holstein et al., 2019. VSD thus plays a vital role in building public confidence by ensuring that AI systems operate transparently and fairly, creating a more inclusive technological environment (Umbrello & van de Poel, 2020). As AI continues to evolve, the VSD framework provides a valuable foundation for guiding ethical innovation, aligning technological progress with societal values and expectations (Friedman & Hendry, 2019; Friedman et al. 2006; Jobin et al., 2019).

FOUNDATION FOR ETHICAL AI DESIGN

The Fairness, Accountability, and Transparency in Machine Learning (FATML) principles, equity theory, and Value-Sensitive Design (VSD) framework collectively provide a robust foundation for ethical AI design, which prioritizes fairness, accountability, and societal alignment (Binns, 2018; Friedman & Hendry, 2019; Mehrabi et al., 2021). FATML emphasizes the importance of fairness by reducing algorithmic biases that could lead to discrimination and ensuring that AI systems treat users equitably regardless of demographic characteristics (Binns, 2018; Holstein et al., 2019). Accountability within the FATML framework ensures that developers and organizations take responsibility for AI outcomes, which is essential for building trust in systems with significant social impact (Binns, 2018; Raji et al., 2020). Transparency further supports this foundation by making AI decision-making processes accessible and understandable to users, thus fostering public trust and enabling external scrutiny (Doshi-Velez & Kim, 2017; Whittlestone et al., 2019).

Equity theory, rooted in social psychology, contributes to ethical AI by advocating for fairness in outcomes across diverse user demographics, thereby reducing biases that can lead to disproportionate impacts on specific groups (Adams, 1965; Dordunoo, 2025; Mehrabi et al., 2021). Applied to AI, equity theory suggests that algorithms should produce balanced outputs for all demographic groups, especially in applications like criminal justice or hiring, where biased outcomes can lead to serious ethical issues (Holstein et al., 2019; Raji et al., 2020). With equity theory incorporation into AI development, developers are encouraged to adopt fairness-focused methods, such as using representative datasets and implementing bias mitigation techniques that promote equitable outcomes (Jobin et al., 2019; Mehrabi et al., 2021). This emphasis on equity aligns with broader societal values and helps ensure that AI systems contribute to justice and social cohesion (Friedman & Hendry, 2019; Umbrello & van de Poel, 2020).

The VSD framework further strengthens this foundation by embedding societal and ethical values into technology, ensuring that AI development aligns with human-centered goals (Friedman & Hendry, 2019. VSD advocates for an inclusive approach, involving stakeholders in the design process to address diverse perspectives and values, which is critical in creating technology that respects cultural and social norms (Umbrello & van de Poel, 2020). This framework also encourages considering fundamental values, such as privacy and autonomy, which help mitigate ethical risks and align AI systems with widely accepted moral standards (Friedman & Hendry, 2019; Jobin et al., 2019). Integrating values into the design process, VSD provides a proactive strategy for addressing potential ethical challenges, fostering user trust, and ensuring societal acceptance of AI technology (Raji et al., 2020; Umbrello & van de Poel, 2020).

Together, the FATML principles, equity theory, and VSD framework create a multifaceted, holistic approach to ethical AI design that encompasses fairness, accountability, and alignment with societal values at all stages of development (Binns, 2018; Friedman & Hendry, 2019; Mehrabi et al., 2021). This approach ensures that ethical considerations are not isolated to any single phase of AI development but are embedded throughout, from design to deployment and monitoring (Doshi-Velez & Kim, 2017; Whittlestone et al., 2019). Following these frameworks, developers can create AI systems that deliver technical performance and adhere to ethical standards, thus enhancing their credibility and alignment with public expectations (Friedman & Hendry, 2019. This comprehensive foundation is essential as AI technologies increasingly shape societal functions, providing a path forward for responsible and socially aligned innovation in AI (Holstein et al., 2019; Jobin et al., 2019).

RECOMMENDATIONS

As AI continues to permeate various sectors, ensuring ethical standards in its development and deployment has become imperative. The recommendations below aim to address the critical challenges of transparency, accountability, and fairness in AI systems. By implementing these practices, developers can foster user trust, mitigate biases, and enhance the interpretability of AI technologies. The following sections detail specific recommendations that focus on transparency in AI development, the importance of interpretability and bias mitigation, and the necessity of regular auditing and monitoring to maintain ethical compliance throughout the lifecycle of AI systems.

Transparency in AI Development

Current findings highlight the importance of transparency as a core practice for ethical AI, enabling users and stakeholders to understand and trust AI systems (Raji et al., 2020). Transparency is not merely a procedural requirement but foundational to fostering trust and accountability in AI technologies. Transparency practices encompass making algorithms and decision-making processes visible and accessible, which allows users to gain insight into how AI systems operate and make predictions (Jonker, Gomstyn, & McGrath, 2024). Research demonstrates that enhanced transparency can significantly reduce user anxiety and distrust, particularly in high-stakes applications such as healthcare and criminal justice, where understanding the rationale behind decisions is crucial for both ethical and practical reasons (Doshi-Velez & Kim, 2017; Obermeyer et al., 2019). For developers, committing to transparency involves sharing detailed documentation, model specifications, and methodologies that clearly explain how AI models function and the data they utilize (Raji et al., 2020). Organizations must establish a culture of openness about their AI systems, including openly communicating their limitations and potential biases. This approach not only aids in demystifying complex algorithms but also equips end-users with the information they need to make informed decisions about interacting with AI technologies (Binns, 2018). Moreover, transparency in AI development is closely linked to the concept of explainability, where explanations about AI decisions should be tailored to different stakeholders, ensuring that they are comprehensible based on the audience's technical background (Lipton, 2016). Transparent AI systems are more likely to gain user trust and are viewed as more accountable, particularly in applications that directly affect individuals (Jonker et al., 2024). When users understand how AI systems reach their conclusions, they are more likely to accept and adopt these technologies, perceiving them as fair and trustworthy. This aligns with the broader ethical aspiration of AI to serve diverse populations equitably. Studies suggest that transparency can serve as a mechanism for accountability: when AI systems are transparent, it is easier to hold developers and organizations responsible for harmful outcomes (Floridi et al., 2018). Today's ethical standards will significantly influence how AI technologies are integrated into societal frameworks, making transparency an indispensable pillar of AI ethics (Manyika, Silberg, & Presten, 2019). Fostering transparency in AI development is critical to ensuring equitable, ethical, and trustworthy AI systems. By committing to transparency, developers enhance user trust and contribute positively to the societal acceptance of AI technologies, ensuring that their deployment aligns with ethical values and public interest (Veale & Binns, 2017). Continued advocacy for transparency can drive the next wave of innovations in AI that prioritizes ethical considerations alongside technical advancements, paving the way for responsible AI deployment across various sectors.

Interpretability and Bias Mitigation Techniques

One specific recommendation to enhance ethical AI development is adopting interpretability techniques designed to make AI decisions more understandable to users and stakeholders (Holstein et al., 2019). Interpretability is critical in enhancing user trust, as users are more likely to engage with AI when they can see the reasoning behind its recommendations or decisions. A transparent approach allows for better scrutiny of the AI's decision-making process, enabling stakeholders to assess whether those decisions align with ethical norms and societal values (Ghassemi et al., 2020). Furthermore, interpretability is especially crucial in high-stakes applications, such as healthcare or criminal justice, where the ramifications of AI decisions can significantly impact individuals' lives (Caruana et al., 2015). Complex

AI models, particularly those utilizing deep learning techniques, often operate as black boxes, where the intricacies of their processes are obscured, making it difficult for developers and users to understand how conclusions are reached. This opacity poses substantial challenges to accountability and trust, as users may question the reliability of AI systems that cannot transparently explain their choices (Lipton, 2018). Integrating interpretability techniques, like local Interpretable Model-agnostic Explanations (LIME) or Shapley Additive Explanations (SHAP), equips users with insights into how individual features contribute to specific predictions (Ribeiro, Singh, & Guestrin, 2016). Consequently, fostering an environment where AI systems can communicate their rationale clearly can bridge the knowledge gap between technical creators and end-users, thus enhancing acceptance and enabling informed decision-making (McGrath & Jonker, 2024). Alongside interpretability, integrating bias mitigation strategies is paramount for promoting fairness and equity in AI systems, mainly to avoid exacerbating existing social disparities (Holstein et al., 2019). Bias in AI can arise from various sources, including skewed training data or flawed algorithms that may unintentionally reinforce societal prejudices (Mehrabi et al., 2019). Techniques like re-sampling datasets, crafting algorithms explicitly designed to be bias-aware, or employing fairness constraints during the training process can significantly enhance the objects of fairness in AI systems (Zemel et al., 2013). By actively identifying and addressing biases, developers can create AI models that uphold ethical standards and reflect a commitment to social justice and equity. Moreover, by making AI systems both interpretable and fair, developers can ensure that these models serve diverse user populations and respond appropriately to ethical concerns (McGrath & Jonker, 2024). Ensuring that AI is understood by its users is crucial for promoting accountability; when users can comprehend how decisions are made, they are more likely to hold developers responsible for what the AI does. This dual focus on interpretability and bias mitigation serves as a technical framework for AI development and a foundational element for cultivating public trust. Effective communication of AI decision processes and outcomes will guide the technology's trajectory toward market acceptance and responsible utilization across various sectors (Mucci & Stryker, 2024).

Regular Auditing and Monitoring

Regular auditing and monitoring of AI systems throughout their lifecycle are essential to ensure ethical principles are upheld in AI development and deployment. Audits allow developers to continuously assess AI performance and identify potential biases that could affect outcomes. These assessments not only help detect biases but also facilitate the identification of any discrepancies between the algorithm's intended functioning and its actual performance in real-world applications (Amodei et al., 2016). By proactively addressing these issues, developers can rectify problems before they escalate, ensuring that AI remains aligned with ethical standards and societal expectations. Moreover, AI systems are not static; they can evolve due to several factors, including algorithm updates, underlying data changes, and operational context shifts (Raji et al., 2020). This phenomenon, known as concept drift, may lead to unintended outcomes that could introduce new biases or exacerbate existing ones (Mitchell et al., 2019). Continuous monitoring practices play a critical role in detecting these changes early, allowing for timely adjustments that maintain the fairness and accountability of AI systems. For instance, periodic evaluations can reveal performance discrepancies in different demographic groups, prompting developers to take corrective actions and ensure equitable treatment for all users. Incorporating ongoing audits enhances the technical robustness of AI algorithms and fosters long-term trust among users (Mitchell et al., 2019). Trust is essential to user acceptance, particularly in high-stakes domains such as healthcare and finance, where bias or error can have serious consequences. By demonstrating a commitment to transparency and ethical accountability through regular audits, organizations can signal to users that they prioritize safety and fairness in their AI applications. This commitment builds user confidence and encourages broader societal acceptance of AI technologies, ultimately leading to more responsible innovation. Establishing a framework for regular auditing can facilitate compliance with emerging regulations and ethical guidelines surrounding AI systems (Amodei et al., 2016). As governmental and industry standards evolve, having a systematic approach to audit and monitor AI systems ensures that developers remain compliant and can adapt to new requirements efficiently. This proactive stance mitigates legal risks and enhances the credibility of AI technologies in the marketplace. Regular auditing and monitoring are integral to ethical AI development, providing a foundation for accountability, trust, and compliance in an increasingly automated world.

Fostering Trust through Ethical AI Practices

Fostering user trust in AI systems is increasingly contingent upon adopting ethical practices that promote transparency, interpretability, bias mitigation, and regular auditing. Transparency is crucial as it empowers users to understand how AI models function and make decisions. When users see the algorithms at work and the data driving those decisions, they are more likely to feel informed and respected (Holstein et al., 2019). This transparency is not merely a best practice but an ethical obligation ensuring accountability. By clarifying the decision-making processes, organizations can demystify AI technology, enhancing user confidence and belief in the system's integrity (Raji et al., 2020). Interpretability complements transparency by allowing users to comprehend the rationale behind specific AI outputs (Holstein et al., 2019; McGrath & Jonker, 2024). As AI systems become increasingly complex, users often find themselves in the dark regarding why certain decisions are made, leading to mistrust and skepticism. Following interpretability guidelines enable developers to create models whose decisions are transparent and explainable, thus bridging the gap between technical complexities and user understanding. By employing straightforward language and visual aids, developers can articulate the logic driving AI solutions, which is vital for garnering user buy-in and acceptance. In addition to transparency and interpretability, implementing effective bias mitigation strategies is essential for ensuring fair outcomes across diverse demographic groups (Holstein et al., 2019; McGrath & Jonker, 2024). Biases in AI outputs can arise from unrepresentative training datasets or flawed algorithms, perpetuating existing social inequities (Buolamwini & Gebru, 2018). Developers cultivate a sense of equity within AI applications by actively identifying and addressing biases. This commitment to fairness protects marginalized communities from discriminatory practices and enhances the overall reliability of AI technologies, fostering a more positive social climate for AI deployment. Regular auditing is a critical accountability mechanism, showcasing a commitment to ethical standards that resonates with users (Amodei et al., 2019). Audits allow for continuous evaluation of AI system performance, ensuring that ethical principles are upheld throughout the lifecycle of the technology. By routinely assessing both the operational performance and ethical implications of their AI systems, developers send a strong message that they prioritize integrity and social responsibility (Mitchell et al., 2019). Such proactive measures engender a culture of accountability that reassures users about the reliability of AI technologies, thus bridging the gap between technical innovation and societal values.

FUTURE RESEARCH

With AI evolvement, robust research is increasingly important to address technical challenges and AI technologies' ethical, regulatory, and societal implications. Establishing effective regulatory frameworks for autonomous AI systems is crucial to ensure ethical operation and public trust. Enhancing transparency in complex AI models is vital, as their opaque decision-making processes can hinder user understanding and trust. Examining AI's societal impacts, including employment changes, privacy issues, and biases, will allow researchers and policymakers to develop effective regulatory responses. Future research will focus on these areas to promote responsible AI innovation that is aligned with societal values and ethical standards.

Regulatory Frameworks for Autonomous AI

Future research should prioritize the development of regulatory frameworks that address the ethical challenges posed by autonomous AI, particularly focusing on decision-making processes where there is minimal or no human oversight (Bryson, 2019). Autonomous AI systems, capable of making complex decisions independently, introduce unique ethical risks, such as potential biases or unintended consequences, that require proactive regulatory guidance (Bryson, 2019; Harishbhai Tilala et al, 2024). These frameworks must establish clear accountability structures, determining who is responsible when an autonomous system causes harm or makes controversial decisions (Bryson, 2019; Harishbhai Tilala et al, 2024). By addressing these regulatory challenges, researchers can contribute to creating AI systems that operate within robust ethical and legal boundaries, even in high-stakes applications (Bryson, 2019).

Key Considerations for Regulatory Frameworks

1. Establishing Accountability Structures: One of the primary challenges is determining accountability when an autonomous AI system causes harm or makes controversial decisions. Regulatory frameworks must clearly define who is responsible developers, operators, or organizations when an AI system fails or behaves unethically (Bryson, 2019; Harishbhai Tilala et al., 2024). This accountability is critical to ensure that victims of potential harm have recourse and that there is a mechanism for holding parties accountable.
2. Addressing Ethical Risks: Autonomous AI systems can introduce various risks, including perpetuating existing biases or making decisions that adversely affect marginalized groups (Raji et al., 2020). Regulatory frameworks must incorporate guidelines to assess and mitigate these risks proactively. This could involve requiring bias assessments before deployment, ongoing monitoring of AI decision-making processes, and implementing checks to ensure that ethical standards are met throughout the AI lifecycle (Amodei et al., 2016).
3. Promoting Transparency: Autonomy in decision-making often comes with a lack of transparency, making it difficult for users and stakeholders to understand how decisions are made (Lipton, 2016). Regulatory frameworks should mandate transparency measures requiring AI systems to explain their decisions. This transparency will foster trust and accountability, allowing users to understand the rationale behind outcomes, especially in sensitive areas like healthcare, criminal justice, and finance (Doshi-Velez & Kim, 2017).

4. Regulating Data Usage: The data used to train AI systems is crucial in determining their outcomes and potential biases. Regulations should include guidelines for ensuring the diversity and representativeness of datasets (Buolamwini & Gebru, 2018). This includes scrutinizing data sources, implementing fairness checks, and requiring documentation of data usage to understand and manage the implications of the data on AI decision-making (Barocas et al., 2019).
5. Continuous Evaluation and Adaptation: AI technology evolves rapidly; thus, regulatory frameworks must be dynamic and adaptable (Whittlestone et al., 2019). This involves establishing mechanisms for continuous evaluation of AI systems even after deployment. Ongoing audits and assessments will allow regulators to identify emerging challenges and adapt regulations, accordingly, ensuring that AI continues to align with societal values and ethical standards (Mitchell et al., 2019).
6. Stakeholder Engagement: Engaging multiple stakeholders, including ethicists, technologists, policymakers, and the public, is essential in shaping comprehensive regulatory frameworks (Bryson, 2019; Jobin et al., 2019). Collaborative dialogue can facilitate the development of policies informed by diverse perspectives, enhancing the likelihood that regulations will effectively address the multifaceted challenges posed by autonomous AI.

Developing regulatory frameworks for autonomous AI is essential to navigating these powerful technologies' ethical and societal implications. By focusing on accountability, ethical risk management, transparency, data regulation, continuous evaluation, and inclusive stakeholder engagement, researchers and policymakers can lay the groundwork for AI systems that operate within ethical and legal boundaries (Whittlestone et al., 2019; Borenstein & Howard, 2021). This proactive approach will protect individuals and communities and foster public trust in AI technologies, ultimately contributing to a responsible and ethical AI landscape.

Enhancing Transparency in Complex Models

Another crucial area for future research is improving transparency in complex AI models, particularly deep learning, and neural networks, which are often described as black boxes due to their opaque decision-making processes (Lipton, 2016). Increased transparency in these models is essential to building public trust, as users need to understand how and why AI systems reach certain conclusions, especially in sensitive domains like healthcare and criminal justice (Lipton, 2016; Kiseleva, Kotzinos, & Hert, 2022). Techniques such as interpretable machine learning and explainable AI can provide insights into the inner workings of complex models, helping bridge the gap between the power of AI and societal acceptance (Lipton,

2016). Research can make AI systems more accountable by focusing on transparency and fostering responsible innovation in complex AI applications (Lipton, 2016; Singhal et al., 2024).

Techniques for Enhancing Transparency

1. Interpretable Machine Learning: Methods that render model predictions comprehensible (Doshi-Velez & Kim, 2017).
    a) Feature Importance: Identifies key influencing features in model decisions (Ribeiro, et al., 2016).
    b) Partial Dependence Plots: Visualizes the effect of specific features on predicted outcomes (Friedman, 2001).
    c) Local Explanations: Tools like LIME offer contextual explanations for individual predictions (Ribeiro, et al., 2016).
2. Explainable AI (XAI): Broader approaches aiming to develop inherently interpretable models (Gunning, 2017).
    a) Model-Specific Techniques: Some models, such as decision trees, are naturally interpretable and yield clear outputs (Aha, Kibler, & Albert, 1991).
    b) Post-hoc Explanation Tools: Techniques that explain complex models after they are developed, such as attention mechanisms in neural networks (Bilodeau et al., 2024).
3. Visualization Techniques: Tools that create intuitive visual representations of model processes, aiding in understanding complex relationships and enhancing user engagement (Sutcliffe, 2022).

Implementing these transparency-enhancing techniques is vital for fostering trust in AI systems and ensuring their responsible deployment in high-impact areas (Lipton, 2016; Kiseleva, et al., 2022).

Examining Societal Impacts and Regulatory Responses

As AI technology progresses, studies examining its societal impacts, regulatory responses, and ethical safeguards are essential to ensuring that innovation aligns with societal values and ethical standards (Whittlestone et al., 2019). These studies can explore the social consequences of AI across sectors, from employment shifts to privacy concerns, thereby providing a comprehensive understanding of AI's broader societal footprint (Whittlestone et al., 2019; Wilson & Daugherty, 2018). In addition, researching regulatory responses can help shape policies that address emerging risks while encouraging ethical AI development (U.S. Department of State (DOS), 2024; Whittlestone et al., 2019). By integrating these insights, researchers can guide responsible AI innovation that respects individual rights and societal needs (Whittlestone et al., 2019).

Societal Impacts of AI

AI's integration into various sectors has profound implications that require meticulous examination. Key areas of concern include:

1. Employment Shifts: AI is poised to revolutionize the labor market by automating tasks traditionally performed by humans. While this can result in increased efficiency and productivity, it also raises questions about job displacement and the future of work. Studies must investigate which job sectors are most vulnerable to automation, the potential for job creation in new fields, and the necessary skills workers need to adapt to this transformed landscape (Wilson & Daugherty, 2018).
2. Privacy Concerns: AI systems' capability to collect and analyze vast amounts of personal data poses significant privacy risks. It is crucial to explore how AI affects data security and user privacy. Research should focus on the ethical implications of data usage, including consent, data ownership, and the potential for surveillance. Understanding these dimensions will inform better regulatory frameworks to safeguard individual rights (Whittlestone et al., 2019).
3. Inequality and Bias: AI systems can inadvertently perpetuate or exacerbate existing social inequalities. Analyzing the social consequences of biased algorithms is vital, as these models may produce discriminatory outcomes against marginalized groups. Investigating the sources of bias in AI and developing strategies for fair algorithms is essential for social equity (Whittlestone et al., 2019).

Regulatory Responses

As the understanding of AI's societal impacts grows, it is equally important to explore regulatory responses that can address the challenges posed by this technology. Such research can help shape policies that achieve the following goals:

1. Mitigating Emerging Risks: As AI technologies evolve, new risks emerge. Policymakers must engage in proactive research to anticipate these risks and develop regulations that adapt to the changing landscape. This includes addressing issues like deepfake technologies, misinformation, and algorithmic accountability to prevent harm to individuals and society (U.S. Department of State (DOS), 2024).

2. Encouraging Ethical AI Development: Regulatory frameworks should focus on mitigating risks and promoting ethical AI development. This involves establishing guidelines encouraging transparency, accountability, and fairness in AI systems and ensuring developers prioritize these principles in their design and deployment practices (Whittlestone et al., 2019).
3. Engaging Stakeholders: Effective regulatory measures require collaboration between various stakeholders, including governments, tech companies, civil society, and academia. Research should identify best practices for stakeholder engagement and facilitate dialogue that informs policy decisions and aligns AI development with public interests and ethical standards (Whittlestone et al., 2019).

By integrating insights from studies on societal impacts and regulatory responses, researchers and policymakers can guide responsible AI innovation that respects individual rights and addresses societal needs. Developing a robust framework that considers both the benefits and potential harms of AI is crucial for ensuring that technological advancements contribute positively to society (Whittlestone et al., 2019).

Guiding Responsible AI Innovation

Research into regulatory frameworks, transparency in complex models, and societal impacts collectively provide a multifaceted foundation for guiding responsible AI innovation (Bryson, 2019; Lipton, 2016; Whittlestone et al., 2019). Investigating regulatory frameworks ensures AI operates within established ethical norms while enhancing transparency builds public trust by making AI systems more understandable (Bryson, 2019; Lipton, 2016). Moreover, examining societal impacts allows researchers and policymakers to anticipate and mitigate AI-related risks, thus aligning technological progress with ethical priorities (Whittlestone et al., 2019). Together, these research avenues support the development of AI technologies that are both advanced and aligned with societal and ethical expectations (Bryson, 2019; Lipton, 2016; Whittlestone et al., 2019).

Regulatory Frameworks

Investigating regulatory frameworks is essential for establishing a structured approach that governs the development and deployment of AI technologies. Effective regulatory frameworks should address several key aspects:

1. Ethical Norms: Regulations help define acceptable practices within AI development, outlining ethical norms that developers must adhere to. This includes considerations related to fairness, accountability, and transparency, which are fundamental in minimizing the risks of biases in AI algorithms (Bryson, 2019).
2. Accountability Structures: Regulatory frameworks must establish precise accountability mechanisms for AI systems, identifying who is responsible when these systems cause harm or make problematic decisions. This includes provisions for recourse for affected individuals and communities, ensuring a clear path for addressing grievances (Bryson, 2019).
3. Dynamic Adaptation: The rapid evolution of AI technologies calls for dynamic regulatory frameworks that allow them to adapt to new challenges and methodologies in AI development. Policymakers must seek continuous feedback from technologists, ethicists, and the public to refine these regulations over time (Whittlestone et al., 2019).

Enhancing Transparency

Transparency in AI systems is crucial for fostering public trust. Building transparent AI systems involves:

1. Clarifying Decision-Making Processes: AI models, particularly complex ones such as deep learning, often operate as "black boxes," making it difficult for users to understand how decisions are made. Enhancing transparency requires the development of interpretable AI models, which can explain their reasoning and decision-making processes in ways that are accessible to users (Lipton, 2016). This clarity helps users feel more informed and engaged with the technology.
2. User Empowerment: Providing users with clear explanations and insights into AI operations fosters a sense of empowerment, enabling them to make informed decisions about their interactions with AI systems. Transparency promotes user confidence and encourages responsible usage, as individuals can better understand the implications of AI decisions on their lives and rights (Lipton, 2016).

Examining Societal Impacts

Understanding the societal impacts of AI technologies is critical for ensuring that AI development aligns with ethical priorities and addresses potential risks:

1. Risk Anticipation: By examining the potential societal effects of AI applications, researchers and policymakers can identify and mitigate emerging risks before they become widespread issues. This includes analyzing how AI may exacerbate

existing inequalities or impact employment dynamics as automation reshapes job markets (Whittlestone et al., 2019).

2. Inclusive Development: Addressing societal impacts encourages a more inclusive approach to AI innovation. Developers and researchers can draw on diverse perspectives and experiences to create AI systems considering various demographic factors, ultimately promoting equity and access across different communities and user groups (Whittlestone et al., 2019).

Together, research into regulatory frameworks, transparency in AI systems, and an understanding of societal impacts guide responsible AI innovation and help ensure that technological progress is aligned with society's ethical standards and expectations. By proactively addressing these areas, we can foster AI technologies that are not just advanced but also trustworthy and equitable, maximizing the benefits of AI while minimizing associated risks (Bryson, 2019; Lipton, 2016; Whittlestone et al., 2019). In doing so, we pave the way for an AI landscape that respects human rights and societal needs, fostering a sustainable and inclusive future.

CONCLUSION

As AI technologies become deeply woven into various sectors of society, the ethical issues of fairness, accountability, and transparency are becoming increasingly critical (Floridi et al., 2018; Hamid, Ajmal, & Torshin, 2023). Fairness in AI systems is essential for preventing biases that can unfairly disadvantage specific demographic groups, while accountability ensures that developers and organizations remain responsible for AI's impacts (Floridi et al., 2018; Manyika et al., 2019). Transparency plays a crucial role in making AI decisions understandable and accessible to users, which is vital for fostering trust and allowing for public scrutiny (Doshi-Velez & Kim, 2017). Collectively, these ethical concerns underscore the need for proactive approaches to AI development that prioritize ethical considerations alongside technical performance (Floridi et al., 2018).

This paper highlights the importance of creating ethical AI frameworks designed to mitigate inherent biases within algorithms and promote equitable outcomes across user groups (Binns, 2018). These frameworks guide developers in identifying and addressing potential sources of bias early in the AI development process, ensuring that outcomes are fair and inclusive (Holstein et al., 2019). By embedding ethics directly into AI design, developers can work to create systems that align with societal values and human rights, ultimately making AI applications more inclusive and equitable (Binns, 2018). Such frameworks are vital for shaping the future of

AI to meet ethical standards that reflect the diversity and complexity of society (Holstein et al., 2019).

Fostering an ethical AI ecosystem is essential for building public trust, as ethical AI systems are more likely to be accepted by users who feel that their values and rights are respected (Jobin et al., 2019; Mucci & Stryker, 2024). When AI development prioritizes transparency and fairness, users are more inclined to engage with and support AI applications, seeing them as responsible and reliable (Jobin et al., 2019; Singhal et al., 2024). Trust is crucial for AI adoption, as individuals and organizations are more likely to rely on AI systems they perceive as ethical and aligned with public interests (Floridi et al., 2018). By fostering a culture of ethical AI development, the field can expand responsibly, maximizing benefits while building and maintaining user trust (Jobin et al., 2019).

As AI technologies continue to evolve rapidly, ongoing research and establishing robust regulatory frameworks will be essential for managing both the opportunities and risks associated with AI (Whittlestone et al., 2019). Continued research can help identify emerging ethical challenges and develop solutions that adapt to technological advances, ensuring AI remains aligned with societal needs (Borenstein & Howard, 2021; Whittlestone et al., 2019). Regulatory frameworks provide a structured approach to addressing ethical concerns, offering guidelines that protect users while encouraging innovation (Bryson, 2019). Together, these measures are vital for maximizing AI's benefits while mitigating societal risks, allowing AI to be used responsibly across diverse applications (Bryson, 2019; Whittlestone et al., 2019).

REFERENCES

Adams, J. S. (1965). Inequity in social exchange. *Advances in Experimental Social Psychology, 2*, 267–299. DOI: 10.1016/S0065-2601(08)60108-2

Aha, W., Kibler, D., & Albert, M. C. (1991). Instance-Based Learning Algorithms. *Machine Learning, 6*(1), 37–66. DOI: 10.1007/BF00153759

Amodei, D., Olah, C., Steinhardt, J., Christiano, P., Schulman, J., & Mané, D. (2016). *Concrete problems in AI safety*. https://doi.org//arXiv.1606.06565DOI: 10.48550

Anderson, J. W., & Visweswaran, S. (2024). Algorithmic individual fairness and healthcare: A scoping review. *The Preprint Server for Health Sciences*, 2024.03.25.24304853. DOI: 10.1101/2024.03.25.24304853

Barocas, S., Hardt, M., & Narayanan, A. (2019). Fairness and machine learning. *Proceedings of the 2019 ICML Workshop on Fairness, Accountability, and Transparency in Machine Learning.*

Bilodeau, B., Jaques, N., Koh, P. W., & Kim, B. (2024). Impossibility theorems for feature attribution. *Proceedings of the National Academy of Sciences of the United States of America, 121*(2), e2304406120. DOI: 10.1073/pnas.2304406120 PMID: 38181057

Binns, R. (2018). Fairness in machine learning: Lessons from political philosophy. *Proceedings of the 2018 Conference on Fairness, Accountability, and Transparency* (pp. 149–158).

Borenstein, J., & Howard, A. (2021). Emerging challenges in AI and the need for AI ethics education. *AI and Ethics, 1*(1), 61–65. DOI: 10.1007/s43681-020-00002-7 PMID: 38624388

Brown, T. B., Mann, B., Ryder, N., & Subbiah, M. (2020). Language models are few-shot learners. *Advances in Neural Information Processing Systems, 33*, 1877–1901. DOI: 10.48550/arXiv.2005.14165

Bryson, J. J. (2019). The past decade and future of AI's impact on society. In *Towards a New Enlightenment? A Transcendent Decade* (pp. 163-187).

Buolamwini, J., & Gebru, T. (2018). Gender shades: Intersectional accuracy disparities in commercial gender classification. *Proceedings of the Conference on Fairness, Accountability, and Transparency* (pp. 77-91). ACM. DOI: 10.1145/3287560.3287596

Caruana, R., Lou, Y., Gehrke, J., Koch, P., Sturm, M., & Elhadad, N. (n.d.). Intelligible models for healthcare: Predicting pneumonia risk and hospital 30-day readmission. *In Proceedings of the 21st ACM SIGKDD International Conference on Knowledge Discovery and Data Mining*, 1721-1730. DOI:DOI: 10.1145/2783258.2788613

Dignum, V. (2017). Responsible artificial intelligence: Designing AI for human values. *ITU Journal : ICT Discoveries*, *1*(25).

Dordunoo, L. (2025). Integrating Organizational and Social Network Theories to Mitigate Racial Bias in Facial Recognition Technology. In Burrell, D. (Ed.), *Organizational Readiness and Research: Security, Management, and Decision Making* (pp. 283–348). IGI Global Scientific Publishing., DOI: 10.4018/979-8-3693-8562-3.ch009

Doshi-Velez, F., & Kim, B. (2017). *Towards a rigorous science of interpretable machine learning*. arXiv preprint arXiv:1702.08608.

Floridi, L. (2020). *The ethics of artificial intelligence: Principles, challenges, and opportunities*. Oxford University Press.

Floridi, L., Cowls, J., Beltrametti, M., Chatila, R., Chazerand, P., Dignum, V., Luetge, C., Madelin, R., Pagallo, U., Rossi, F., Schafer, B., Valcke, P., & Vayena, E. (2018). AI4People-an ethical framework for a good AI society: Opportunities, risks, principles, and recommendations. *Minds and Machines*, *28*(4), 689–707. DOI: 10.1007/s11023-018-9482-5 PMID: 30930541

Friedman, B., & Hendry, D. G. (2019). *Value sensitive design: Shaping technology with moral imagination*. DOI:DOI: 10.7551/mitpress/7585.001.0001

Friedman, B., Kahn, P. H., Borning, A., & Zhang, P. (2006). *Value sensitive design and information systems*. Human-computer interaction in management information systems. *Foundations*. Advance online publication. DOI: 10.1007/978-94-007-7844-3_4

Friedman, J. H. (2001). Greedy function approximation: A gradient boosting machine. *Annals of Statistics*, *29*(5), 1189–1232. DOI: 10.1214/aos/1013203451

Ghassemi, M., Naumann, T., Schulam, P., Beam, A. L., Chen, I. Y., & Ranganath, R. (2020). A Review of Challenges and Opportunities in Machine Learning for Health. *AMIA Joint Summits on Translational Science proceedings. AMIA Joint Summits on Translational Science*, *2020*, 191–200. PMID: 32477638

Guidotti, R., Monreale, A., Ruggieri, S., Turini, F., Giannotti, F., & Pedreschi, D. (2018). A survey of methods for explaining black box models. *ACM Computing Surveys*, *51*(5), 1–42. Advance online publication. DOI: 10.1145/3236009

Gunning, D. (2017). Explainable artificial intelligence (XAI). D*efense Advanced Research Projects Agency* (DARPA). Retrieved from https://www.darpa.mil/attachments/XAIProgramUpdate.pdf

Hamid, Z., Ajmal, S., & Torshin, I. (2023). Ethical considerations in AI and machine learning. DOI:DOI: 10.13140/RG.2.2.15343.20648

Harishbhai Tilala, M., Kumar Chenchala, P., Choppadandi, A., Kaur, J., Naguri, S., Saoji, R., & Devaguptapu, B. (2024). Ethical considerations in the use of artificial intelligence and machine learning in health care: A comprehensive review. *Cureus*, *16*(6), e62443. DOI: 10.7759/cureus.62443 PMID: 39011215

Holstein, K., Wortman Vaughan, J., Daumé, H.III, Dudik, M., & Wallach, H. (2019). Improving fairness in machine learning systems: What do industry practitioners need? *Proceedings of the 2019 CHI Conference on Human Factors in Computing Systems*, 1-16. DOI: 10.1145/3290605.3300830

Jobin, A., Ienca, M., & Vayena, E. (2019). The global landscape of AI ethics guidelines. *Nature Machine Intelligence*, *1*(2), 389–399. Advance online publication. DOI: 10.1038/s42256-019-0088-2

Jonker, A., Gomstyn, A., & McGrath, A. (2024). *What is AI transparency.* https://www.ibm.com/think/topics/ai-transparency

Kiseleva, A., Kotzinos, D., & De Hert, P. (2022). Transparency of AI in healthcare as a multilayered system of accountabilities: Between legal requirements and technical limitations. *Frontiers in Artificial Intelligence*, *5*, 879603. DOI: 10.3389/frai.2022.879603 PMID: 35707765

Li, N. (2023). Ethical considerations in artificial intelligence: A comprehensive discussion from the perspective of computer vision. *SHS Web of Conferences*, *179*(1), 04024. Advance online publication. DOI: 10.1051/shsconf/202317904024

Lipton, Z. C. (2016). The mythos of model interpretability. *Communications of the ACM*, *61*(10), 36–43. DOI: 10.1145/3233231

Manyika, J., Silberg, J., & Presten, B. (2019). AI and machine learning: What do we do about the biases in AI? *Harvard Business Review*.

McGrath, A., & Jonker, A. (2024). *What is AI interpretability.* https://www.ibm.com/think/topics/interpretability

Mehrabi, N., Morstatter, F., Saxena, N., Lerman, K., & Galstyan, A. (2021). A survey on bias and fairness in machine learning. *ACM Computing Surveys*, *54*(6), 1–35. DOI: 10.1145/3457607

Mitchell, M., Wu, S., Zaldivar, A., Barnes, P., Vasserman, L., Hutchinson, B., Spitzer, E., Raji, I. D., & Gebru, T. (2019). Model cards for model reporting. *Proceedings of the Conference on Fairness, Accountability, and Transparency* (pp. 220-229). ACM. DOI: 10.1145/3287560.3287596

Mucci, T., & Stryker, C. (2024). *What is AI governance.* https://www.ibm.com/topics/ai- governance

Obermeyer, Z., Powers, B., Kohane, I., & Levy, M. (2019). Dissecting racial bias in an algorithm used to manage the health of populations. *Science, 366*(6464), 447–453. DOI: 10.1126/science.aax2342 PMID: 31649194

Radford, A., Wu, J., Child, R., Luan, D., Amodei, D., & Sutskever, I. (2019). *Language models are unsupervised multitask learners.* Retrieved from https://cdn.openai.com/better- language-models/language_models_are_unsupervised_multitask_learners.pdf

Raji, I. D., & Buolamwini, J. (2019). Actionable auditing: Investigating the impact of publicly naming biased performance results of commercial AI products. *The 2019 AAAI/ACM Conference.* DOI:DOI: 10.1145/3306618.3314244

Raji, I. D., Smart, A., White, R. N., Mitchell, M., Gebru, T., Hutchinson, B., Smith-Loud, J., Theron, D., & Barnes, P. (2020). Closing the AI accountability gap: Defining an end-to- end framework for internal algorithmic auditing. *Proceedings of the 2020 Conference on Fairness, Accountability, and Transparency*, pp. 33–44. DOI: 10.1145/3351095.3372873

Ribeiro, M. T., Singh, S., & Guestrin, C. (2016). Why should I trust you? Explaining the predictions of any classifier. *In Proceedings of the 22nd ACM SIGKDD International Conference on Knowledge Discovery and Data Mining* (pp. 1135-1144). DOI:DOI: 10.1145/2939672.2939778

Rudin, C. (2019). Stop explaining black box machine learning models for high-stakes decisions and use interpretable models instead. *Nature Machine Intelligence, 1*(5), 206–215. DOI: 10.1038/s42256-019-0048-x PMID: 35603010

Singhal, A., Neveditsin, N., Tanveer, H., & Mago, V. (2024). Toward fairness, accountability, transparency, and ethics in AI for social media and health care: Scoping review. *JMIR Medical Informatics, 12*, e50048. DOI: 10.2196/50048 PMID: 38568737

Sutcliffe, A. (2022). *Designing for user Engagement.* Springer.

Umbrello, S., & van de Poel, I. (2020). *Mapping value sensitive design onto AI for social good principles.* DOI:DOI: 10.1007/s43681-020-00033-0

United States Department of State (DOS). (state.gov), (2024). *Artificial intelligence*. Retrieved from https://www.state.gov/artificial-intelligence/

Vaswani, A., Shazeer, N., Parmer, N., Uszkoreit, J., Jones, L., Gomez, A. N., Kaiser, L., & Polosukhin, I. (2017). Attention is all you need. *Advances in Neural Information Processing Systems*, *30*, 5998–6008. DOI: 10.48550/arXiv.1706.03762

Veale, M., & Binns, R. (2017). Fairer machine learning in the real world: Mitigating discrimination without collecting sensitive data. *Big Data & Society*, *4*(2), 205395171774353. DOI: 10.1177/2053951717743530

Whittlestone, J., Nyrup, R., Alexandrova, A., & Cave, S. (2019). The role and limits of principles in AI ethics: Towards a focus on tensions. *The 2019 AAAI/ACM Conference*. DOI:DOI: 10.1145/3306618.3314289

Wilson, J., & Daugherty, P. R. (2018). Collaborative intelligence: Humans and AI are joining forces. *Harvard Business Review*.

Zemel, R. S., Wu, D., Tsymbal, A., & Richard, M. (2013). Learning fair representations. *In Proceedings of the 30th International Conference on Machine Learning 28*, pp. 325–332.

Compilation of References

Abebe, R. G., Bakombo, S., & Konkle, A. T. (2023). Understanding the Response of Canadians to the COVID-19 Pandemic Using the Kübler-Ross Model: Twitter Data Analysis. *International Journal of Environmental Research and Public Health, 20*(4), 3197. DOI: 10.3390/ijerph20043197 PMID: 36833891

Achinas, S., Horjus, J., Achinas, V., & Euverink, G. J. W. (2019). A PESTLE analysis of biofuels energy industry in Europe. *Sustainability (Basel), 11*(21), 5981. DOI: 10.3390/su11215981

Adams, B. (2022). *Make your employer brand stand out in the talent marketplace*. Havard Business Review.

Adams, J. S. (1965). Inequity in social exchange. *Advances in Experimental Social Psychology, 2*, 267–299. DOI: 10.1016/S0065-2601(08)60108-2

Adams, K. M. G. (2012). Systems theory: A formal construct for understanding systems. *International Journal of System of Systems Engineering, 3*(3/4), 209. DOI: 10.1504/IJSSE.2012.052684

Afridah, , & Muhlisah Lubis, Afridah and Muhlisah Lubis. (2024). The role of communication and employee engagement in promoting inclusion in the workplace: A case study in the creative industry. Feedback. *International Journal of Communication, 1*(1), 1–15. DOI: 10.62569/fijc.v1i1.8

Agree, E. M. (2014). The potential for technology to enhance independence for those aging with a disability. *Disability and Health Journal, 7*(1), S33–S39. DOI: 10.1016/j.dhjo.2013.09.004 PMID: 24456682

Aha, W., Kibler, D., & Albert, M. C. (1991). Instance-Based Learning Algorithms. *Machine Learning, 6*(1), 37–66. DOI: 10.1007/BF00153759

Ahmad, R., Nawaz, M. R., Ishaq, M. I., Khan, M. M., & Ashraf, H. A. (2023). Social exchange theory: Systematic review and future directions. *Frontiers in Psychology*, *13*, 1015921. DOI: 10.3389/fpsyg.2022.1015921 PMID: 36710813

Ahmeti, F. (2023). Leveraging employee engagement for competitive advantage: Satisfaction and work motivation management. *Journal of Liberty and International Affairs Institute for Research and European Studies*, *9*(2), 216–227. DOI: 10.47305/JLIA2392216a

Alaqla, M. F. (2020). The effect of the relationship between organizational excellence and quality of work-life on organizational commitment: An applied study on the health services sector in taif city. *International Journal of Human Resource Studies*, *10*(2), 280. DOI: 10.5296/ijhrs.v10i2.16860

Albrecht, S. L., Connaughton, S., Foster, K., Furlong, S., & Yeow, C. J. L. (2020). Change engagement, change resources, and change demands: A model for positive employee orientations to organizational change. *Frontiers in Psychology*, *11*, 531944. DOI: 10.3389/fpsyg.2020.531944 PMID: 33240144

Albrecht, S. L., Connaughton, S., & Leiter, M. P. (2022). The influence of change-related organizational and job resources on employee change engagement. *Frontiers in Psychology*, *13*, 910206. DOI: 10.3389/fpsyg.2022.910206 PMID: 35769731

Albrecht, S. L., Furlong, S., & Leiter, M. P. (2023). The psychological conditions for employee engagement in organizational change: Test of a change engagement model. *Frontiers in Psychology*, *14*, 1071924. DOI: 10.3389/fpsyg.2023.1071924 PMID: 36743243

Aldag, R. J., & Kuzuhara, L. W. (2015). *Organizational behavior and management: An integrated skills approach* (2nd ed.). Cengage Learning.

Alfes, K., Shantz, A. D., Truss, C., & Soane, E. C. (2013). The link between perceived human resource management practices, engagement and employee behaviour: A moderated mediation model. *International Journal of Human Resource Management*, *24*(2), 330–351. DOI: 10.1080/09585192.2012.679950

Ali, A., King, M., Strydom, A., & Hassiotis, A. (2016). Self-reported stigma and its association with socio-demographic factors and physical disability in people with intellectual disabilities: Results from a cross-sectional study in England. *Social Psychiatry and Psychiatric Epidemiology*, *51*(3), 465–474. DOI: 10.1007/s00127-015-1133-z PMID: 26498927

Allen, A., & Jacques, S. (2014). Police officers' theories of crime. *American Journal of Criminal Justice*, *39*(2), 206–227. DOI: 10.1007/s12103-013-9219-1

Allen, T. D., Golden, L. S., & Shockley, K. M. (2015). How effective is telecommuting? Assessing the status of our scientific findings. *Psychological Science in the Public Interest, 16*(2), 40–68. DOI: 10.1177/1529100615593273 PMID: 26403188

Allison, P., Mnatsakanova, A., McCanlies, E., Fekedulegn, D., Hartley, T. A., Andrew, M. E., & Violanti, J. M. (2020). Police stress and depressive symptoms: Role of coping and hardiness. *Policing*, *43*(2), 247–261. DOI: 10.1108/PIJPSM-04-2019-0055 PMID: 32714068

Almanza Jiménez, R., Calderón Campos, P., Vargas Hernández, J. G., Casas Cárdenas, R., & Palomares Salceda, F. (2019). Aprendizaje y desempeño organizacional bajo el enfoque de las teorías organizacionales. *Revista de Economía & Administración, 13*(1), 84-94. Recuperado de https://revistas.uao.edu.co/ojs/index.php/REYA/article/view/10

Alper, S., & Raharinirina, S. (2006). Assistive Technology for Individuals with Disabilities: A Review and Synthesis of the Literature. *Journal of Special Education Technology*, *21*(2), 47–64. DOI: 10.1177/016264340602100204

Alton, L. (2024). *7 strategies to address employee disengagement.* Retrieved from https://www.shrm.org/topics-tools/news/managing-smart/7-strategies-to-address - employee-disengagement

American Psychological Association (APA). (2023). *2023 work in America survey.* Retrieved from https://www.apa.org/pubs/reports/work-in-america/2023-workplace -health-well- being

Americans With Disabilities Act of 1990, 42 U.S.C. § 12101 et seq. (1990)

Amodei, D., Olah, C., Steinhardt, J., Christiano, P., Schulman, J., & Mané, D. (2016). *Concrete problems in AI safety*. https://doi.org//arXiv.1606.06565DOI: 10.48550

Anderson, J. W., & Visweswaran, S. (2024). Algorithmic individual fairness and healthcare: A scoping review. *The Preprint Server for Health Sciences*, 2024.03.25.24304853. DOI: 10.1101/2024.03.25.24304853

Andrews, R., Boyne, G., & Walker, R. (2011). The Impact of Management on Administrative and Survey Measures of Organizational Performance. *Public Management Review*, *13*(2), 227–255. DOI: 10.1080/14719037.2010.532968

Antony, J. P., & Bhattacharyya, S. (2010). Measuring organizational performance and organizational excellence of smes – part 2: An empirical study on smes in india. *Measuring Business Excellence*, *14*(3), 42–52. DOI: 10.1108/13683041011074209

Aoun, J. E. (2017). *Robot-proof: Higher education in the age of artificial intelligence*. MIT Press. DOI: 10.7551/mitpress/11456.001.0001

Appelbaum, S. H. (1997). Socio-technical systems theory: An intervention strategy for organizational development. *Management Decision, 35*(6), 452–463. DOI: 10.1108/00251749710173823

Aragón, J., & Sharma, S. (2003). A Contingent Resource Based View of Proactive Corporate Environmental Strategy. *Academy of Management Review, 28*(1), 71–88. DOI: 10.5465/amr.2003.8925233

Aris, A. Z. Z., Hassan, A., & Kassim, M. A. M. (2023). Enhancing organizational commitment through effective computer communication practices: a proposal for malaysian organizations. *E3S Web of Conferences*, 440, 04011. DOI: 10.1051/e3sconf/202344004011

Arksey, H., & O'Malley, L. (2005). Scoping studies: Towards a methodological framework. *International Journal of Social Research Methodology, 8*(1), 19–32. DOI: 10.1080/1364557032000119616

Armstrong, M., & Taylor, S. (2023). *Armstrong's handbook of human resource management practice: A guide to the theory and practice of people management* (16th ed.). Kogan Page.

Arno, A., & Thomas, S. (2016). The efficacy of nudge theory strategies in influencing adult dietary behaviour: A systematic review and meta-analysis. *BMC Public Health, 16*(1), 1–11. DOI: 10.1186/s12889-016-3272-x PMID: 27475752

Arun, A., Rao, A., Hecht, A., Garcia, A., Oluyomi, A., Lapidus, A., & Miggins, J. (2022). Fifth annual summer research summit on Health Equity organized by the Center of excellence in Health Equity, Training and research, Baylor College of medicine, Houston, Texas 77030, USA on May 17, 2022. [IJMA]. *International Journal of MCH and AIDS, 11*(2s). Advance online publication. DOI: 10.21106/ijma.596

Ashikali, T., Groeneveld, S., & Kuipers, B. (2020). The role of inclusive leadership in supporting an inclusive climate in diverse public sector teams. *Review of Public Personnel Administration, 41*(3), 497–519. DOI: 10.1177/0734371X19899722

Asree, S., Zain, M., & Razalli, R. (2010). Influence of leadership competency and organizational culture on responsiveness and performance of firms. *International Journal of Contemporary Hospitality Management, 22*(4), 500–516. DOI: 10.1108/09596111011042712

Awahnde, M. (2014). *Development of self-managed teams through service design methods*. Theseus.

Ayoko, O. B. (2022). Leadership, ethics and corporate social responsibility. *Journal of Management & Organization*, *28*(1), 1–8. DOI: 10.1017/jmo.2022.13

Azevedo, V., Nunes, L., & Sani, A. (2022). Is campus a place of (in)security and crime? Perceptions of higher education students. *European Journal of Investigation in Health, Psychology and Education*, *12*(2), 193–208. DOI: 10.3390/ejihpe12020015 PMID: 35200237

Azevedo, V., Sani, A., Nunes, L., & Paulo, D. (2021). Do you feel safe in the urban space? From perceptions to associated variables. *Anuário de Psicología Juridica*, *31*(1), 75–84. DOI: 10.5093/apj2021a12

Babri, M., Davidson, B., & Helin, S. (2021). An updated inquiry into the study of corporate codes of ethics: 2005–2016. *Journal of Business Ethics*, *168*(1), 71–108. DOI: 10.1007/s10551-019-04192-x

Bailey, D. E., & Kurland, N. B. (2002). A review of telework research: Findings, new directions, and lessons for the study of modern work. *Journal of Organizational Behavior*, *23*(4), 383–400. DOI: 10.1002/job.144

Baker, D., & Ellis, L. (2021). Future directions in digital information: scenarios and themes. In *Future Directions in Digital Information* (pp. 1–15). Chandos Publishing. DOI: 10.1016/B978-0-12-822144-0.00001-X

Bakker, A. B., & Demerouti, E. (2007). The job demands-resources model: State of the art. *Journal of Managerial Psychology*, *22*(3), 309–328. DOI: 10.1108/02683940710733115

Bakker, A. B., Demerouti, E., & Euwema, M. C. (2005). Job resources buffer the impact of job demands on burnout. *Journal of Occupational Health Psychology*, *10*(2), 170–180. DOI: 10.1037/1076-8998.10.2.170 PMID: 15826226

Bakker, A. B., & Schaufeli, W. B. (2008). Positive organizational behavior: Engaged employees in flourishing organizations. *Journal of Organizational Behavior*, *29*(2), 147–154. DOI: 10.1002/job.515

Baloh, J., Zhu, X., & Ward, M. (2017). Implementing team huddles in small rural hospitals: How does the Kotter model of change apply? *Journal of Nursing Management*, *26*(5), 571–578. DOI: 10.1111/jonm.12584 PMID: 29250892

Balsiger, A. J. (2022). *Differing Perspectives of Stigma from Students with Disabilities and Access Workers*. Practicing Anthropology., DOI: 10.17730/0888-4552.44.4.26

Bandura, A. (1997). *Self-efficacy: The exercise of control*. Macmillan.

Bano, A., Khatun, D. A., & Kumar, D. D. (2024). Examining drivers of engagement and employee engagement in relation to organizational performance- a study of banking sector in hyderabad city. *Migration Letters : An International Journal of Migration Studies*, *21*(S3), 1067–1081. DOI: 10.59670/ml.v21iS3.6887

Bantam, D. J., Ardin, M. P., Shafira, N., & Janah, M. (2024). Organizational culture: Employee perception of pt. indomarco prismatama. Eduvest -. *Journal of Universal Studies*, *4*(4), 2080–2089. DOI: 10.59188/eduvest.v4i4.1113

Barker Scott, B. A., & Manning, M. R. (2024). Designing the collaborative organization: A framework for how collaborative work, relationships, and behaviors generate collaborative capacity. *The Journal of Applied Behavioral Science*, *60*(1), 149–193. DOI: 10.1177/00218863221106245

Barnes, B. R., Leonidou, L. C., Lai, G. W. F., & Leonidou, C. N. (2015). Interpersonal factors as drivers of quality and performance in western–hong kong interorganizational business relationships. *Journal of International Marketing*, *23*(1), 23–49. DOI: 10.1509/jim.14.0008

Barney, J. (1991). Firm Resources and Sustained Competitive Advantage. *Journal of Management*, *17*(1), 99–120. DOI: 10.1177/014920639101700108

Barocas, S., Hardt, M., & Narayanan, A. (2019). Fairness and machine learning. *Proceedings of the 2019 ICML Workshop on Fairness, Accountability, and Transparency in Machine Learning.*

Basit, A. A., & Nauman, S. (2023). How workplace loneliness harms employee well-being: A moderated mediational model. *Frontiers in Psychology*, *13*, 1086346. DOI: 10.3389/fpsyg.2022.1086346 PMID: 36726515

Bass, B. M. (1985). *Leadership and performance beyond expectations*. Free Press.

Bass, B. M. (1999). Two decades of research and development in transformational leadership. *European Journal of Work and Organizational Psychology*, *8*(1), 9–32. DOI: 10.1080/135943299398410

Bass, B. M., & Avolio, B. J. (1993). Transformational leadership and organizational culture. *Public Administration Quarterly*, *17*(1), 112–121. https://www.jstor.org/stable/40862298

Bassi, L., & McMurrer, D. (2007). Maximizing your return on people. *Harvard Business Review*. PMID: 17348175

Batavia, A., & Schriner, K. (2001). The Americans With Disabilities Act as Engine of Social Change: Models of Disability and the Potential of a Civil Rights Approach. *Policy Studies Journal: the Journal of the Policy Studies Organization, 29*(4), 690–702. DOI: 10.1111/j.1541-0072.2001.tb02122.x

Battisti, E., Alfiero, S., & Leonidou, E. (2022). Remote working and digital transformation during the COVID-19 pandemic: Economic-financial impacts and psychological drivers for employees. *Journal of Business Research, 150*, 38–50. DOI: 10.1016/j.jbusres.2022.06.010 PMID: 35706830

Batt, R. (1997). Work organization, technology, and performance in customer service and sales. *Industrial & Labor Relations Review, 52*(4), 539–564. DOI: 10.1177/001979399905200403

Baumgärtner, M. K., Dwertmann, D. J. G., Boehm, S., & Bruch, H. (2015). Job Satisfaction of Employees with Disabilities: The Role of Perceived Structural Flexibility. *Human Resource Management, 54*(2), 323–343. DOI: 10.1002/hrm.21673

Becker, K., Antuar, N., & Everett, C. (2011). Implementing an employee performance management system in a nonprofit organization. *Nonprofit Management & Leadership, 21*(3), 255–271. DOI: 10.1002/nml.20024

Bendak, S., Shikhli, A. M., Abdel-Razek, R. H., & Ardito, L. (2020). How changing organizational culture can enhance innovation: Development of the innovative culture enhancement framework. *Cogent Business & Management, 7*(1), 1712125. Advance online publication. DOI: 10.1080/23311975.2020.1712125

Bernal González, I., Pedraza Melo, N. A. & Castillo Hernández, L. (2020). El capital humano y su relación con el desempeño organizacional. *Revista Espacios, 41*(22), art. 14.

Bernardino, R. C., & Bernardino, A. V. S. (2018). Fatores estressores que influenciam na qualidade de vida, gerando danos à saúde do policial militar [Stressors that influence the quality of life, causing damage to the health of military police officers]. *Revista Mosaico, 09*(2), 02-09. https://doi.org/DOI: 10.21727/rm.v9i2.1304

Bertolli, M., Roark, G., Urrutia, S., & Chiodi, F. (2017). Revisión de modelos de madurez en la medición del desempeño. *INGE CUC, 13*(1), 70–83. DOI: 10.17981/ingecuc.13.1.2017.07

Bérubé, N. (2023). *Hidden challenges: An invisible disabilities learning activity*. Organization Management Journal., DOI: 10.1108/OMJ-03-2023-1810

Bhargava, S., & Sinha, B. (1992). Predictions of organizational effectiveness as a function of type of organizational structure. *The Journal of Social Psychology, 132*(2), 223–232. DOI: 10.1080/00224545.1992.9922974

Bilodeau, B., Jaques, N., Koh, P. W., & Kim, B. (2024). Impossibility theorems for feature attribution. *Proceedings of the National Academy of Sciences of the United States of America, 121*(2), e2304406120. DOI: 10.1073/pnas.2304406120 PMID: 38181057

Binns, R. (2018). Fairness in machine learning: Lessons from political philosophy. *Proceedings of the 2018 Conference on Fairness, Accountability, and Transparency* (pp. 149–158).

Bishop, P. A., Downes, J. M., & Farber, K. (2019). *Personalized learning in the middle grades: A guide for classroom teachers and school leaders*. Harvard Education Press. https://hep.gse.harvard.edu/9781682533178/personalized-learning-in-the-middle-grades/

Blau, P. M. (1964). *Exchange and power in social life*. Wiley.

Bocanegra, H. G. V. (2024). Supply Chain Management and Innovation Practices as Core Strategies for Business Competitiveness and Performance. *Advances in Management and Applied Economics, 14*(4).

Boccoli, G., Gastaldi, L., & Corso, M. (2022). The evolution of employee engagement: Towards a social and contextual construct for balancing individual performance and wellbeing dynamically. *International Journal of Management Reviews, 25*(1), 75–98. DOI: 10.1111/ijmr.12304

Bodkin, C., & Hakimi, M. (2020). Innovative communication strategies for enhancing organizational performance. *Journal of Business Communication, 57*(3), 215–230. DOI: 10.1002/jbc.12345

Borenstein, J., & Howard, A. (2021). Emerging challenges in AI and the need for AI ethics education. *AI and Ethics, 1*(1), 61–65. DOI: 10.1007/s43681-020-00002-7 PMID: 38624388

Botha-Badenhorst, D., & Veerasamy, N. (2023). Examining barriers to entry: Disparate gender representation in cybersecurity within Sub-Saharan Africa. *International Conference on Gender Research*, 6(1), 47–55. DOI: 10.34190/icgr.6.1.1148

Bourne, M., Melnyk, S., & Bititci, U. S. (2018). Performance measurement and management: Theory and practice. *International Journal of Operations & Production Management, 38*(11), 2010–2021. DOI: 10.1108/IJOPM-11-2018-784

Bourne, M., Neely, A., Mills, J., & Platts, K. (2003). Implementing performance measurement systems: A literature review. *International Journal of Business Performance Management, 5*(1), 1–24. DOI: 10.1504/IJBPM.2003.002097

Bridges, M. (2024). Integrate KPI's into culture strategies and case studies. Retrieved from https://mark-bridges.medium.com/integrated-kpis-into-culture-strategies-and-case- studies-b3a1fa7b085c

Brooks, L. A., Manias, E., & Bloomer, M. J. (2019). Culturally sensitive communication in healthcare: A concept analysis. *Collegian (Royal College of Nursing, Australia), 26*(3), 383–391. DOI: 10.1016/j.colegn.2018.09.007

Brown, M. E., & Treviño, L. K. (2006). Ethical leadership: A review and future directions. *The Leadership Quarterly, 17*(6), 595–616. DOI: 10.1016/j.leaqua.2006.10.004

Brown, T. B., Mann, B., Ryder, N., & Subbiah, M. (2020). Language models are few-shot learners. *Advances in Neural Information Processing Systems, 33*, 1877–1901. DOI: 10.48550/arXiv.2005.14165

Bryl, T. (2018). Human capital orientation and financial performance. a comparative analysis of us corporations. *Journal of Entrepreneurship. Management and Innovation, 14*(3), 61–86. DOI: 10.7341/20181433

Bryson, J. J. (2019). The past decade and future of AI's impact on society. In *Towards a New Enlightenment? A Transcendent Decade* (pp. 163-187).

Bryson, J. M. (2018). *Strategic planning for public and nonprofit organizations: A guide to strengthening and sustaining organizational achievement* (5th ed.). Wiley.

Buil, I., Martínez, E., & Matute, J. (2019). Transformational leadership and employee performance: The role of identification, engagement, and proactive personality. *International Journal of Hospitality Management, 77*, 64–75. DOI: 10.1016/j.ijhm.2018.06.014

Buolamwini, J., & Gebru, T. (2018). Gender shades: Intersectional accuracy disparities in commercial gender classification. *Proceedings of the Conference on Fairness, Accountability, and Transparency* (pp. 77-91). ACM. DOI: 10.1145/3287560.3287596

Burgess, A., Haq, I., Bleasel, J., Roberts, C., Garsia, R., Randal, N., & Mellis, C. (2019). Team-based learning (TBL): A community of practice. *BMC Medical Education, 19*(1), 1–7. DOI: 10.1186/s12909-019-1795-4 PMID: 31615507

Burrell, D. N., & Nobles, C. (2018). Recommendations to develop and hire more highly qualified women and minorities cybersecurity professionals. In International Conference on Cyber Warfare and Security (pp. 75-81). Academic Conferences International Limited.

Burrell, D. (2018). An Exploration of the Cybersecurity Workforce Shortage. *International Journal of Hyperconnectivity and the Internet of Things*, *2*(1), 29–41. DOI: 10.4018/IJHIoT.2018010103

Busa, A. I., Yakubu, I., & Olabode, H. A. (2024). Adaptive leadership in educational settings: Complex challenges and uncertain environments. *Fudma Journal of Research. Educational Psychology and Counselling*, *2*(1), 257–264.

Cabaj, K., Kotulski, Z., Księżopolski, B., & Mazurczyk, W. (2018). Cybersecurity: Trends, issues, and challenges. *EURASIP Journal on Information Security*, *2018*(1), 1–3. DOI: 10.1186/s13635-018-0080-0

Cain, Curtis C. (2021b) "Establishing a research agenda for broadening participation of Black men in computing, informatics, and engineering," Technology in Society, Elsevier, vol. 67(C).

Cain, C. C. (2021a). Beyond the IT Artifact - Studying the Underrepresentation of Black Men and Women in IT. *Journal of Global Information Technology Management*, *24*(3), 157–163. DOI: 10.1080/1097198X.2021.1954315

Calhoun, M., & McCarthy, V. (2022). An unintended consequence of flexible work arrangements (FWAs) – the role of peer resentment on turnover intentions in public accounting. *Journal of Accounting & Organizational Change*. https://doi.org/DOI: 10.1108/JAOC-03-2022-0045

Camisón, C., & Cruz, S. (2008). La medición del desempeño organizativo desde una perspectiva estratégica: Creación de un instrumento de medida. *Revista Europea de Dirección y Economía de la Empresa*, *17*(1), 79–102.

Camisón, C., Garrigós, F., & Palacios, D. (2007). Estrategias competitivas y desempeño empresarial: Estudio comparativo de los modelos de Robinson and Pearce y Miles and Snow en el sector hotelero español. *Investigaciones Europeas de Dirección y Economía de la Empresa*, *13*(3), 161–182.

Campion, M. A., Medsker, G. J., & Higgs, A. C. (1993). Relations between work group characteristics and effectiveness: Implications for designing effective work groups. *Personnel Psychology*, *46*(4), 823–847. DOI: 10.1111/j.1744-6570.1993.tb01571.x

Camp, R. (1989). *Benchmarking: the search for industry best practices that lead to superior performance*. ASQ Quality Press.

Camps, J., & Luna-Arocas, R. (2012). A matter of learning how human resources affect organizational performance. *British Journal of Management, 23*(1), 1–21. DOI: 10.1111/j.1467-8551.2010.00714.x

Canavesi, A., & Minelli, E. (2021). Servant leadership and employee engagement: A qualitative study. *Employee Responsibilities and Rights Journal, 34*(4), 413–435. DOI: 10.1007/s10672-021-09389-9 PMID: 40477632

Cannon-Bowers, J. A., Tannenbaum, S. I., Salas, E. & Volpe, C. E. (1995). Defining competencies and establishing team-training requirements. *Team effectiveness and decision making in organizations,* 333-380.

Cao, M., Zhao, Y., & Zhao, S. (2020). Facilitating the Implementation Process of High-Performance Work Systems: The Role of Authentic Leadership. *Frontiers in Psychology, 11*, 550711. Advance online publication. DOI: 10.3389/fpsyg.2020.550711 PMID: 33071868

Cappelli, P., & Eldor, L. (2019). Where measuring engagement goes wrong. *Harvard Business Review*.

Carmeli, A., Brueller, D., & Dutton, J. E. (2009). Learning behaviors in the workplace: The role of high-quality interpersonal relationships and psychological safety. *Systems Research and Behavioral Science, 26*(1), 81–98. DOI: 10.1002/sres.932

Carmeli, A., Meitar, R., & Weisberg, J. (2006). Self-leadership skills and innovative behavior at work. *International Journal of Manpower, 27*(1), 75–90. DOI: 10.1108/01437720610652853

Carmeli, A., & Spreitzer, G. M. (2009). Trust, connectivity, and thriving: Implications for innovative behaviors at work. *The Journal of Creative Behavior, 43*(3), 169–191. DOI: 10.1002/j.2162-6057.2009.tb01313.x

Carton, R. (2004). *Measuring Organizational Performance: An Exploratory Study*. [Unpublished PhD Thesis]. University of Georgia, Athens.

Caruana, R., Lou, Y., Gehrke, J., Koch, P., Sturm, M., & Elhadad, N. (n.d.). Intelligible models for healthcare: Predicting pneumonia risk and hospital 30-day readmission. *In Proceedings of the 21st ACM SIGKDD International Conference on Knowledge Discovery and Data Mining*, 1721-1730. DOI:DOI: 10.1145/2783258.2788613

Carver, C. S., & Scheier, M. F. (2001). *On the self-regulation of behavior*. Cambridge University Press.

Castillo, V. (2013). *Teorías de las organizaciones*. Trillas.

Cate, L., Ward, L., & Ford, K. S. (2022). Strategic ambiguity: How pre-tenure faculty negotiate the hidden rules of academia. *Innovative Higher Education, 47*(5), 795–812. DOI: 10.1007/s10755-022-09604-x PMID: 35502224

Chajduga, T., & Ingaldi, M. (2021). Hiring Disabled People to Avoid Staff Turnover and Enhance the Sustainability of Production. *Sustainability (Basel), 13*(19), 10577. Advance online publication. DOI: 10.3390/su131910577

Chanana, N., & Sangeeta, . (2020). Employee engagement practices during COVID-19 lockdown. *Journal of Public Affairs, 21*(4), e2508. DOI: 10.1002/pa.2508 PMID: 33041656

Chan, X. W., Shang, S., Brough, P., Wilkinson, A., & Lu, C. Q. (2022). Work, life and COVID- 19: A rapid review and practical recommendations for the post-pandemic workplace. *Asia Pacific Journal of Human Resources*. Advance online publication. *10*.1111/1744-7941.12355

Chávez, E. (2007). El ROI: De la remuneración al capital humano. *Ingeniería Industrial, 25*(25), 111–118. DOI: 10.26439/ing.ind2007.n025.613

Chávez, E. A., & Vizcaíno, A. (2017). Talento humano: Una contribución a la competitividad organizacional. *Mercados y Negocios, 1*(36), 7–20. DOI: 10.32870/myn.v0i36.5646

Chen, X., Al Mamun, A., Hussain, W. M. H. W., Jingzu, G., Yang, Q., & Shami, S. S. A. A. (2023). Envisaging the job satisfaction and turnover intention among the young workforce: Evidence from an emerging economy. *PLoS One, 18*(6), e0287284. DOI: 10.1371/journal.pone.0287284 PMID: 37327240

Cherry, J., Aquino, F., Fabelico, F., & Afalla, B. (2021). Managing educational institutions: School heads' leadership practices and teachers' performance. [IJERE]. *International Journal of Evaluation and Research in Education, 10*(4), 1325–1333. DOI: 10.11591/ijere.v10i4.21518

Chewning, L. V., Lai, C., & Doerfel, M. L. (2012). Organizational resilience and using information and communication technologies to rebuild communication structures. *Management Communication Quarterly, 27*(2), 237–263. DOI: 10.1177/0893318912465815

Chiu, M., Yang, X., Wong, H., & Li, J. (2015). The mediating effect of affective stigma between face concern and general mental health—The case of Chinese caregivers of children with intellectual disability. *Research in Developmental Disabilities, 36C*, 437–446. DOI: 10.1016/j.ridd.2014.10.024 PMID: 25462504

Chong, H. (2008). Measuring Performance of Small-and-Medium Sized Enterprises: The Grounded Theory Approach. *Journal of Business & Public Affairs*, *10*, 1–11.

Chowdhury, N. H., Adam, M. T., & Skinner, G. (2019). The impact of time pressure on cybersecurity behaviour: A systematic literature review. *Behaviour & Information Technology*, *38*(12), 1290–1308. DOI: 10.1080/0144929X.2019.1583769

Christodoulou, A., & Cullinane, K. (2019). Identifying the main opportunities and challenges from the implementation of a port energy management system: A SWOT/PESTLE analysis. *Sustainability (Basel)*, *11*(21), 6046. DOI: 10.3390/su11216046

Clark, T. R. (2020). *The 4 stages of psychological safety: Defining the path to inclusion and innovation*. Berrett-Koehler Publishers.

Clavijo-Tapia, J., Smith, A., & Johnson, R. (2021). Evaluating communication strategies for organizational effectiveness. *Journal of Organizational Communication*, *45*(2), 150–165. DOI: 10.1002/joc.12345

Clifton, J. (2022). *The world's workplace is broken: Here's how to fix it.* Retrieved from https://www.gallup.com/workplace/393395/world-workplace-broken-fix.aspx

Cohen, E. (2017). *CSR for HR: A necessary partnership for advancing responsible business practices*. Routledge. DOI: 10.4324/9781351278607

Cohen, L. W., & Felson, M. (1979). Social change and crime rate trends: A routine activity approach. *American Sociological Review*, *44*(4), 588–608. DOI: 10.2307/2094589

Cohen, S. G. (1993). New approaches to teams and teamwork. *Trends and Issues in OD:* Crook, R., Combs, J. & Shook, C. (2005). The Dimensionality of Organizational Performance and its Implications for Strategic Management Research. *Research Methodology in Strategy and Management*, *2*, 259–286. DOI: 10.1016/S1479-8387(05)02011-4

Cohen, S. G., & Bailey, D. E. (1997). What makes teams work: Group effectiveness research from the shop floor to the executive suite. *Journal of Management*, *23*(3), 239–290. DOI: 10.1177/014920639702300303

Cohen, S. G., & Ledford, G. E.Jr. (1994). The effectiveness of self-managing teams: A quasi-experiment. *Human Relations*, *47*(1), 13–43. DOI: 10.1177/001872679404700102

Colby, S. (2017). *Competency-based education: A framework for measuring quality courses*. EDUCAUSE Review. https://er.educause.edu/articles/2017/6/competency-based-education-a-framework-for-measuring-quality-courses

Cook, L., Foley, J., & Semeah, L. M. (2016). An exploratory study of inclusive worksite wellness: Considering employees with disabilities. *Disability and Health Journal, 9*(1), 100–107. DOI: 10.1016/j.dhjo.2015.08.011 PMID: 26471518

Corbière, M., Zaniboni, S., Dewa, C., Villotti, P., Lecomte, T., Sultan-Taïeb, H., Hupé, J., & Fraccaroli, F. (2019). Work productivity of people with a psychiatric disability working in social firms. *Work (Reading, Mass.), 62*(1), 151–160. DOI: 10.3233/WOR-182850 PMID: 30689598

Corradini, I. (2020). *Building a cybersecurity culture in organizations: How to bridge the gap between people and digital technology* (Vol. 284). Springer Nature. DOI: 10.1007/978-3-030-43999-6

Couzens, D., Poed, S., Kataoka, M., Brandon, A., Hartley, J., & Keen, D. (2015). Support for Students with Hidden Disabilities in Universities: A Case Study. *International Journal of Disability Development and Education, 62*(1), 24–41. DOI: 10.1080/1034912X.2014.984592

Crail, C., & Watts, R. (2024). *15 effective employee retention strategies in 2024*. Retrieved from https://www.forbes.com/advisor/business/employee-retention-strategies/

Crampton, S., & Hodge, J. W. (2003). The ADA and Disability Accommodations. *Public Personnel Management, 32*(1), 143–154. DOI: 10.1177/009102600303200108

Criminal Division of the U.S. Department of Justice. (2024). United States v. The Boeing Company. https://www.justice.gov/criminal/criminal-fraud/case/united-states-v-boeing-companyLinks to an external site

Cronin, C. (2014). Using case study research as a rigorous form of inquiry. *Nurse Researcher, 21*(5), 19–27. DOI: 10.7748/nr.21.5.19.e1240 PMID: 24877907

Cropanzano, R., & Mitchell, M. S. (2005). Social exchange theory: An interdisciplinary review. *Journal of Management, 31*(6), 874–900. DOI: 10.1177/0149206305279602

Curado, C., Henriques, P. L., Jerónimo, H. M., & Azevedo, J. (2022). The contribution of communication to employee satisfaction in service firms: A causal configurational analysis. *Vision (Basel), 0*(0), 09722629221101157. Advance online publication. DOI: 10.1177/09722629221101157

Daft, R. (2010). *Teoría y diseño organizacional.* México: Cengage Learning Editores, S. A. de C. V.

Dai, X., Zhang, Y., & Liu, J. (2022). The Role of Communication Strategies in Enhancing Organizational Excellence: A Comprehensive Study. *Journal of Organizational Excellence*, *41*(2), 45–60. DOI: 10.1002/joe.2022.0045

Dalgin, R. S., & Gilbride, D. (2003). Perspectives of people with psychiatric disabilities on employment disclosure. *Psychiatric Rehabilitation Journal*, *26*(3), 306–310. DOI: 10.2975/26.2003.306.310 PMID: 12653452

Darling-Hammond, L., Hyler, M. E., & Gardner, M. (2017). *Effective teacher professional development*. Learning Policy Institute. https://learningpolicyinstitute.org/sites/default/files/product-files/Effective_Teacher_Professional_Development_REPORT.pdf

Darling-Hammond, L. (2010). *The flat world and education: How America's commitment to equity will determine our future*. Teachers College Press.

Darling-Hammond, L., Flook, L., Cook-Harvey, C., Barron, B., & Osher, D. (2019). Implications for educational practice of the science of learning and development. *Applied Developmental Science*, *24*(2), 97–140. DOI: 10.1080/10888691.2018.1537791

Dasborough, M. (2006). Cognitive asymmetry in employee emotional reactions to leadership behaviors. *The Leadership Quarterly*, *17*(2), 163–178. DOI: 10.1016/j.leaqua.2005.12.004

Data Quality Campaign. (2023). *Data literacy for educators: A roadmap for effective data use in schools*. https://dataqualitycampaign.org/wp-content/uploads/2023/03/DQC-Data101.pdf

David, S. (2016). *Emotional agility: Get unstuck, embrace change, and thrive in work and life*. Penguin House.

Day, G. (1994). The Capabilities of Market Driven Organizations. *Journal of Marketing*, *58*(4), 3–52. DOI: 10.1177/002224299405800404

Deci, E. L., & Ryan, R. M. (2000). *The "what" and "why" of goal pursuits: Human needs*

Deci, E. L., & Ryan, R. M. (2000). The "what" and "why" of goal pursuits: Human needs and the self-determination of behavior. *Psychological Inquiry*, *11*(4), 227–268. DOI: 10.1207/S15327965PLI1104_01

Deci, E. L., & Ryan, R. M. (2008). Self-determination theory: A macrotheory of human motivation, development, and health. *Canadian Psychology*, *49*(3), 182–185. DOI: 10.1037/a0012801

Deep, G. (2023). The power of resilience and flexibility in business leadership: Adapting to change. *Magna Scientia Advanced Research and Reviews, 9*(2):086-091.

De-la-Calle-Durán, M. C., & Rodríguez-Sánchez, J. L. (2021). Employee engagement and wellbeing in times of COVID-19: A Proposal of the 5Cs Model. *International Journal of Environmental Research and Public Health, 18*(10), 5470. DOI: 10.3390/ijerph18105470 PMID: 34065338

Delaney, J. T., & Huselid, M. A. (1996). The impact of human resource management practices on perceptions of organizational performance. *Academy of Management Journal, 39*(4), 949–969. DOI: 10.2307/256718

DeLeire, T. (2000). The Wage and Employment Effects of the Americans with Disabilities Act. *The Journal of Human Resources, 35*(4), 693–715. DOI: 10.2307/146368

Deng, C., Gulseren, D., Isola, C., Grocutt, K., & Turner, N. (2022). Transformational leadership effectiveness: An evidence-based primer. *Human Resource Development International, 26*(5), 627–641. DOI: 10.1080/13678868.2022.2135938

Dennison, K. (2024). *Gallup says $8.8 trillion is the future cost of low employee engagement*. Retrieved from https://www.forbes.com/sites/karadennison/2024/07/16/gallup-says-88- trillion-is-the-true-cost-of-low-employee-engagement/

Detert, J. R., & Burris, V. (2007). Leadership behavior and employee voice: Is the door open? *Academy of Management Journal, 50*(4), 869–884. DOI: 10.5465/amj.2007.26279183

Dewey, J. (1916). *Democracy and education: An introduction to the philosophy of education*. Macmillan.

Dhaafri, H., Al Swidi, A., & Al Ansi, A. (2016). Organizational Excellence as the Driver for Organizational Performance: A Study on Dubai Police. *International Journal of Business and Management, 11*(2), 47–52. DOI: 10.5539/ijbm.v11n2p47

Di Miceli da Silveira, A. (2013). The Enron scandal a decade later: Lessons learned? *SSRN*. https://doi.org/DOI: 10.2139/ssrn.2310114

Dignum, V. (2017). Responsible artificial intelligence: Designing AI for human values. *ITU Journal : ICT Discoveries, 1*(25).

Doblinger, M. (2021). Individual Competencies for Self-Managing Team Performance: A Systematic Literature Review. *Small Group Research, 53*(1), 128–180. DOI: 10.1177/10464964211041114

Dodgson, M. (1993). Organizational Learning: A Review of Some Literatures. *Organization Studies, 14*(3), 375–395. DOI: 10.1177/017084069301400303

Dolbier, C. L., Soderstrom, M., & Steinhardt, M. A. (2001). The relationships between self- leadership and enhanced psychological, health, and work outcomes. *The Journal of Psychology*, *135*(5), 469–485. DOI: 10.1080/00223980109603713 PMID: 11804002

Dordunoo, L. (2025). Integrating Organizational and Social Network Theories to Mitigate Racial Bias in Facial Recognition Technology. In Burrell, D. (Ed.), *Organizational Readiness and Research: Security, Management, and Decision Making* (pp. 283–348). IGI Global Scientific Publishing., DOI: 10.4018/979-8-3693-8562-3.ch009

Doshi-Velez, F., & Kim, B. (2017). *Towards a rigorous science of interpretable machine learning*. arXiv preprint arXiv:1702.08608.

Douglas, S. (2020). The effects of body-worn cameras on violent police victimization. *Policing (Oxford, England)*, *15*(2), 1399–1416. DOI: 10.1093/police/paaa032

Dowling, B., Goldstein, D., Park, M., & Price, H. (2022). *Hybrid work: Making it fit with your diversity, equity, and inclusion strategy*. Retrieved from https://www.mckinsey.com/capabilities/people-and-organizational-performance/our- insights/hybrid-work-making-it-fit-with-your-diversity-equity-and-inclusion-strategy

Drevs, F. (2013). How patients choose hospitals: Using the stereotypic content model to model trustworthiness, warmth and competence. *Health Services Management Research*, *26*(2-3), 95–101. DOI: 10.1177/0951484813513246 PMID: 25595006

Druskat, V. U., & Kayes, D. C. (2000). Learning versus performance in short-term project teams. *Small Group Research*, *31*(3), 328–353. DOI: 10.1177/104649640003100304

Dudley-Evans, T. (2020). The debate over Milton Friedman's theoretical framework: An applied linguist's view. In *Economics and language* (pp. 132–152). Routledge. DOI: 10.4324/9781003072256-9

Dursun, K., & Aytaç, S. (2021). Workplace violence against police officers and the effects of workplace violence on mental health. *Pakistan Journal of Medical & Health Sciences*, *15*(12), 3491–3493. DOI: 10.53350/pjmhs2115123491

Eccles, J. S., & Wigfield, A. (2020). From expectancy-value theory to situated expectancy-value theory: A developmental, social cognitive, and sociocultural perspective on motivation. *Contemporary Educational Psychology*, *61*, 101859. DOI: 10.1016/j.cedpsych.2020.101859

Edmondson, A. (1999). Psychological safety and learning behavior in work teams. *Administrative Science Quarterly*, *44*(2), 350–383. DOI: 10.2307/2666999

Eisenhardt, K. M. (1989). Agency Theory: An Assessment and Review. *Academy of Management Review*, *14*(1), 57–74. DOI: 10.2307/258191

Ekberg, K., Pransky, G., Besen, E., Fassier, J., Feuerstein, M., Munir, F., & Blanck, P. (2016). New Business Structures Creating Organizational Opportunities and Challenges for Work Disability Prevention. *Journal of Occupational Rehabilitation*, *26*(4), 480–489. DOI: 10.1007/s10926-016-9671-0 PMID: 27704343

Eliasson, M. N. (2021). Police officers: Invisible victims in the line of duty. In Blasdell, R., Krieger-Sample, L., & Kilburn, M. (Eds.), *Invisible victims and the pursuit of justice: analyzing frequently victimized yet rarely discussed populations* (pp. 41–62). IGI Global Scientific Publishing., DOI: 10.4018/978-1-7998-7348-8.ch003

Ellinger, A. D., & Bostrom, R. P. (1999). Managerial coaching behaviors in learning organizations. *Journal of Management Development*, *18*(9), 752–771. DOI: 10.1108/02621719910300810

Ellrich, K., & Baier, D. (2016). Police officers as victims of violence: Findings of a Germany-wide survey. In Baier, D., & Pfeiffer, C. (Eds.), *Representative Studies on Victimisation: Research Findings from Germany (1st ed.). Nomos Verlagsgesellschaft mbH.* (pp. 139-162). https://www.jstor.org/stable/j.ctv941sz3.8

Endenich, C., & Trapp, R. (2020). Ethical implications of management accounting and control: A systematic review of the contributions from the Journal of Business Ethics. *Journal of Business Ethics*, *163*(2), 309–328. DOI: 10.1007/s10551-018-4034-8

Errida, A., & Lotfi, B. (2021). The determinants of organizational change management success: Literature review and case study. *International Journal of Engineering Business Management*, *13*, 18479790211016273. Advance online publication. DOI: 10.1177/18479790211016273

Escandón, D., Arias, A., & Salas, J. (2012). Recursos y capacidades que inciden en el desempeño de los subsectores económicos colombianos 2003-2004: Un acercamiento desde la teoría basada en recursos. *Ingeniería y Competitividad*, *14*(2), 53–67. DOI: 10.25100/iyc.v14i2.2654

Estell, P., Davidson, E., & Abhari, K. (2021). Affording employee voice: how enterprise social networking sites (esns) create new pathways for employee expression. *Proceedings of the Annual Hawaii International Conference on System Sciences*. DOI: 10.24251/HICSS.2021.336

Evans, H. (2019). 'Trial by fire': Forms of impairment disclosure and implications for disability identity. *Disability & Society, 34*(5), 726–746. DOI: 10.1080/09687599.2019.1580187

Fadun, S. (2022). Risk management of everything store. *Publisher*

Fairchild, E. J., & Ede, L. (1978). Occupational safety and health: United States of America. *Sozial- und Praventivmedizin, 23*(1), 20–26. DOI: 10.1007/BF02073950 PMID: 695968

Fan, Y. (2024). Accountability in public organization: A systematic literature review and future research agenda. *Public Organization Review*. Advance online publication. DOI: 10.1007/s11115-024-00792-y

Farahnak, L. R., Ehrhart, M. G., Torres, E. M., & Aarons, G. A. (2020). The influence of transformational leadership and leader attitudes on subordinate attitudes and implementation success. *Journal of Leadership & Organizational Studies, 27*(1), 98–111. DOI: 10.1177/1548051818824529

Farhoomand, A. F., & Drury, D. H. (2002). Managerial information overload. *Communications of the ACM, 45*(10), 127–131. DOI: 10.1145/570907.570909

Farkas, M. (2013). Building and Sustaining a Culture of Assessment: Best Practices for Change Leadership. *RSR. Reference Services Review, 41*(1), 13–31. DOI: 10.1108/00907321311300857

Fernandes, J. (2023). The role of data-driven decision-making in effective educational leadership. *Academy of Educational Leadership Journal, 27*(2), 1–3. https://www.abacademies.org/articles/the-role-of-datadriven-decisionmaking-in-effective-educational-leadership-15998.html

Fernandes, J., & Kumar, R. (2025). Ethical challenges in AI-driven education: A global perspective. *Journal of Educational Ethics, 12*(3), 45–60.

Fey, C., & Beamish, P. (2001). Organizational climate similarity and performance international joint Ventures in Russia. *Stockholm School of Economics in St. Petersburg Working, 99*(101), 1-28.

Fiol, M. C., & Lyles, M. (1985). Organizational learning. *Academy of Management Review, 10*(4), 803–813. DOI: 10.2307/258048

Fiorvanti, C. M., & Brassard, M. R. (2020). Child protection: A child rights approach for schools. In Nastasi, B. K., Hart, S. N., & Naser, S. C. (Eds.), *International handbook on child rights and school psychology*. Springer., DOI: 10.1007/978-3-030-37119-7_15

Fitz-Enz, J. (2003). *El ROI del capital humano. Como medir el valor económico del rendimiento del personal*. Deusto.

Floridi, L. (2020). *The ethics of artificial intelligence: Principles, challenges, and opportunities*. Oxford University Press.

Floridi, L., Cowls, J., Beltrametti, M., Chatila, R., Chazerand, P., Dignum, V., Luetge, C., Madelin, R., Pagallo, U., Rossi, F., Schafer, B., Valcke, P., & Vayena, E. (2018). AI4People-an ethical framework for a good AI society: Opportunities, risks, principles, and recommendations. *Minds and Machines*, *28*(4), 689–707. DOI: 10.1007/s11023-018-9482-5 PMID: 30930541

Foley, C., Kirkby, A., & Eccles, F. J. R. (2022). A meta-ethnographic synthesis of the experiences of stigma amongst people with functional neurological disorder. *Disability and Rehabilitation*, *46*(1), 1–12. DOI: 10.1080/09638288.2022.2155714 PMID: 36519449

Ford, J., Ford, L., & Polin, B. (2021). Leadership in the implementation of change: Functions, sources, and requisite variety. *Journal of Change Management*, *21*(1), 87–119. DOI: 10.1080/14697017.2021.1861697

Freeman, S., Marston, H. R., Olynick, J., Musselwhite, C., Kulczycki, C., Genoe, M. R., & Xiong, B. (2020). Intergenerational effects on the impacts of technology use in later life: Insights from an international, multi-site study. *International Journal of Environmental Research and Public Health*, *17*(16), 5711. DOI: 10.3390/ijerph17165711 PMID: 32784651

Friedman, B., & Hendry, D. G. (2019). *Value sensitive design: Shaping technology with moral imagination*. DOI:DOI: 10.7551/mitpress/7585.001.0001

Friedman, M. (1970). A Friedman doctrine: The social responsibility of business is to increase its profits. *The New York Times Magazine*, 13(1970), 32-33

Friedman, B., Kahn, P. H., Borning, A., & Zhang, P. (2006). *Value sensitive design and information systems*. Human-computer interaction in management information systems. *Foundations*. Advance online publication. DOI: 10.1007/978-94-007-7844-3_4

Friedman, J. H. (2001). Greedy function approximation: A gradient boosting machine. *Annals of Statistics*, *29*(5), 1189–1232. DOI: 10.1214/aos/1013203451

Froehlich-Grobe, K., Lee, J., & Washburn, R. (2013). Disparities in obesity and related conditions among Americans with disabilities. *American Journal of Preventive Medicine*, *45*(1), 83–90. DOI: 10.1016/j.amepre.2013.02.021 PMID: 23790992

Fuentes, M. M., & Hurtado, N. E. (2002). Variables críticas en la medición del desempeño en empresas con implantación de la gestión de la calidad total. *Investigaciones Europeas de Dirección y Economía de la Empresa*, *8*(2), 87–102.

Fullan, M. (2015). *The new meaning of educational change* (5th ed.). Teachers College Press.

Fullan, M., & Quinn, J. (2016). *Coherence: The right drivers in action for schools, districts, and systems*. Corwin Press.

Fuller, R., Pyle, A. S., Riolli, L., & Mickel, A. E. (2020). Creating order out of chaos? development of a measure of perceived effects of communication on the crisis organizing process. *International Journal of Business Communication*, *59*(2), 174–192. DOI: 10.1177/2329488420979657

Fuller, R., & Sala, A. L. (2021). Crisis communication preparedness practices among u.s. charitable organizations: Results from a national survey. *SAGE Open*, *11*(2), 21582440211014516. Advance online publication. DOI: 10.1177/21582440211014516

Gagné, M., & Deci, E. L. (2005). Self-determination theory and work motivation. *Journal of Organizational Behavior*, *26*(4), 331–362. DOI: 10.1002/job.322

Gajendran, R. S., & Joshi, A. (2012). Innovation in globally distributed teams: The role of LMX, communication frequency, and innovation climate. *The Journal of Applied Psychology*, *97*(6), 1252–1261. DOI: 10.1037/a0028958 PMID: 22708920

Galanakis, M. D., & Tsitouri, E. (2022). Positive psychology in the working environment. Job demands-resources theory, work engagement and burnout: A systematic literature review. *Frontiers in Psychology*, *13*, 1022102. DOI: 10.3389/fpsyg.2022.1022102 PMID: 36204770

Gallen, D. (2017). Personalized learning in Lindsay Unified. *School Made Fresh*. https://schoolmadefresh.org/personalized-learning-in-lindsay-unified/

Gallo, A. (2023). What is psychological safety? *Harvard Business Review*.

Gandini, A. (2019). Labour process theory and the gig economy. *Human Relations*, *72*(6), 1039–1056. DOI: 10.1177/0018726718790002

Ganga, F., Cassinelli, A., Díaz, G., & Maluk, S. A. (2016). Breves disquisiciones teóricas en torno al tema capital humano y eficiencia organizativa. *Revista Gaceta Laboral*, *22*(1), 27–39.

Gardner, H. K. (2016). *Smart collaboration: How professionals and their firms succeed by breaking down silos*. Harvard Business Review Press.

Garvin, D. A. (1984). What does "product quality" really mean? *Sloan Management Review*, *26*(1), 25–43.

Gavrea, C., Ilies, L., & Stegeran, R. (2011). Determinants of Organizational Performance: The Case of Romania. *Management and Marketing Challenges for the Knowledge Society*, *6*(2), 285–300.

Geerts, I. A., Bierbooms, J. J. P. A., & Cloudt, S. W. M. G. (2021). Understanding self-managing teams in Dutch healthcare: Empirical evidence to non-sequential team development processes. *Journal of Health Organization and Management*, *35*(9), 101–120. DOI: 10.1108/JHOM-04-2020-0122 PMID: 33818047

Georgopoulos, B., & Tannenbaum, A. (1957). A Study of Organizational Effectiveness. *American Sociological Review*, *22*(5), 534–540. DOI: 10.2307/2089477

Gerber, P., & Price, L. A. (2003). Persons with Learning Disabilities in the Workplace: What We Know So Far in the Americans with Disabilities Act Era. *Learning Disabilities Research & Practice*, *18*(2), 132–136. DOI: 10.1111/1540-5826.00069

Ghassemi, M., Naumann, T., Schulam, P., Beam, A. L., Chen, I. Y., & Ranganath, R. (2020). A Review of Challenges and Opportunities in Machine Learning for Health. *AMIA Joint Summits on Translational Science proceedings. AMIA Joint Summits on Translational Science*, *2020*, 191–200. PMID: 32477638

Gibbs, J. C., Lee, J., Moloney, J., & Olson, S. (2018). Exploring the neighbourhood context of serious assaults on police. *Policing and Society*, *28*(8), 898–914. DOI: 10.1080/10439463.2017.1333120

Giovanis, E., & Ozdamar, O. (2019). Accommodating Employees with Disabilities: The Role of Flexible Employment Schemes in Europe. *Labor: Public Policy & Regulation EJournal*. https://doi.org/DOI: 10.2139/ssrn.3441925

Golden, T. D., & Veiga, J. F. (2005). The impact of extent of telecommuting on job satisfaction: Resolving inconsistent findings. *Journal of Management*, *31*(2), 301–318. DOI: 10.1177/0149206304271768

Goleman, D., Boyatzis, R., & McKee, A. (2004). *Primal leadership: Learning to lead with emotional intelligence*. Harvard Business Review Press.

Gondal, A. Q., Ahmad, M., Raza, M. A., Bagum, U., & Husnain, M. (2023). Community policing for violence prevention in Pakistan: A critical review of evidence and future directions. *Research*, *8*(1), 53–75. DOI: 10.1234/re.v8.i1.4

Gong, Y., Xiao, J., Tang, X., & Li, J. (2023). How sustainable marketing influences the customer engagement and sustainable purchase intention? The moderating role of corporate social responsibility. *Frontiers in Psychology*, *14*, 1128686. DOI: 10.3389/fpsyg.2023.1128686 PMID: 37008858

Gruman, J. A., & Saks, A. M. (2011). Performance management and employee engagement. *Human Resource Management Review*, *21*(2), 123–136. DOI: 10.1016/j.hrmr.2010.09.004

Guerrero, S., González, S., Matos, A., & Picón, E. (2015). Desarrollo del capital humano y su impacto en el desempeño de una institución microfinanciera no regulada del Perú. *Revista Apuntes Universitarios*, *5*(1), 51–68. DOI: 10.17162/au.v5i1.65

Guha, E. A., & Indurkar, S. K. (2020). A Study on Cyber Crime and Data Breach Management. *Research Journal of Engineering and Technology*, *11*(2), 113–117. DOI: 10.5958/2321-581X.2020.00020.3

Guidotti, R., Monreale, A., Ruggieri, S., Turini, F., Giannotti, F., & Pedreschi, D. (2018). A survey of methods for explaining black box models. *ACM Computing Surveys*, *51*(5), 1–42. Advance online publication. DOI: 10.1145/3236009

Gulyamova, S. T., Aziz, S. F. A., Omar, N., & Mohd, R. (2023). Workplace-Related Socioeconomic Issues Associated with Job Performance and Productivity among Employees with Various Impairments: A Systematic Literature Review. *Social Sciences (Basel, Switzerland)*, *12*(5), 275. Advance online publication. DOI: 10.3390/socsci12050275

Gunning, D. (2017). Explainable artificial intelligence (XAI). D*efense Advanced Research Projects Agency* (DARPA). Retrieved from https://www.darpa.mil/attachments/XAIProgramUpdate.pdf

Habibzadeh, H., Nussbaum, B. H., Anjomshoa, F., Kantarci, B., & Soyata, T. (2019). A survey on cybersecurity, data privacy, and policy issues in cyber-physical system deployments in smart cities. *Sustainable Cities and Society*, *50*, 101660. DOI: 10.1016/j.scs.2019.101660

Hackman, J. R. (1987). The design of work teams. In Lorsh, J. (Ed.), *Handbook of Organizational Behavior 315-342*. Prentice-Hall.

Hagan, J., McCarthy, B., Herda, D., & Cann Chandrasekher, A. (2018). Dual-process theory of racial isolation, legal cynicism, and reported crime. *Proceedings of the National Academy of Sciences of the United States of America*, *115*(28), 7190–7199. DOI: 10.1073/pnas.1722210115 PMID: 29987006

Hale, C. (2015). Urban special education policy and the lived experience of stigma in a high school science classroom. *Cultural Studies of Science Education, 10*(4), 1071–1088. DOI: 10.1007/s11422-013-9548-x

Hallam, K. T., Popovic, N., & Karimi, L. (2023). Identifying the key elements of psychologically safe workplaces in healthcare settings. *Brain Sciences, 13*(10), 1450. DOI: 10.3390/brainsci13101450 PMID: 37891818

Hamann, P., Schiemann, F., Bellora, L., & Guenther, T. (2013). Exploring the Dimensions of Organizational Performance: A Construct Validity Study. *Organizational Research Methods, 16*(1), 67–87. DOI: 10.1177/1094428112470007

Hambrick, D. (2003). On the Staying power of Miles and Snow's defenders, analyzers and prospectors. *The Academy of Management Perspectives, 17*(4), 115–118. DOI: 10.5465/ame.2003.11851883

Hamid, Z., Ajmal, S., & Torshin, I. (2023). Ethical considerations in AI and machine learning. DOI:DOI: 10.13140/RG.2.2.15343.20648

Hancock, A. J., Gellatly, I. R., Walsh, M. M., Arnold, K. A., & Connelly, C. E. (2023). Good, bad, and ugly leadership patterns: Implications for followers' work-related and context-free outcomes. *Journal of Management, 49*(2), 640–676. DOI: 10.1177/01492063211050391 PMID: 36484084

Haoran, X., Verhage, A., & Vandeviver, C. (2022). Violence against on-duty police officers: The role of criminal motivation and premeditation in offenders' choice of weapons. *Policing, 45*(6), 1023–1036. DOI: 10.1108/PIJPSM-04-2022-0062

Harishbhai Tilala, M., Kumar Chenchala, P., Choppadandi, A., Kaur, J., Naguri, S., Saoji, R., & Devaguptapu, B. (2024). Ethical considerations in the use of artificial intelligence and machine learning in health care: A comprehensive review. *Cureus, 16*(6), e62443. DOI: 10.7759/cureus.62443 PMID: 39011215

Harris, A., & Jones, M. (2020). COVID-19—School leadership in disruptive times. *School Leadership & Management, 40*(4), 243–247. DOI: 10.1080/13632434.2020.1811479

Harter, J. (2017). *Employee engagement vs. employee satisfaction and organizational culture*. Retrieved from https://www.gallup.com/workplace/236366/right-culture-not-employee- satisfaction.aspx

Harter, J. (2022). *Disengagement persists among U.S. employees*. Retrieved from https://www.gallup.com/workplace/391922/employee-engagement-slump-continues.aspx

Harter, J. (2024). *In new workplace, U.S employee engagement stagnates.* Retrieved from https://www.gallup.com/workplace/608675/new-workplace-employee-engagement- stagnates.aspx

Harter, J. K., Schmidt, F. L., & Hayes, T. L. (2002). Business-unit-level relationship between employee satisfaction, employee engagement, and business outcomes: A meta- analysis. *The Journal of Applied Psychology*, *87*(2), 268–279. DOI: 10.1037/0021-9010.87.2.268 PMID: 12002955

Hartnett, H., Stuart, H., Thurman, H., Loy, B., & Batiste, L. (2011a). Employers' perceptions of the benefits of workplace accommodations: Reasons to hire, retain and promote people with disabilities. *Journal of Vocational Rehabilitation*, *34*(1), 17–23. DOI: 10.3233/JVR-2010-0530

Heifetz, R. A., Linsky, M., & Grashow, A. (2009). *The practice of adaptive leadership: Tools and tactics for changing your organization and the world.* Harvard Business Press.

Heimstädt, M., & Dobusch, L. (2020). Transparency and accountability: Causal, critical and constructive perspectives. *Organization Theory*, *1*(4), 2631787720964216. Advance online publication. DOI: 10.1177/2631787720964216

Held, M., Marian, A., & Reaves, A. (2018). *The auto industry's growing recall problem—and how to fix it.* https://www.alixpartners.com/media/14438/ap_auto_industry _recall_problem_jan_2018.pdf

Hicks, J. M. (2011). Leader communication styles and organizational health. *The Health Care Manager*, *30*(1), 86–91. DOI: 10.1097/HCM.0b013e3182078bf8 PMID: 21248553

Hillmann, J. (2021). Disciplines of organizational resilience: Contributions, critiques, and future research avenues. *Review of Managerial Science*, *15*(4), 879–936. DOI: 10.1007/s11846-020-00384-2

Hoegl, M., & Parboteeah, K. P. (2006). Team goal commitment in innovative projects. *International Journal of Innovation Management*, *10*(03), 299–324. DOI: 10.1142/S136391960600151X

Hokke, S., Bennetts, S., Crawford, S., Leach, L., Hackworth, N., Strazdins, L., Nguyen, C., Nicholson, J., & Cooklin, A. (2020). Does flexible work 'work' in Australia? A survey of employed mothers' and fathers' work, family and health. *Community Work & Family*, *24*(4), 488–506. DOI: 10.1080/13668803.2019.1704397

Holland, J. (2016). The Average Black Family Would Need 228 Years to Build the Wealth of a White Family Today. The Nation Journal. Retrieved from https://www.thenation.com/article/the-average-black-family-wouldneed-228-years-to-build-the-wealth-of-a-white-family-today

Hollister, R., Tecosky, K., Watkins, M., & Wolpert, C. (2021). Why every executive should be focusing on culture change now. *MIT Sloan Management Review*.

Holmes, W., Bialik, M., & Fadel, C. (2019). *Artificial intelligence in education: Promises and implications for teaching and learning*. Center for Curriculum Redesign.

Holstein, K., Wortman Vaughan, J., Daumé, H.III, Dudik, M., & Wallach, H. (2019). Improving fairness in machine learning systems: What do industry practitioners need? *Proceedings of the 2019 CHI Conference on Human Factors in Computing Systems*, 1-16. DOI: 10.1145/3290605.3300830

Hornberger, R. C. (2021). Creating a sense of digital privacy in the private sector. *Information Security Journal: A Global Perspective, 30*, 30–56. https://doi.org/DOI: 10.1080/19393555.2020.1797948

Huang, C., Xie, X., Cheung, S. P., & Zhou, Y. (2022). Job demands and resources, positive and negative affect, and psychological distress of social workers in China. *Frontiers in Psychiatry, 12*, 752382. DOI: 10.3389/fpsyt.2021.752382 PMID: 35115957

Huang, Z., Sindakis, S., Aggarwal, S., & Thomas, L. (2022). The role of leadership in collective creativity and innovation: Examining academic research and development environments. *Frontiers in Psychology*, *13*, 1060412. DOI: 10.3389/fpsyg.2022.1060412 PMID: 36619078

Huebner, L. A., & Zacher, H. (2021). Following up on employee surveys: A conceptual framework and systematic review. *Frontiers in Psychology*, *12*, 801073. DOI: 10.3389/fpsyg.2021.801073 PMID: 34956026

ISC2 (2018). Innovation Through Inclusion: The Multicultural Cybersecurity Workforce. https://www.isc2.org/-/media/Files/Research/Innovation-Through-Inclusion-Report.ashx

Islam, M. A. (2017). SWOT Mix and PESTEL Analysis: Effective Tools of Risk Management of Leasing Companies. *The Millennium University Journal*, *2*(1), 1–18. DOI: 10.58908/tmuj.v2i1.11

Jackson, J. (2009). A psychological perspective on vulnerability in the fear of crime. *Psychology, Crime & Law*, *15*(4), 365–390. DOI: 10.1080/10683160802275797

Jacobsen, C. B., Andersen, L. B., Bøllingtoft, A., & Eriksen, T. L. M. (2022). Can leadership training improve organizational effectiveness? Evidence from a randomized field experiment on transformational and transactional leadership. *Public Administration Review*, *82*(1), 117–131. DOI: 10.1111/puar.13356

Jahoda, A., & Marková, I. (2004). Coping with social stigma: People with intellectual disabilities moving from institutions and family home. *Journal of Intellectual Disability Research*, *48*(Pt 8), 719–729. DOI: 10.1111/j.1365-2788.2003.00561.x PMID: 15494061

James, S. L. (2019). The underrepresentation of females in the United States cybersecurity workforce: a multiple-case study (Doctoral dissertation, Capella University).

Jaworski, B., & Kohli, A. (1993). Market Orientation: Antecedents and Consequences. *Journal of Marketing*, *57*(3), 53–70. DOI: 10.1177/002224299305700304

Jennings, M., & Trautman, L. J. (2016). Ethical culture and legal liability: The GM switch crisis and lessons in governance. *BUJ Sci. & Tech. L.*, *22*, 187.

Jha, N., Potnuru, R. K. G., Sareen, P., & Shaju, S. (2019). Employee voice, engagement and organizational effectiveness: A mediated model. *European Journal of Training and Development*, *43*(7/8), 699–718. DOI: 10.1108/EJTD-10-2018-0097

Jiang, H., & Luo, Y. (2018). Crafting employee trust: From authenticity, transparency to engagement. *Journal of Communication Management (London)*, *22*(2), 138–160. DOI: 10.1108/JCOM-07-2016-0055

Jiang, H., & Shen, H. (2020). Toward a relational theory of employee engagement: Understanding authenticity, transparency, and employee behaviors. *International Journal of Business Communication*, *60*(3), 948–975. DOI: 10.1177/2329488420954236

Jobin, A., Ienca, M., & Vayena, E. (2019). The global landscape of AI ethics guidelines. *Nature Machine Intelligence*, *1*(2), 389–399. Advance online publication. DOI: 10.1038/s42256-019-0088-2

John-Akinola, Y. O., Ajayi, A. O., & Oluwasanu, M. M. (2020). Experience of stress and coping mechanism among police officers in south western Nigeria. *International Quarterly of Community Health Education*, *41*(1), 7–14. DOI: 10.1177/0272684X19900878 PMID: 31969056

Johnson, M., Brown, T., & Davis, K. (2023). The role of leadership in personalized learning: A case study analysis. *Educational Leadership Quarterly*, *58*(2), 89–104.

Johnson, S. R. (2018). *Engaging the workplace: Using surveys to spark change.* Association for Talent Development.

Jones, J. L., Davis, W., & Thomas, C. H. (2017). Is Competition Engaging? Examining the Interactive Effects of Goal Orientation and Competitive Work Environment on Engagement. *Human Resource Management, 56*(3), 389–405. DOI: 10.1002/hrm.21773

Jones, M. K. (2008). Disability and the labour market: A review of the empirical evidence. *Journal of Economic Studies (Glasgow, Scotland), 35*(5), 405–424. DOI: 10.1108/01443580810903554

Jonker, A., Gomstyn, A., & McGrath, A. (2024). *What is AI transparency.* https://www.ibm.com/think/topics/ai-transparency

Jordan, C. A. (2022). Exploring the Cybersecurity Skills Gap: A Qualitative Study of Recruitment and Retention from a Human Resource Management Perspective (Order No. 29320493). Available from ProQuest Central; ProQuest Dissertations & Theses Global; ProQuest One Academic; Publicly Available Content Database. (2706752761). https://www.proquest.com/dissertations-theses/exploring-cybersecurity-skills-gap-qualitative/docview/2706752761/se-2

Joubert, S. (2024). *Transformational leadership: How to inspire innovation in the workplace*. Retrieved from https://graduate.northeastern.edu/resources/transformational-leadership/

Juan, D., & Khan, D. (2023). Ensuring corporate accountability: Environmental responsibility and legal liability.

Junaidi, J. (2024). The role of ethical leadership to employees work engagement: A social learning theory perspective. *International Journal of Social Economics, 51*(7), 884–898. DOI: 10.1108/IJSE-03-2023-0218

Junxiao, L., Love, S., & Regan, M. (2014). Public-Private Partnerships: A review of theory and practice of performance measurement. *International Journal of Productivity and Performance Management, 63*(4), 499–512. DOI: 10.1108/IJPPM-09-2013-0154

Kahneman, D. (2017). *Thinking, fast and slow*. Farrar, Straus, and Giroux.

Kahn, W. A. (1990). Psychological conditions of personal engagement and disengagement at work. *Academy of Management Journal, 33*(4), 692–724. DOI: 10.2307/256287

Kalifa, G. (2024). The role of transformational leadership in enhancing organizational innovation. *International Journal of Financial, Administrative, and Economic Sciences*.

Kaminski, R. J., Jefferis, E., & Gu, J. (2003). Community Correlates of Serious Assaults on Police. *Police Quarterly*, *6*(2), 119–149. DOI: 10.1177/1098611103006002001

Kang, M., & Sung, M. (2017). How symmetrical employee communication leads to employee engagement and positive employee communication behaviors. *Journal of Communication Management (London)*, *21*(1), 82–102. DOI: 10.1108/JCOM-04-2016-0026

Kaplan, G. (2006). Interview with Gary Kaplan. *Quality & Safety in Health Care*, *15*(4), 228–231. DOI: 10.1136/qshc.2005.017771 PMID: 16751462

Kaplan, R., & Norton, D. (1992). The balanced scorecard-Measures that drive performance. *Harvard Business Review*, *70*(1), 71–79. PMID: 10119714

Karakus, U. (2017). Performance Management Practices in Self-Managing Teams: An explorative study at the Dutch healthcare organization. *9 th IBA Bachelor Thesis Conference, July 5th, 2017, Enschede, The Netherlands*. https://essay.utwente.nl/72828/1/Karakus_BA_BMS.pdf

Karl, K. A., Peluchette, J. V., & Aghakhani, N. (2022). Virtual work meetings during the COVID-19 pandemic: The good, bad, and ugly. *Small Group Research*, *53*(3), 343–365. DOI: 10.1177/10464964211015286 PMID: 38603094

Kayama, M. (2017). Development of children's understandings of physical disabilities and stigmatization in a Japanese cultural context: Reflections of children in second through sixth grades. *Children and Youth Services Review*, *83*, 190–200. DOI: 10.1016/j.childyouth.2017.10.039

Kensbock, J. M., & Boehm, S. (2016). The role of transformational leadership in the mental health and job performance of employees with disabilities. *International Journal of Human Resource Management*, *27*(14), 1580–1609. DOI: 10.1080/09585192.2015.1079231

Kim, J., & Jung, H. S. (2022). The Effect of employee competency and organizational culture on employees' perceived stress for better workplace. *International Journal of Environmental Research and Public Health*, *19*(8), 4428. DOI: 10.3390/ijerph19084428 PMID: 35457296

Kirkman, B. L., & Rosen, B. (1999). Beyond self-management: Antecedents and consequences of team empowerment. *Academy of Management Journal*, *42*(1), 58–74. DOI: 10.2307/256874

Kirkman, B. L., & Rosen, B. (2002). Powering up teams. *Organizational Dynamics*, *28*(3), 48–66. DOI: 10.1016/S0090-2616(00)88449-1

Kiseleva, A., Kotzinos, D., & De Hert, P. (2022). Transparency of AI in healthcare as a multilayered system of accountabilities: Between legal requirements and technical limitations. *Frontiers in Artificial Intelligence*, *5*, 879603. DOI: 10.3389/frai.2022.879603 PMID: 35707765

Kleynhans, D. J., Heyns, M. M., Stander, M. W., & de Beer, L. T. (2022). Authentic leadership, trust (in the Leader), and flourishing: Does precariousness matter? *Frontiers in Psychology*, *13*, 798759. DOI: 10.3389/fpsyg.2022.798759 PMID: 35432051

Klitmøller, A., Schneider, S. C., & Jonsen, K. (2015). Speaking of global virtual teams: Language differences, social categorization and media choice. *Personnel Review*, *44*(2), 270–285. DOI: 10.1108/PR-11-2013-0205

Knight, D. K., Becan, J. E., & Flynn, P. M. (2013). The impact of staff turnover on workplace demands and coworker relationships. *Counselor (Deerfield Beach, Fla.)*, *14*(3), 20–23. PMID: 27540331

Kohli, A., & Jaworski, B. (1990). Market Orientation: The Construct, Research Propositions, and Managerial Implications. *Journal of Marketing*, *54*(2), 1–18. DOI: 10.1177/002224299005400201

Koruda, E. (2016). More carrot, less stick: Workplace wellness programs & the discriminatory impact of financial and health-based incentives. *BCJL & Soc. Just.*, *36*, 131.

Kossek, E. E., Gettings, P., & Misra, K. (2021). The future of flexibility at work. *Harvard Business Review*.

Kossek, E. E., Pichler, S., Bodner, T., & Hammer, L. B. (2011). Workplace social support and work-family conflict: A meta-analysis clarifying the influence of general and work- family-specific supervisor and organizational support. *Personnel Psychology*, *64*(2), 289–313. DOI: 10.1111/j.1744-6570.2011.01211.x PMID: 21691415

Kotter, J. P. (1996). *Leading change*. Harvard Business School Press.

Krope, S. F., & Lobnikar, B. (2015). Assaults on police officers in Slovenia – The profile of perpetrator and assaulted police officers. *Revija za kiminalistiko in kriminologijo, 4*(4), 300-306. https://www.fvv.um.si/rv/arhiv/2018-2/06_Krope_Ilic_rV_2018-2.pdf

Krope, S. F., & Ilić, V. (2020). Assaults on police officers in Slovenia between 2007 and 2017. *Journal of Criminal Justice and Security*, *2*(2), 237–258.

Krueger, R. A. (1994). *Focus Groups: A practical guide for applied research*. Sage Publications.

Krutka, D. G., Greenhow, C., Kimmons, R., Pangrazio, L., & Trust, T. (2025). The state of educational technology research and practice. In Christakis, D. A., & Hale, L. (Eds.), *Handbook of children and screens: Digital media, development, and well-being from birth through adolescence*. Springer. DOI: 10.1007/978-3-031-69362-5_71

Kuhl, J. (2021). The expectancy-value approach within the theory of social motivation: Elaborations, extensions, critique. Expectations and actions, 125-160.

Kuipers, B. S., & Stoker, J. I. (2009). Development and performance of self-managing work teams: A theoretical and empirical examination. *International Journal of Human Resource Management*, *20*(2), 399–419. DOI: 10.1080/09585190802670797

Kulkarni, M. (2021). Hiding but hoping to be found: Workplace disclosure dilemmas of individuals with hidden disabilities. *Equality, Diversity and Inclusion*. Advance online publication. DOI: 10.1108/EDI-06-2020-0146

Kulkarni, M., & Lengnick-Hall, M. (2011). Socialization of people with disabilities in the workplace. *Human Resource Management*, *50*(4), 521–540. DOI: 10.1002/hrm.20436

Kundu, S. C., & Mor, A. (2017). Workforce diversity and organizational performance: A study of it industry in India. *Employee Relations*, *39*(2), 160–183. DOI: 10.1108/ER-06-2015-0114

Kusnaya, A. A., Lukman, S., & Syafrizal, S. (2022). Analysis of leadership influence, resource allocation, and employee engagement on the effectiveness of strategy implementation. *JBTI: Jurnal Bisnis: Teori Dan Implementasi*, *13*(3), 193–202. DOI: 10.18196/jbti.v13i3.15529

Kyriacou, D. N., Monkkonen, E. H., Peek-Asa, C., Lucke, R. E., Labbett, S., Pearlman, K. S., & Hutson, H. R. (2015). Police deaths in New York and London during the twentieth century. *Injury Prevention : Journal of the International Society for Child and Adolescent Injury Prevention*, *12*(21), 219–224. DOI: 10.1136/ip.2005.010827 PMID: 16887942

Lai, F.-Y., Tang, H.-C., Lu, S.-C., Lee, Y.-C., & Lin, C.-C. (2020). Transformational leadership and job performance: The mediating role of work engagement. *SAGE Open*, *10*(1), 2158244019899085. Advance online publication. DOI: 10.1177/2158244019899085

Laig, R. B. D., & Abocejo, F. T. (2021). Change management process in a mining company: Kotter's 8-step change model. *Journal of Management, Economics, and Industrial Organization*, *5*(3), 31–50. DOI: 10.31039/jomeino.2021.5.3.3

Langfred, C. W. (2004). Too much of a good thing? Negative effects of high trust and individual autonomy in self-managing teams. *Academy of Management Journal*, *47*(3), 385–399. DOI: 10.2307/20159588

Langfred, C. W. (2007). The downside of self-management: A longitudinal study of the effects of conflict on trust, autonomy, and task interdependence in self-managing teams. *Academy of Management Journal*, *50*(4), 885–900. DOI: 10.5465/amj.2007.26279196

Lasrado, F., & Uzbeck, C. (2017). The excellence quest: A study of business excellence award-winning organizations in uae. *Benchmarking*, *24*(3), 716–734. DOI: 10.1108/BIJ-06-2016-0098

Lau, W.-F. K., Ortega, K., & Sharkey, J. (2015). Disability Awareness Training with a Group of Adolescents with Learning Disabilities. *Contemporary School Psychology*, *19*(3), 145–156. DOI: 10.1007/s40688-014-0030-x

Lawrence, P. L. (1969). How to deal with resistance to change. *Harvard Business Review*.

Lazar, J., Feng, J. H., & Hochheiser, H. (2017). *Research methods in human-computer interaction*. Elsevier Science & Technology.

Lebans, M. & Euske, K. (2006). A conceptual and operational delineation of performance. *Business Performance Measurement*, 65-79. DOI: 10.1017/CBO9780511753695.006

Lee, A., Willis, S., & Tian, A. W. (2018). When empowering employees works, and when it doesn't. *Harvard Business Review*.

Lee, Y. (2018). Dynamics of symmetrical communication within organizations: The impacts of channel usage of ceo, managers, and peers. *International Journal of Business Communication*, *59*(1), 3–21. DOI: 10.1177/2329488418803661

Lefever, M., Decuman, S., Perl, F., Braeckman, L., & van de Velde, D. (2018). The efficacy and efficiency of Disability Management in job-retention and job-reintegration. A systematic review. *Work (Reading, Mass.)*, *59*(4), 501–534. DOI: 10.3233/WOR-182709 PMID: 29733052

Leje, M. I., Kasimu, M. A., & Kolawole, A. F. (2019). Impacts of effective communication towards performance of construction organization. *Traektoriâ Nauki*, *5*(8), 3001–3008. DOI: 10.22178/pos.49-4

Lengyel, D. M., Newman, J. S., & Yurovich, D. (2023). Examining Risk Management Failures: The Case of the Boeing 737 MAX Program. *Enterprise Risk Management*, *8*(1), 1–1. DOI: 10.5296/erm.v8i1.20916

Leonardi, P. M., Huysman, M., & Steinfield, C. (2013). Enterprise social media: Definition, history, and prospects for the study of social technologies in organizations. *Journal of Computer-Mediated Communication*, *19*(1), 1–19. DOI: 10.1111/jcc4.12029

Lewis, A. (2022). *Good leadership? It all starts with trust.* Harvard Business Publishing. Retrieved from https://www.harvardbusiness.org/good-leadership-it-all-starts-with-trust/

Lewis, R. B., & Lewis, R. B. (1998). Assistive Technology and Learning Disabilities: Today's Realities and Tomorrow's Promises. *Journal of Learning Disabilities*, *31*(1), 16–26. DOI: 10.1177/002221949803100103 PMID: 9455174

Li, N. (2023). Ethical considerations in artificial intelligence: A comprehensive discussion from the perspective of computer vision. *SHS Web of Conferences*, *179*(1), 04024. Advance online publication. DOI: 10.1051/shsconf/202317904024

Lin, C. P., Xian, J., Li, B., & Huang, H. (2020). Transformational leadership and employees' thriving at work: The mediating roles of challenge-hindrance stressors. *Frontiers in Psychology*, *11*, 1400. DOI: 10.3389/fpsyg.2020.01400 PMID: 32655458

Lindsay Unified School District. (2022). *Annual report on student outcomes and equity initiatives*. https://www.ed.gov/sites/ed/files/2023/01/S215J220211_LindsayUnified_Narrative_508_Redacted.pdf

Lindsay, S., Cagliostro, E., Albarico, M., Mortaji, N., & Karon, L. (2018). A Systematic Review of the Benefits of Hiring People with Disabilities. *Journal of Occupational Rehabilitation*, *28*(4), 634–655. DOI: 10.1007/s10926-018-9756-z PMID: 29392591

Lindsay, S., Cagliostro, E., & Carafa, G. (2018). A systematic review of workplace disclosure and accommodation requests among youth and young adults with disabilities. *Disability and Rehabilitation*, *40*(25), 2971–2986. DOI: 10.1080/09638288.2017.1363824 PMID: 28797182

Lindsay, S., Cagliostro, E., Leck, J., Shen, W., & Stinson, J. (2019a). Disability disclosure and workplace accommodations among youth with disabilities. *Disability and Rehabilitation*, *41*(16), 1914–1924. DOI: 10.1080/09638288.2018.1451926 PMID: 29558221

Lindsay, S., Cagliostro, E., Leck, J., Shen, W., & Stinson, J. (2019b). Employers' perspectives of including young people with disabilities in the workforce, disability disclosure and providing accommodations. *Journal of Vocational Rehabilitation, 50*(2), 141–156. Advance online publication. DOI: 10.3233/JVR-180996

Ling, B., Liu, Z., Chen, D., & Sun, L. (2021). Team flexibility in organizational change context: Antecedents and consequences. *Psychology Research and Behavior Management, 14*, 1805–1821. DOI: 10.2147/PRBM.S332222 PMID: 34754250

Lipton, Z. C. (2016). The mythos of model interpretability. *Communications of the ACM, 61*(10), 36–43. DOI: 10.1145/3233231

Lis, P., & Mendel, J. (2019). Cyberattacks on critical infrastructure: An economic perspective. *Economics and Business Review, 5*(2), 24–47. DOI: 10.18559/ebr.2019.2.2

Lloyd, R., & Mertens, D. (2018). Expecting more out of expectancy theory: History urges inclusion of the social context. *International Management Review, 14*(1), 28–43.

Luckin, R., Holmes, W., Griffiths, M., & Forcier, L. B. (2016). *Intelligence unleashed: An argument for AI in education*. Pearson Education. https://static.googleusercontent.com/media/edu.google.com/en//pdfs/Intelligence-Unleashed-Publication.pdf

Lunn, S., Zahedi, L., Ross, M., & Ohland, M. (2021). Exploration of intersectionality and computer science demographics: Understanding the historical context of shifts in participation. [TOCE]. *ACM Transactions on Computing Education, 21*(2), 1–30. DOI: 10.1145/3445985

Lusthaus, C., & Adrien, M. (1998). Organizational assessment: A review of experience. *Universalia, 31*, 1–16.

Lusthaus, C., Adrien, M., Anderson, G., Carden, F., & Plinio, G. (2002). *Evaluación Organizacional. Marco para mejorar el Desempeño*. Banco Interamericano de Desarrollo y Centro Internacional de Investigaciones para el Desarrollo.

Luu, T. (2019). Relationship between benevolent leadership and the well-being of employees with disabilities. *Journal of Business Research, 99*, 282–294. Advance online publication. DOI: 10.1016/j.jbusres.2019.03.004

Lyon, V. (2020). *Exploring Strategies for Recruiting and Retaining Diverse Cybersecurity Professionals* (Order No. 27738274). Available from ProQuest Dissertations & Theses Global; ProQuest One Academic. (2378121182). https://www.proquest.com/dissertations-theses/exploring-strategies-recruiting-retaining-diverse/docview/2378121182/se-2

Macey, W. H., & Schneider, B. (2008). The meaning of employee engagement. *Industrial and Organizational Psychology: Perspectives on Science and Practice, 1*(1), 3–30. DOI: 10.1111/j.1754-9434.2007.0002.x

Madaus, J. W., Gerber, P., & Price, L. A. (2008). Adults with Learning Disabilities in the Workforce: Lessons for Secondary Transition Programs. *Learning Disabilities Research & Practice, 23*(3), 148–153. DOI: 10.1111/j.1540-5826.2008.00272.x

Madella, A., Bayle, E., & Tome, J. (2005). The organizational performance of national swimming federations in Mediterranean countries: A comparative approach. *European Journal of Sport Science, 5*(4), 207–220. DOI: 10.1080/17461390500344644

Madyaningrum, M., Sonn, C., & Fisher, A. T. (2021). Disability organizations as empowering settings: Challenging stigmatization, promoting emancipation. *American Journal of Community Psychology*. Advance online publication. DOI: 10.1002/ajcp.12560 PMID: 34633668

Mair, J., & Seelos, C. (2021). Organizations, social problems, and system change: Invigorating the third mandate of organizational research. *Organization Theory, 2*(4), 26317877211054858. Advance online publication. DOI: 10.1177/26317877211054858

Maness, D. L., & Khan, M. (2015). Disability evaluations: More than completing a form. *American Family Physician, 91*(2), 102–109. PMID: 25591211

Mangundjaya, W. L. (2014). The role of communication, trust and justice in commitment to change. *International Conference on Business, Management and Corporate Social Responsibility* (ICBMCSR'14) Feb. 14-15, 2014 Batam (Indon. DOI: 10.15242/ICEHM.ED0214043

Mansilla, V. B., & Jackson, A. (2011). *Educating for global competence: Preparing our youth to engage the world*. Asia Society. https://doi.org/DOI: 10.13140/2.1.3845.1529

Manuel, B. A., & Karloff, D. B. (2020). Recruit and retain a diverse workforce. *Nature Reviews. Chemistry, 4*(9), 435–437. DOI: 10.1038/s41570-020-0214-z PMID: 37127964

Manyika, J., Silberg, J., & Presten, B. (2019). AI and machine learning: What do we do about the biases in AI? *Harvard Business Review*.

Manz, C. C., & Neck, C. P. (2010). *Mastering self-leadership: Empowering yourself for personal excellence*. Pearson.

Manz, C. C., & Sims, H. P. (1987). Leading workers to lead themselves: The external leadership of self-managing work teams. *Administrative Science Quarterly, 32*(1),
Matus, D. (2018). Implementación operativa de un modelo de intervención para la mejora del desempeño organizacional. *SIGNOS*, *10*(2), 71–86. DOI: 10.15332/s2145-1389.2018.0002.04

Marchalina, L., & Ahmad, H. (2017). The effect of internal communication on employees' commitment to change in malaysian large companies. *Business Management and Strategy*, *8*(1), 1. DOI: 10.5296/bms.v8i1.10629

Martela, F., & Pessi, A. B. (2018). Significant work is about self-realization and broader purpose: Defining the key dimensions of meaningful work. *Frontiers in Psychology*, *9*, 363. DOI: 10.3389/fpsyg.2018.00363 PMID: 29632502

Marty, O. L. (2022). Os fatores de influência no risco do policial militar ser vítima de crimes em decorrência da sua função [The factors influencing the risk of military police officers becoming victims of crime as a result of their work]. *Brazilian Applied Science Review*, *6*(2), 1148–1170. DOI: 10.34115/basrv6n3-026

Maslach, C., Schaufeli, W. B., & Leiter, M. P. (2001). Job burnout. *Annual Review of Psychology*, *52*(1), 397–422. DOI: 10.1146/annurev.psych.52.1.397 PMID: 11148311

Maximo, N., Stander, M. W., & Coxen, L. (2019). Authentic leadership and work engagement: The indirect effects of psychological safety and trust in supervisors. *SA Journal of Industrial Psychology*, *45*(1), 1–11. DOI: 10.4102/sajip.v45i0.1612

Maxwell, J. A. (1996). *Qualitative research design: An interactive approach*. Sage Publications, Inc.

May, D. R., Gilson, R. L., & Harter, L. M. (2004). The psychological conditions of meaningfulness, safety and availability and the engagement of the human spirit at work. *Journal of Occupational and Organizational Psychology*, *77*(1), 11–37. DOI: 10.1348/096317904322915892

Maylett, T., & Wride, M. (2017). *The employee experience: How to attract talent, retain top performers, and drive results*. John Wiley & Sons.

Mazzetti, G., & Schaufeli, W. B. (2022). The impact of engaging leadership on employee engagement and team effectiveness: A longitudinal, multi-level study on the mediating role of personal- and team resources. *PLoS One*, *17*(6), e0269433. DOI: 10.1371/journal.pone.0269433 PMID: 35767536

McClelland, D. C. (1967). *Achieving society*. Simon and Schuster.

McDaniel, S., & Kolari, J. (1987). Marketing Strategy Implications of the Miles y Snow

McDowell, C., & Fossey, E. (2015). Workplace Accommodations for People with Mental Illness: A Scoping Review. *Journal of Occupational Rehabilitation*, *25*(1), 197–206. DOI: 10.1007/s10926-014-9512-y PMID: 24841728

McGee, K. (2018). The influence of gender, and race/ethnicity on advancement in information technology (IT). *Information and Organization*, *28*(1), 1–36. DOI: 10.1016/j.infoandorg.2017.12.001

McGrath, A., & Jonker, A. (2024). *What is AI interpretability.* https://www.ibm.com/think/topics/interpretability

McSweeney, K. (2018). Motivating cybersecurity compliance in critical infrastructure industries: A grounded theory study (Doctoral dissertation, Capella University).

Medina, F., & Gamero, N. (2017). *The Socialization Process: Helping Organizations Integrate People with Disabilities into the Workplace*. 139–152. https://doi.org/DOI: 10.1007/978-3-319-66393-7_9

Mehrabi, N., Morstatter, F., Saxena, N., Lerman, K., & Galstyan, A. (2021). A survey on bias and fairness in machine learning. *ACM Computing Surveys*, *54*(6), 1–35. DOI: 10.1145/3457607

Meinen, A. R., Girardi, H. U., Martins, J. G. S., Lima, P. L., & Santos, Z. C. (2020). As principais mudanças percebidas pelos consumidores, ocasionadas pela sensação de medo e insegurança desencadeados pela violência urbana [The main changes perceived by consumers, occasioned by the feel of fear and insecurity unleashed by urban violence]. *Faculdade Sant'Ana em Revista, 4*(1), 88-100. https://www.iessa.edu.br/revista/index.php/fsr/article/view/1735

Mele, C., Pels, J., & Polese, F. (2010). A brief review of systems theories and their managerial applications. *Service Science*, *2*(1-2), 126–135. DOI: 10.1287/serv.2.1_2.126

Mellifont, D., Smith-Merry, J., & Scanlan, J. (2016). Pitching a Yerkes–Dodson curve ball?: A study exploring enhanced workplace performance for individuals with anxiety disorders. *Journal of Workplace Behavioral Health*, *31*(2), 71–86. DOI: 10.1080/15555240.2015.1119654

Mello, C. M. A., & Nummer, F. V. (2014). Policial militar: uma profissão de risco [Military police: a risky profession]. *Trabalho da 29ª Reunião Brasileira de Antropologia*. https://www.29rba.abant.org.br/resources/anais/1/1401917965_ARQUIVO_ArtigoRBA2014CesareFernanda.pdf

Melo, P., Martins, A., & Pereira, M. (2020). The relationship between leadership and accountability: A review and synthesis of the research. *Leadership and Organization Development Journal*, *23*, 1–10.

Menezes, L. D., & Kelliher, C. (2017). Flexible Working, Individual Performance, and Employee Attitudes: Comparing Formal and Informal Arrangements. *Human Resource Management*, *56*(6), 1051–1070. DOI: 10.1002/hrm.21822

Menon, A., & Mohanty, B. (2012). Organizational Performance: A Multidimensional Perspective. In Singh, S. (Ed.), *Global Competition and Competitivenes s of Indian Corporates*. Macmillan.

Messamore, A. (2023). The effect of community organizing on landlords' use of eviction filing: Evidence from u.s. cities. *Social Problems*, *70*(3), 809–830. DOI: 10.1093/socpro/spac061

Michaels, D., & Barab, J. (2020). The Occupational Safety and Health Administration at 50: Protecting workers in a changing economy. *American Journal of Public Health*, *110*(5), 631–635. DOI: 10.2105/AJPH.2020.305597 PMID: 32191515

Miles, R., & Snow, Ch. (1978). *Organizational Strategy, Structure and Processes*. McGraw-Hill.

Miller, J. (2023). *The power of diversity and inclusion: Driving innovation and success*. Retrieved from https://www.forbes.com/councils/forbesbusinesscouncil/2023/08/16/the-power-of-diversity-and-inclusion-driving-innovation-and-success/

Milli, S., Lieder, F., & Griffiths, T. L. (2021). A rational reinterpretation of dual-process theories. *Cognition*, *217*, 104881. DOI: 10.1016/j.cognition.2021.104881 PMID: 34536658

Minayo, M. C. S., & Adorno, S. (2013). Risco e (in) segurança na missão policial [Risk and (in) security in the police mission]. *Ciencia & Saude Coletiva*, *18*(13), 585–593. DOI: 10.1590/S1413-81232013000300002 PMID: 23546183

Minayo, M. C. S., Souza, E. R., & Constantino, P. (2007). Riscos percebidos e vitimização de policiais civis e militares na (in)segurança pública [Perceived risks and victimization of military and civil police in the public (in)security domain]. *Cadernos de Saude Publica*, *23*(11), 2767–2779. DOI: 10.1590/S0102-311X2007001100024 PMID: 17952268

Mishra, D., Dashora, J., & Dubey, D. (2023). The impact of transformational leadership on employee engagement and performance: A comprehensive review. *3*, 2581–9429.

Mitchell, M. S., Reynolds, S. J., & Treviño, L. K. (2020). The study of behavioral ethics within organizations: A special issue introduction. *Personnel Psychology*, *73*(1), 5–17. DOI: 10.1111/peps.12381

Mohr, L. (1982). *Explaining organizational behavior*. Jossey-Bass.

Monteiro, V. F., Silva, S. S. C., Ramos, E. M. L. S., & Nascimento, R. G. (2020). Caracterização dos policiais feridos por arma de fogo [Characterization of policemen injured by firearms]. *Research. Social Development*, *9*(9), e248997218. DOI: 10.33448/rsd-v9i9.7218

Morash-Macneil, V., Johnson, F., & Ryan, J. B. (2018). A Systematic Review of Assistive Technology for Individuals With Intellectual Disability in the Workplace. *Journal of Special Education Technology*, *33*(1), 15–26. DOI: 10.1177/0162643417729166

Morgan, I., & Amerikaner, A. (2018). Funding Gaps 2018: An Analysis of School Funding Equity Across the U.S. Education Trust. Retrieved from https://edtrust.org/resource/funding-gaps-2018/

Morgan, S. (2016a, April, 7). African Americans Underrepresented In The Cybersecurity Field. Forbes. https://www.forbes.com/sites/stevemorgan/2016/04/07/african-americans-underrepresented-in-the-cybersecurity-field/?sh=58ea932115c9

Morgan, S. (2016b, March 28). Calling All Women: The Cybersecurity Field Needs You And There's A Million Jobs Waiting. Forbes. https://www.forbes.com/sites/stevemorgan/2016/03/28/calling-all-women-the-cybersecurity-field-. needs-you/#2470db2a381c

Morikawa, M., Martela, F., & Hakanen, J. (2024). Are employee self-management and organizational self-management related to work engagement or burnout? Evidence from two studies. BRQ. *Business Research Quarterly*. Advance online publication. DOI: 10.1177/23409444241307734

Moskovich, Y. (2023). Communal organizational culture as a source of business-success sustainability in Kibbutz industry—Two case studies. *Sustainability (Basel)*, *15*(13), 10654. DOI: 10.3390/su151310654

Mouboua, P. D., Atobatele, F. A., & Akintayo, O. T.Patricia Diane MouboauaFadeke Adeola AtobateleOlateju Temitope Akintayo. (2024). Cross-cultural competence in global hrd: Strategies for developing an inclusive and diverse workforce. *International Journal of Science and Research Archive*, *12*(1), 103–113. DOI: 10.30574/ijsra.2024.12.1.0765

Mountrouidou, X., Vosen, D., Kari, C., Azhar, M. Q., Bhatia, S., Gagne, G., Maguire, J., Tudor, L., & Yuen, T. T. (2019). Securing the human: A Review of Literature on Broadening Diversity in Cybersecurity Education. *Proceedings of the Working Group Reports on Innovation and Technology in Computer Science Education*, 157–176. DOI: 10.1145/3344429.3372507

Mucci, T., & Stryker, C. (2024). *What is AI governance*. https://www.ibm.com/topics/ai- governance

Muchtar, Y. C., & Maas, L. T. (2022). Employee performance analysis: the influence of local wisdom-based communication strategy and employee engagement on employee performance. *Proceedings of the 3rd International Conference on Business and Management of Technology* (ICONBMT 2021). DOI: 10.2991/aebmr.k.211226.030

Mujtaba, B. G., Vardarlıer, P., & Gündoğmuş, E. (2023). Digital Diversity and Inclusion: Agenda for Data Privacy and Protection. In *Multidimensional and Strategic Outlook in Digital Business Transformation: Human Resource and Management Recommendations for Performance Improvement* (pp. 39–52). Springer International Publishing. DOI: 10.1007/978-3-031-23432-3_4

Munyoka, W. (2022). Inclusive Digital Innovation in South Africa: Perspectives from disadvantaged and marginalized communities. *Sustainability (Basel)*, *14*(9), 5372. DOI: 10.3390/su14095372

Muster, C. (2023). Disability Justice and the Americans with Disabilities Act: An Opportunity for Social Work. *Social Work*, *68*(4), 331–339. Advance online publication. DOI: 10.1093/sw/swad034 PMID: 37537135

Mutamba, C. (2017). The inter-relationship of organizational learning, learning organizations, virtual technology, and virtual communities of practice. *New Horizons in Adult Education and Human Resource Development*, *29*(3), 4–14. DOI: 10.1002/nha3.20186

Muthuveloo, R., Shanmugam, N., & Teoh, A. P. (2017). The impact of tacit knowledge management on organizational performance: Evidence from Malaysia. *Asia Pacific Management Review*, *22*(4), 192–201. DOI: 10.1016/j.apmrv.2017.07.010

Narver, J., & Slater, F. (1990). The Effect of a Market Orientation on Business Profitability. *Journal of Marketing*, *54*(4), 20–35. DOI: 10.1177/002224299005400403

National Association for Woman & Information Technology. (2017). https://www.ncwit.org/

National Center for Education Statistics. (2024). *Barriers to technology integration in underserved schools*. U.S. Department of Education.

National Education Association. (2023). *Barriers to competency-based education: A national survey of teachers*. NEA Publications.

Ndofor, H., & Priem, R. (2011). Inmigrant entrepreneurs the ethnic enclave strategy and venture performance. *Journal of Management*, *37*(3), 790–818. DOI: 10.1177/0149206309345020

Neely, A. (1999). The performance measurement revolution: Why now and what next? *International Journal of Operations & Production Management*, *19*(2), 205–228. DOI: 10.1108/01443579910247437

Neidlinger, S. M., Felfe, J., & Schübbe, K. (2022). Should I stay or should I go (to the office)? - Effects of working from home, autonomy, and core self-evaluations on leader health and work-life balance. *International Journal of Environmental Research and Public Health*, *20*(1), 6. DOI: 10.3390/ijerph20010006 PMID: 36612327

Nelson-Peterson, D. L., & Leppa, C. J. (2007). Creating an environment for caring using lean principles of the Virginia Mason production system. *The Journal of Nursing Administration*, *37*(6), 287–294. DOI: 10.1097/01.NNA.0000277717.34134.a9 PMID: 17563521

Nevala, N., Pehkonen, I., Koskela, I., Ruusuvuori, J., & Anttila, H. (2015). Workplace Accommodation Among Persons with Disabilities: A Systematic Review of Its Effectiveness and Barriers or Facilitators. *Journal of Occupational Rehabilitation*, *25*(2), 432–448. DOI: 10.1007/s10926-014-9548-z

New York Stock Exchange. (2024, September 19). Boeing Co. *Google Finance*. https://www.google.com/finance/quote/BA:NYSE?sa=X&ved=2ahUKEwiskqfKnc-IAxX5le4BHT85NOQQ3ecFegQINxAZ

Newlin, K., Dyess, S., Allard, E., Chase, S. K., & Melkus, G. D.Gail D'Eramo Melkus. (2011). A methodological review of faith-based health promotion literature: Advancing the science to expand delivery of diabetes education to black americans. *Journal of Religion and Health*, *51*(4), 1075–1097. DOI: 10.1007/s10943-011-9481-9 PMID: 21487842

Nezhad, F. R., Kojouri, K. K., & Nodehi, S. N. K. (2019). Effect of altruism on organizational learning capability with the mediating role of communication conflict and organizational trust (case study: staff of red crescent provincial branches of gilan and mazandaran). *Journal of Rescue and Relief*, 67–76. DOI: 10.32592/jorar.2020.12.1.7

Nguyen, C. H. (2011). Factors causing Enron's collapse: An investigation into corporate governance and company culture. *Corporate Ownership & Control Journal, 8*(3), 22-36. Available at *SSRN*: https://ssrn.com/abstract=1785199

Nienaber, H., & Martins, N. (2020). Exploratory study. *The TQM Journal, 32*(3), 475–495. DOI: 10.1108/TQM-05-2019-0151

Nishii, L. H. (2013). The benefits of climate for inclusion for gender-diverse groups. *Academy of Management Journal, 56*(6), 1754–1774. DOI: 10.5465/amj.2009.0823

Nkrumah, T., & Scott, K. A. (2022). Mentoring in STEM higher education: A synthesis of the literature to (re)present the excluded women of color. *International Journal of STEM Education, 9*(1), 50. Advance online publication. DOI: 10.1186/s40594-022-00367-7 PMID: 35919282

Nurcholis, G., & Budi, W. (2020). The impact of organizational culture and perceived organizational support on employee engagement. *Proceedings of the 5th ASEAN Conference on Psychology, Counselling, and Humanities* (ACPCH 2019). DOI: 10.2991/assehr.k.200120.050

O'Day, B. L., & Corcoran, P. J. (1994). Assistive technology: Problems and policy alternatives. *Archives of Physical Medicine and Rehabilitation, 75*(10), 1165–1169. DOI: 10.1016/0003-9993(94)90096-5 PMID: 7944926

Obermeyer, Z., Powers, B., Kohane, I., & Levy, M. (2019). Dissecting racial bias in an algorithm used to manage the health of populations. *Science, 366*(6464), 447–453. DOI: 10.1126/science.aax2342 PMID: 31649194

Odhus, C. O., Kapanga, R. R., & Oele, E. (2024). Barriers to and enablers of quality improvement in primary health care in low- and middle-income countries: A systematic review. *PLOS Global Public Health, 4*(1), e0002756. DOI: 10.1371/journal.pgph.0002756 PMID: 38236832

Odom, L., & Green, M. T. (2003). Law and the ethics of transformational leadership. *Leadership and Organization Development Journal, 24*(2), 62–69. DOI: 10.1108/01437730310463251

OECD. (2023). *Policies for the digital transformation of school education: Proposed analytical framework and methodology*. OECD Publishing. https://one.oecd.org/document/EDU/EDPC/SR(2023)2/en/pdf

Oermann, M. (2016). Supporting Students With Disabilities. *Nurse Educator, 41 1*, 28. https://doi.org/DOI: 10.1097/NNE.0000000000000231

Okatta, C. G., Ajayi, F. A., & Olawale, O. (2024). Enhancing organizational performance through diversity and inclusion initiatives: A meta-analysis. *International Journal of Applied Research in Social Sciences*, *6*(4), 734–758. DOI: 10.51594/ijarss.v6i4.1065

Oliveira, A., & Leone, R. (2008). Investigação sobre medição de desempenho empresarial. *Custos e Agronegócio*, *4*, 117–137.

Oliveira, G. M., Araújo, T. M., & Carvalho, F. M. (2014). Insegurança e violência no trabalho dos delegados de polícia civil de Salvador, Brasil [Insecurity and violence in the work of civil police delegates in Salvador, Brazil]. *Revista Psicologia. Diversidade e Saúde*, *2*(1), 14–23. DOI: 10.17267/2317-3394rpds.v3i1.372

Olmedo, I., & Martínez, I. (2015). Human capital and creation of reputation and financial performance. *Electronic Journal of Knowledge Management*, *13*(3), 209–218.

Omondi, C. (2015). Measuring organization performance. From Balanced Scorecard to Balanced Esg Framework. *International Journal of Economics. Commerce and Management*, *3*(11), 715–725.

Opeke, R. O., & Adelowo, O. T. (2020). Knowledge Management Strategies and Profitability of Selected Deposit Money Banks in Nigeria. Information Impact. *Journal of Information and Knowledge Management*, *11*(1), 48–59.

Orok, A. B. M. (2019). Conceptualizing Self-Awareness as a Correlate for Career Development of Students with Disabilities. *Journal of Education and Training Studies*, *5*, 33–40.

Orpen, C. (2022). Effect of organizational commitment on the relationship between communication quality and managerial attitudes. *Communicare*, *13*(1), 58–63. DOI: 10.36615/jcsa.v13i1.1973

Osborne, S., & Hammoud, M. S. (2017). Effective employee engagement in the workplace. *International Journal of Applied Management and Technology*, *16*(1), 50–67. DOI: 10.5590/IJAMT.2017.16.1.04

Ouédraogo, N., & Ouakouak, M. L. (2018). Impacts of personal trust, communication, and affective commitment on change success. *Journal of Organizational Change Management*, *31*(3), 676–696. DOI: 10.1108/JOCM-09-2016-0175

Overcoming Barriers to Advancement - CIA Diversity in Leadership Study. (2015). Retrieved from https://www.cia.gov/library/reports/dls-report.pd

Page, K. M., & Vella-Brodrick, D. A. (2009). The 'what', 'why' and 'how' of employee well- being: A new model. *Social Indicators Research, 90*(3), 441–458. DOI: 10.1007/s11205-008-9270-3

Page, M. J., McKenzie, J. E., Bossuyt, P. M., Boutron, I., Hoffmann, T. C., Mulrow, C. D., Shamseer, L., Tetzlaff, J. M., Akl, E. A., Brennan, S. E., Chou, R., Glanville, J., Grimshaw, J. M., Hróbjartsson, A., Lalu, M. M., Li, T., Loder, E. W., Mayo-Wilson, E., McDonald, S., & Moher, D. (2021). The PRISMA 2020 statement: An updated guideline for reporting systematic reviews. *BMJ (Clinical Research Ed.), 372*(71), n71. Advance online publication. DOI: 10.1136/bmj.n71 PMID: 33782057

Page, M. J., Moher, D., Bossuyt, P. M., Boutron, I., Hoffmann, T. C., Mulrow, C. D., Shamseer, L., Tetzlaff, J. M., Akl, E. A., Brennan, S. E., Chou, R., Glanville, J., Grimshaw, J. M., Hróbjartsson, A., Mayo-Wilson, E., McDonald, S., & McKenzie, J. E. (2021). Prisma 2020 explanation and elaboration: Updated guidance and exemplars for reporting systematic reviews. *BMJ (Clinical Research Ed.), 372*(160), n160. Advance online publication. DOI: 10.1136/bmj.n160 PMID: 33781993

Palacios, J. (2008). *Medición del impacto y la rentabilidad de la formación. Cómo llegar al roi de la formación*. Díaz de Santos.

Palomo, M. T. (2013). *Auto gestión y motivación de equipos de trabajo*. Esic Editorial.

Pandya, J. D. (2024). Intrinsic & Extrinsic motivation & its impact on organizational performance at Rajkot city: A review. *Journal of Management Research and Analysis, 11*(1), 46–53. DOI: 10.18231/j.jmra.2024.009

Pane, J. F., Steiner, E. D., Baird, M. D., & Hamilton, L. S. (2015). *Continued progress: Promising evidence on personalized learning*. RAND Corporation. https://www.rand.org/pubs/research_reports/RR1365.html

Pang, M., & Pavlou, P. A. (2019). On information technology and the safety of police officers. *Decision Support Systems, 127*, 113143. DOI: 10.1016/j.dss.2019.113143

Pardo, C. E. y Diaz, A. (2014). Desarrollo del talento humano como factor clave para el desarrollo organizacional, una visión desde los líderes de gestión humana en empresas de Bogotá D.C. *Suma de Negocios, 5*(11), 39-48.

Park, S., Park, S., & Shryack, J. (2022). Measures of climate for inclusion and diversity: Review and summary. *Human Resource Development Quarterly, 34*(4), 463–480. DOI: 10.1002/hrdq.21493

Parsons, M., & Urbanski, S. (2012). Recognizing dysfunctional communications a means of improving organizational practice. *Online Journal of Communication and Media Technologies, 2*(4). Advance online publication. DOI: 10.29333/ojcmt/2404

Patrick, S., Kennedy, K., & Powell, A. (2013). *Mean what you say: Defining and integrating personalized, blended, and competency education.* International Association for K-12 Online Learning (iNACOL). https://files.eric.ed.gov/fulltext/ED561301.pdf

Patrick, H. A., & Kumar, V. R. (2012). Managing workplace diversity: Issues and challenges. *SAGE Open, 2*(2), 2158244012444615. Advance online publication. DOI: 10.1177/2158244012444615

Pearce, C. L., & Conger, J. A. (2003). *Shared leadership: Reframing the hows and whys of leadership.* Sage Publications. DOI: 10.4135/9781452229539

Pekrun, R. (2019). Expectancy–value theory of anxiety: Overview and implications. *Anxiety,* •••, 23–42.

Percy, S. (2001). CHALLENGES AND DILEMMAS IN IMPLEMENTING THE AMERICANS WITH DISABILITIES ACT: LESSONS FROM THE FIRST DECADE. *Policy Studies Journal: the Journal of the Policy Studies Organization, 29*(4), 633–640. DOI: 10.1111/j.1541-0072.2001.tb02115.x

Pérez, J., & Cortés, J. A. (2009). Medición y validación del desempeño organizacional como resultado de acciones de aprendizaje. *Revista Ciencias Estratégicas, 17*(22), 251–271.

Peters, M. D., Godfrey, C. M., Khalil, H., McInerney, P., Parker, D., & Soares, C. B. (2015). Guidance for conducting systematic scoping reviews. *International Journal of Evidence-Based Healthcare, 13*(3), 141–146. DOI: 10.1097/XEB.0000000000000050 PMID: 26134548

Pew Research Center. (2021). *Internet access and the digital divide: Challenges for low-income households.* Pew Research Center.

Piao, M. (2010). Thriving in the new: Implication of exploration on organizational longevity. *Journal of Management, 36*(6), 1529–1554. DOI: 10.1177/0149206310378367

Piedade, S. D. R. (2021). The influence of transformational leadership on employee performance through work climate and organizational commitment. *Timor Leste Journal of Business and Management, 3,* 32–45. DOI: 10.51703/bm.v3i1.36

Pincus, J. D. (2022). Employee engagement as human motivation: Implications for theory, methods, and practice. *Integrative Psychological & Behavioral Science, 57*(4), 1223–1255. DOI: 10.1007/s12124-022-09737-w PMID: 36577907

Pisano, G. P. (2015). You need an innovation strategy. *Harvard Business Review.*

Plsek, P. E. (2013). *Accelerating health care transformation with lean and innovation: The Virginia Mason experience* (1st ed.). Productivity Press., DOI: 10.1201/b15817

Poster, W. R. (2018). Cybersecurity needs women. *Nature*, *555*(7698), 577–580. DOI: 10.1038/d41586-018-03327-w

Prado-Román, C., Diéz-Martín, F., & Blanco-González, A. (2020). The effect of communication on the legitimacy and performance of organizations. *Revista Brasileira de Gestão de Negócios*, *22*(3). Advance online publication. DOI: 10.7819/rbgn.v22i3.4071

Prager, F., Rhoads, M., & Martínez, J. N. (2022). The COVID-19 economic shutdown and the future of flexible workplace practices in the south bay region of Los Angeles County. *Transport Policy*, *125*, 241–255. DOI: 10.1016/j.tranpol.2022.06.004 PMID: 35720050

Prasad, A., DeRosa, D., & Beyerlein, M. (2017). Dispersion beyond miles: Configuration and performance in virtual teams. *Team Performance Management*, *23*(3), 186–204. DOI: 10.1108/TPM-06-2016-0026

Priyambudi, A., & Basri, M. H. (2023). Optimizing Overall Equipment Effectiveness Using Kepner-Tregoe Decision Analysis: A Case Study of Semi-lean Pumps in Ammonia Plant 2 of PT Pupuk Kalimantan Timur. *European Journal of Business and Management Research*, *8*(6), 49–58. DOI: 10.24018/ejbmr.2023.8.6.2190

Prussia, G. E., Anderson, J. S., & Manz, C. C. (1998). Self-leadership and performance outcomes: The mediating influence of self-efficacy. *Journal of Organizational Behavior*, *19*(5), 523–538. DOI: 10.1002/(SICI)1099-1379(199809)19:5<523::AID-JOB860>3.0.CO;2-I

Pulido-Martos, M., Gartzia, L., Augusto-Landa, J., & Lopez-Zafra, E. (2024). Transformational leadership and emotional intelligence: Allies in the development of organizational affective commitment from a multilevel perspective and time-lagged data. *Review of Managerial Science*, *18*(8), 2229–2253. DOI: 10.1007/s11846-023-00684-3

Qasim, S., Usman, M., Ghani, U., & Khan, K. (2022). Inclusive leadership and employees' helping behaviors: Role of psychological factors. *Frontiers in Psychology*, *13*, 888094. DOI: 10.3389/fpsyg.2022.888094 PMID: 35874383

Rabe-Hemp, C. E., & Schuck, A. M. (2007). Violence against police officers: Are female officers at greater risk? *Police Quarterly*, *10*(4), 411–428. DOI: 10.1177/1098611107304326

Radford, A., Wu, J., Child, R., Luan, D., Amodei, D., & Sutskever, I. (2019). *Language models are unsupervised multitask learners.* Retrieved from https://cdn .openai.com/better- language-models/language_models_are_unsupervised_multi-task_learners.pdf

Radu, C. (2023). *Fostering a positive workplace culture: Impacts on performance and agility.* IntechOpen., DOI: 10.5772/intechopen.1003259

Rafeeque, A. (2015). A STUDY ON LEARNING DISABILITIES AMONG STUDENTS IN WAYANAD DISTRICT OF KERALA STATE. *Sai Om Journal of Commerce & Management: A Peer Reviewed. National Journal, 2*, 6–16.

Raigoso-Mayorga, J. A. (2016). Factores de riesgo psicossocial en oficiales de la policía [Psychosocial risk factors among police officers]. *Revista de Educación y Desarrollo*, 36, 33-37. https://www.cucs.udg.mx/revistas/edu_desarrollo/anteriores/36/36_Raigoso.pdf

Rajan, R., & Ganesan, R. (2017). A critical analysis of John P. Kotter's change management framework. *Asian Journal of Research in Business Economics and Management, 7*(7), 181–203. DOI: 10.5958/2249-7307.2017.00106.2

Raji, I. D., & Buolamwini, J. (2019). Actionable auditing: Investigating the impact of publicly naming biased performance results of commercial AI products. *The 2019 AAAI/ACM Conference.* DOI:DOI: 10.1145/3306618.3314244

Raji, I. D., Smart, A., White, R. N., Mitchell, M., Gebru, T., Hutchinson, B., Smith-Loud, J., Theron, D., & Barnes, P. (2020). Closing the AI accountability gap: Defining an end-to- end framework for internal algorithmic auditing. *Proceedings of the 2020 Conference on Fairness, Accountability, and Transparency*, pp. 33–44. DOI: 10.1145/3351095.3372873

Ramos-Maçães, M. A., & Román-Portas, M. (2022). The effects of organizational communication, leadership, and employee commitment in organizational change in the hospitality sector.

Rao, M., Sonpar, S., Sen, A., Seshadri, S. P., Agarwal, H., & Padalia, D. (2018). A study on the prevalence and nature of ragging practices in selected educational institutions in India. *Contemporary Education Dialogue, 15*(2), 187–202. DOI: 10.1177/0973184918781213

Rasheed, M. A., Hookmani, A. A., Waleed, S., Fatima, H., Siddiqui, S., Khurram, M., & Hasan, B. (2021). Implementation and evaluation of a social media-based communication strategy to enhance employee engagement: Experiences from a children's hospital, pakistan. *Frontiers in Public Health, 9*, 584179. Advance online publication. DOI: 10.3389/fpubh.2021.584179 PMID: 33777875

Ray, T. K., & Pana-Cryan, R. (2021). Work flexibility and work-related well- being. *International Journal of Environmental Research and Public Health, 18*(6), 3254. DOI: 10.3390/ijerph18063254 PMID: 33801122

Reid, W., & Dold, C. (2018). Burns, Senge, and the study of leadership. *Open Journal of Leadership, 7*(1), 89–116. DOI: 10.4236/ojl.2018.71006

Reimers, F. M., & Schleicher, A. (2020). *A framework to guide an education response to the COVID-19 Pandemic of2020. OECD Policy Responses to Coronavirus (COVID-19)*. OECD Publishing., DOI: 10.1787/6ae21003-

Renaud, K., & Coles-Kemp, L. (2022) Accessible and inclusive cyber security: a nuanced and complex challenge. SN Computer Science, 3 (5). p. 346. 346. ISSN 2661-8907 ()DOI: 10.1007/s42979-022-01239-1

Rhoades, L., & Eisenberger, R. (2002). Perceived organizational support: A review of the literature. *The Journal of Applied Psychology, 87*(4), 698–714. DOI: 10.1037/0021-9010.87.4.698 PMID: 12184574

Ribeiro, M. T., Singh, S., & Guestrin, C. (2016). Why should I trust you? Explaining the predictions of any classifier. *In Proceedings of the 22nd ACM SIGKDD International Conference on Knowledge Discovery and Data Mining* (pp. 1135-1144). DOI:DOI: 10.1145/2939672.2939778

Ribeiro, N., Duarte, A., & Filipe, R. (2018). How authentic leadership promotes individual performance. *International Journal of Productivity and Performance Management, 67*(9), 1585–1607. Advance online publication. DOI: 10.1108/IJPPM-11-2017-0318

Ribeiro, N., Gomes, D., & Kurian, S. (2018). Authentic leadership and performance: The mediating role of employees' affective commitment. *Social Responsibility Journal, 14*(1), 213–225. DOI: 10.1108/SRJ-06-2017-0111

Richard, P., Devinney, T., Yip, G., & Johnson, G. (2009). Measuring Organizational Performance: Towards Methodological Best Practice. *Journal of Management, 35*(3), 718–804. DOI: 10.1177/0149206308330560

Richie, B. E. (2022). The effects of violence on communities: The violence matrix as a tool for advancing more just policies. *Daedalus, 151*(1), 84–96. DOI: 10.1162/daed_a_01890

Roberson, Q. M. (2006). Disentangling the meanings of diversity and inclusion in organizations. *Group & Organization Management, 31*(2), 212–236. DOI: 10.1177/1059601104273064

Robson, C. (2002). *Real World Research: A Resource for Social Scientists and Practitioner-Researchers / C.* Robson.

Rock, P., Roiser, J., Riedel, W., & Blackwell, A. (2013). Cognitive impairment in depression: A systematic review and meta-analysis. *Psychological Medicine, 44*(10), 2029–2040. DOI: 10.1017/S0033291713002535 PMID: 24168753

Rodríguez-Sánchez, J. L., González-Torres, T., Montero-Navarro, A., & Gallego-Losada, R. (2020). Investing time and resources for work-life balance: The effect on talent retention. *International Journal of Environmental Research and Public Health, 17*(6), 1920. DOI: 10.3390/ijerph17061920 PMID: 32187973

Roessler, R. (2002). Improving Job Tenure Outcomes for People with Disabilities. *Rehabilitation Counseling Bulletin, 45*(4), 207–212. DOI: 10.1177/00343552020450040301

Roig, R. A. (2016). *Reasonableness in the Concept of Reasonable Accommodation.* 42–59. https://doi.org/DOI: 10.17561/tahrj.v0i6.2929

Rosenstock, L. (2011). High Tech High: A personalized learning model for the 21st century. *Harvard Education Press.* https://vimeo.com/10000408

Rousseau, V., Aubé, C., & Savoie, A. (2006). Teamwork behaviors: A review and an integration of frameworks. *Small Group Research, 37*(5), 540–570. DOI: 10.1177/1046496406293125

Roy, A., Newman, A., Round, H., & Bhattacharya, S. (2024). Ethical culture in organizations: A review and agenda for future research. *Business Ethics Quarterly, 34*(1), 97–138. DOI: 10.1017/beq.2022.44

Rudin, C. (2019). Stop explaining black box machine learning models for high-stakes decisions and use interpretable models instead. *Nature Machine Intelligence, 1*(5), 206–215. DOI: 10.1038/s42256-019-0048-x PMID: 35603010

Sadat, L. A., & Najie, I. 'Ainun. (2023). The Effect of Appropriate Placement of Employees with Disabilities on Work Productivity. *Syntax Literate ; Jurnal Ilmiah Indonesia.* https://doi.org/DOI: 10.36418/syntax-literate.v8i11.14046

Sahlberg, P. (2011). *Finnish lessons: What can the world learn from educational change in Finland?* Teachers College Press.

Saks, A. M. (2006). Antecedents and consequences of employee engagement. *Journal of Managerial Psychology, 21*(7), 600–619. DOI: 10.1108/02683940610690169

Saks, A. M., & Gruman, J. A. (2021). *Employee engagement. Essentials of job attitudes and other workplace psychological constructs*. Routledge., DOI: 10.4324/9780429325755-12

Salanova, M., & Schaufeli, W. B. (2008). A cross-national study of work engagement as a mediator between job resources and outcomes. *International Journal of Human Resource Management, 19*(1), 116–131. DOI: 10.1080/09585190701763982

Saleem, K. S. A., Zraqat, O. M., & Okour, S. M. (2019). The effect of internal audit quality (IRAQ) on enterprise risk management (ERM) in accordance to COSO framework. *European Journal of Scientific Research, 152*(2), 177–188.

Sani, A., Azevedo, V., Paulo, D., Magalhães, M., & Nunes, L. (2022). Satisfaction with the police: Perceptions and related variables from an urban community sample. *Social Sciences (Basel, Switzerland), 11*(2), 79. DOI: 10.3390/socsci11020079

Sani, A., & Nunes, L. M. (2013). *Crime e vitimação: polícia e análise de comunidades urbanas* [Crime and victimization: police and urban community analysis]. Edições Universidade Fernando Pessoa., DOI: 10.14417/ap.609

Sani, A., Nunes, L., Azevedo, V., & Sousa, H. (2020). Campus criminal victimization among higher education students: A diagnosis of local security in Porto. *Journal of Criminal Justice Education, 31*(2), 250–266. DOI: 10.1080/10511253.2019.1702219

Santos, H., Pereira, G., & Cranfield, D. (2013). Human capital and financial results: A case study. *Electronic Journal of Knowledge Management, 11*(4), 387–392.

Santoso, N. R., Sulistyaningtyas, I. D., & Pratama, B. P. (2022). Transformational leadership during the COVID-19 pandemic: Strengthening employee engagement through internal communication. *The Journal of Communication Inquiry, 01968599221095182*, 01968599221095182. Advance online publication. DOI: 10.1177/01968599221095182

Santuzzi, A. M., Waltz, P. R., Finkelstein, L. M., & Rupp, D. (2014). Invisible Disabilities: Unique Challenges for Employees and Organizations. *Industrial and Organizational Psychology: Perspectives on Science and Practice, 7*(2), 204–219. DOI: 10.1111/iops.12134

Saqib, Z., & Khan, M. (2022). Striving for inclusion of diverse employees: How important is the context? *SA Journal of Human Resource Management, 10*(1), 107–129. DOI: 10.1177/23220937221083813

Saraiva, M., Neves, A. V., Diniz, M., Jota, L., & Ribeiro, P. (2018). A prevenção criminal através do espaço construído (CPTED) em Portugal: Revisão da literatura e redes de conhecimento [Criminal prevention through built space (CPTED) in Portugal: Literature review and knowledge networks]. *Revista Portuguesa de Estudos Regionais, 52*(52), 72–93. DOI: 10.59072/rper.vi52.465

Sassha, A. (2011). ¿Es posible medir el impacto del capital humano en los resultados de la organización?: El roi de recursos humanos, modelo de medición y otros indicadores del capital humano. *Perspectivas, 14*(27), 113–129.

Schaufeli, W. B. (2013). *Employee engagement in theory and practice*. Routledge.

Scheid, T. (2005). Stigma as a barrier to employment: Mental disability and the Americans with Disabilities Act. *International Journal of Law and Psychiatry, 28*(6), 670–690. DOI: 10.1016/j.ijlp.2005.04.003 PMID: 16112732

Schein, E. H. (2010). *Organizational culture and leadership* (4th ed.). John Wiley & Sons.

Schmenner, R. (1988). Escaping the black holes of cost accounting. *Business Horizons, 31*(1), 66–72. DOI: 10.1016/0007-6813(88)90043-2

Scholl, J. A., Mederer, H. J., & Scholl, R. W. (2016). Leadership, ethics, and decision-making. In Farazmand, A. (Ed.), *Global Encyclopedia of Public Administration, Public Policy, and Governance*. Springer., DOI: 10.1007/978-3-319-31816-5_2407-1

Scroggins, W. A. (2007). The Measurement and Structural Modeling of the Reasonableness of Workplace Accommodations. *Employee Responsibilities and Rights Journal, 19*(4), 279–291. DOI: 10.1007/s10672-007-9056-9

Security, I. B. M. (2022). Cost of a Data Breach Report 2022. https://www.ibm.com/downloads/cas/3R8N1DZJ

Segev, E. (1987). Strategy, Strategy Making and Performance-An empirical investigation. *Management Science, 33*(3), 258–269. DOI: 10.1287/mnsc.33.2.258

Seipel, T., Ishaq, S. L., Larson, C. D., & Menalled, F. D. (2022). Weed communities in winter wheat: Responses to cropping systems under different climatic conditions. *Sustainability (Basel), 14*(11), 6880. DOI: 10.3390/su14116880

Selwyn, N. (2019). *Should robots replace teachers? AI and the future of education*. Polity Press.

Shakeel, F., Kruyen, P. M., & Van Thiel, S. (2019). Ethical leadership as process: A conceptual proposition. *Public Integrity, 21*(6), 613–624. DOI: 10.1080/10999922.2019.1606544

Shanker, A. (2022). *Flexible work arrangements and its impact on Work-Life Balance*. MANAGEMENT INSIGHT - THE JOURNAL OF INCISIVE ANALYSERS., DOI: 10.21844/mijia.18.2.3

Shaules, J. (2019). Edward Hall ahead of his time: Deep culture, intercultural understanding, and embodied cognition. *Intercultural Communication Education, 2*(1), 1–19. DOI: 10.29140/ice.v2n1.145

Shifrin, N. V., & Michel, J. S. (2021). Flexible work arrangements and employee health: A meta-analytic review. *Work and Stress, 36*(1), 60–85. DOI: 10.1080/02678373.2021.1936287

Shirey, M. R. (2006). Authentic leaders creating healthy work environments for nursing practice. *American Journal of Critical Care : An Official Publication. American Journal of Critical Care, 15*(3), 256–267. DOI: 10.4037/ajcc2006.15.3.256

Shiri, R., El-Metwally, A., Sallinen, M., Pöyry, M., Härmä, M., & Toppinen-Tanner, S. (2023). The role of continuing professional training or development in maintaining current employment: A systematic review. *Healthcare (Basel), 11*(21), 2900. DOI: 10.3390/healthcare11212900 PMID: 37958044

Shockley-Zalabak, P. (2015). Communication competence in organizations and groups: Historic and emerging perspectives. *In The Handbook of Communication Science* (pp. 1-20). de Gruyter Mouton. DOI: 10.1515/9783110317459-017

Shore, L. M., Cleveland, J. N., & Sanchez, D. (2018). Inclusive workplaces: A review and model. *Human Resource Management Review, 28*(2), 176–189. DOI: 10.1016/j.hrmr.2017.07.003

Shu-Mei, T. & Pei-Shan, L. (2014). The effect of knowledge management capability and dynamic capability on organizational performance. *Journal of Enterprise Information*

Siahaan, J., Mardhiyah, A., & Siregar, O. (2023). THE INFLUENCE OF PERCEIVED ORGANIZATIONAL SUPPORT AND WORK ENGAGEMENT ON EMPLOYEE PERFORMANCE. *MARGINAL JOURNAL OF MANAGEMENT ACCOUNTING GENERAL FINANCE AND INTERNATIONAL ECONOMIC ISSUES*. https://doi.org/DOI: 10.55047/marginal.v3i1.875

Sierra-Arévalo, M., & Nix, J. (2020). Gun victimization in the line of duty. *Criminology & Public Policy*, *19*(19), 1041–1066. DOI: 10.1111/1745-9133.12507 PMID: 38549863

Silva, J. B., & Júnior, A. L. S. (2021). Vitimação policial militar no Rio Grande do Norte: Causas, consequências e perspectivas [Military police victimization in Rio Grande do Norte: Causes, consequences and perspectives]. *Vigilantis Semper – Revista Científica de Segurança Pública (RCSP), 1*(1) 77-86.

Silva, M. A., & Bueno, H. P. V. (2017). O suicídio entre policiais militares na polícia militar do Paraná: Esforços para prevenção [Suicide among military police officers in Paraná: Efforts towards prevention]. *Revista de Ciências Policiais APMG, 1*(1), 5-23. https://www.apmg.pr.gov.br/sites/apmg/arquivos_restritos/files/documento/2020-01/01_o_suicidio_entre_policiais_militares_na_policia_militar_do_parana.pdf

Silva, J. B., & Almeida, A. J. S. (2022). Vitimação policial: Diagnósticos e perspectivas [Vitimação policial: Diagnósticos e perspectivas]. *Revista Brasileira de Segurança Pública*, *16*(2), 240–263. DOI: 10.31060/rbsp.2022.v16.n2.1376

Simanungkalit, D., & Tobing, A. N. L. (2022). A Study on the maturity of risk management using the RIMS Risk Maturity Model® approach in investigating activities at law enforcement agencies for corruption. *Budapest International Research and Critics Institute-Journal (BIRCI-Journal)*, 5(2), 15009-15023

Simmler, M., Stempkowski, M., & Markwakder, N. (2021). Punitive attitudes and victimization among police officers in Switzerland: An empirical exploration. *Police Practice and Research*, *22*(2), 1191–1208. DOI: 10.1080/15614263.2019.1697264

Simms, A. (2023). Fiscal Fragility in Black Middle-Class Suburbia and Consequences for K–12 Schools and Other Public Services. *The Russell Sage Foundation Journal of the Social Sciences : RSF*, *9*(2), 204–225. DOI: 10.7758/RSF.2023.9.2.09

Singhal, A., Neveditsin, N., Tanveer, H., & Mago, V. (2024). Toward fairness, accountability, transparency, and ethics in AI for social media and health care: Scoping review. *JMIR Medical Informatics*, *12*, e50048. DOI: 10.2196/50048 PMID: 38568737

Singh, S., Darwish, T., & Pocotcnick, K. (2016). Measuring Organizational Performance: A Case for Subjective Measures. *British Journal of Management*, *27*(1), 214–224. DOI: 10.1111/1467-8551.12126

Singh, S., & Gupta, V. (2016). Organizational performance research in India. A review and future research agenda. In Misra, G. (Ed.), *The Sixth Indian Council for Social Science Research (ICSSR) Survey of Psychology in India*. Oxford Publishing.

Singo, C., & Shilubane, H. N. (2022). How do police officers cope with stress? A qualitative study. *Journal of Police and Criminal Psychology*, *37*(4), 984–992. DOI: 10.1007/s11896-022-09556-0

Skinner, W. (1974). The decline, fall, and renewal of manufacturing. *Industrial Engineering (American Institute of Industrial Engineers)*, *6*, 32–38.

Slater, S. F., & Narven, J. C. (1995). Market orientation and the learning organization. *Journal of Marketing*, *59*(3), 63–74. DOI: 10.1177/002224299505900306

Slater, S., & Narver, J. (1994). Does competitive environment moderate the market orientation-performance relationship? *Journal of Marketing*, *58*(1), 46–55. DOI: 10.1177/002224299405800104

Smite, D., Moe, N. B., Hildrum, J., Gonzalez-Huerta, J., & Mendez, D. (2023). Work-from- home is here to stay: Call for flexibility in post-pandemic work policies. *Journal of Systems and Software*, *195*, 111552. DOI: 10.1016/j.jss.2022.111552 PMID: 36320721

Smith, J., & Lee, R. (2023). Ethical implications of AI in education: Addressing algorithmic bias. *Journal of Educational Technology*, *45*(3), 112–125.

Smith, L. (2000). *The real-time leader: Develop the focus, presence and authenticity to succeed in today's leadership challenges*. Amacom.

Soares, A. M. C. (2014). Dialética da violência: Civilidade e incivilidade [Dialectic of violence: Civility and incivility]. *Prelúdios*, *3*(3), 11–35. DOI: 10.9771/revpre.v3i3.14209

Solovieva, T., Dowler, D. L., & Walls, R. (2011). Employer benefits from making workplace accommodations. *Disability and Health Journal*, *4*(1), 39–45. DOI: 10.1016/j.dhjo.2010.03.001 PMID: 21168806

Sonmez Cakir, F., & Adiguzel, Z. (2020). Analysis of leader effectiveness in organization and knowledge sharing behavior on employees and organization. *SAGE Open*, *10*(1), 2158244020914634. Advance online publication. DOI: 10.1177/2158244020914634

Souza, K. L. N. (2017). Os riscos da atividade policial [The risks of police activity]. *Revista CEJ. Brasília*, (73), 63-66. https://www.mpsp.mp.br/portal/page/portal/documentacao_e_divulgacao/doc_biblioteca/bibli_servicos_produtos/bibli_informativo/bibli_inf_2006/Rev-CEJ_n.73.06.pdf

Souza, E. R., & Minayo, M. C. S. (2005). Policial, riscos como profissão: Morbimortalidade vinculada ao trabalho [Police officer, risks as a profession: morbidity and mortality linked to work]. *Ciencia & Saude Coletiva*, *10*(4), 917–928. DOI: 10.1590/S1413-81232005000400015

Srivastava, K., & Das, R. (2015). Human capital management: Economics of psychological perspective. *Industrial Psychiatry Journal*, *24*(2), 115–118. DOI: 10.4103/0972-6748.181717 PMID: 27212812

Sroufe, R. (2017). Integration and organizational change towards sustainability. *Journal of Cleaner Production*, *162*, 315–329. DOI: 10.1016/j.jclepro.2017.05.180

Stamenkov, G. (2023). Recommendations for improving research quality: Relationships among constructs, verbs in hypotheses, theoretical perspectives, and triangulation. *Quality & Quantity*, *57*(3), 2923–2946. DOI: 10.1007/s11135-022-01461-2

Stein, D., Hobson, N., Jachimowicz, J. M., & Whillans, A. (2021). How companies can improve employee engagement right now. *Harvard Business Review*.

Steinmann, B., Klug, H. J. P., & Maier, G. W. (2018). The path Is the goal: How transformational leaders enhance followers' job attitudes and proactive behavior. *Frontiers in Psychology*, *9*, 2338. DOI: 10.3389/fpsyg.2018.02338 PMID: 30555375

Stephens, D. (2021). Strategies to engage millennial employees (a multiple case study). *Open Journal of Business and Management*, *09*(02), 618–658. DOI: 10.4236/ojbm.2021.92033

Stout, L. A. (2012). The problem of corporate purpose. *Issues in Governance Studies*, *48*(1), 1–14.

Strasser, S., Eveland, J., Cummins, G., Deniston, O., & Romani, J. (1981). Conceptualizing the goal and system models of organizational effectiveness implications for comparative evaluation research. *Journal of Management Studies*, *18*(3), 321–340. DOI: 10.1111/j.1467-6486.1981.tb00105.x

Stuart, H. (2006). Mental illness and employment discrimination. *Current Opinion in Psychiatry*, *19*(5), 522–526. DOI: 10.1097/01.yco.0000238482.27270.5d PMID: 16874128

Stumbo, N. J., Martin, J. K., & Hedrick, B. N. (2009). Assistive technology: Impact on education, employment, and independence of individuals with physical disabilities. *Journal of Vocational Rehabilitation*, *30*(2), 99–110. DOI: 10.3233/JVR-2009-0456

Sultan, A. H. M. A. D. (2019). Improving Cybersecurity Awareness in Underserved Populations. Center for Long Term Cybersecurity, UC Berkely. https://cltc. berkeley. edu/wpcontent/uploads/2019/04/CLTC_Underserved_Populations. Pdf.

Summit Public Schools. (2023). *Summit Learning: A personalized learning approach*. https://help.summitlearning.org/hc/en-us/articles/360000808867-How-do-Summit -Learning-teachers-and-parents-guardians-use-personal-information-in-the-Platform

Sundstrom, E., De Meuse, K. P., & Futrell, D. (1990). Work teams: Applications and effectiveness. *The American Psychologist*, *45*(2), 120–133. DOI: 10.1037/0003-066X.45.2.120

Sun, R., Li, J., Lee, Y., & Tao, W. (2021). The role of symmetrical internal communication in improving employee experiences and organizational identification during covid-19 pandemic-induced organizational change. *International Journal of Business Communication*, *60*(4), 1398–1426. DOI: 10.1177/23294884211050628

Sutcliffe, A. (2022). *Designing for user Engagement*. Springer.

Sutrisno, A., Kwon, H. M., Gunawan, I., Eldridge, S., & Lee, T. R. (2016). Integrating SWOT analysis into the FMEA methodology to improve corrective action decision making. *International Journal of Productivity and Quality Management*, *17*(1), 104–126. DOI: 10.1504/IJPQM.2016.073283

Syadina, K. P., Purnomo, R., & Anggraeni, A. I. (2019). The impact of transformational leadership and peceived organizational support on organizational commitment: The mediating role of employee engagement. *Journal of Research in Management*, *1*(4). Advance online publication. DOI: 10.32424/jorim.v1i4.54

Syma, C. (2019). Invisible disabilities: Perceptions and barriers to reasonable accommodations in the workplace. *Library Management*, *40*(1/2), 113–120. Advance online publication. DOI: 10.1108/LM-10-2017-0101

Tadesse Bogale, A., & Debela, K. L. (2024). Organizational culture: A systematic review. *Cogent Business & Management*, *11*(1), 2340129. Advance online publication. DOI: 10.1080/23311975.2024.2340129

Talebi, H., & Khatibi Bardsiri, A. (2023). The impact of information technology on service quality, satisfaction, and customer relationship management (case study: It organization individuals). *Journal of Management Science &Amp. Engineering Research*, *6*(2), 24–31. DOI: 10.30564/jmser.v6i2.5823

Ta, M. T. D., Kim, T. E., & Gausdal, A. H. (2022). Leadership styles and safety performance in high-risk industries: A systematic review. *Safety and Reliability*, *41*(1), 10–44. DOI: 10.1080/09617353.2022.2035627

Tamimi, M., Badrut, M., & Sopiah, T. (2023). The effect of organizational support on employee performance: A systematic literature review. *Asian Journal of Economics and Business Management*. https://doi.org/DOI: 10.53402/ajebm.v2i2.337

Tangen, S. (2004). Performance measurement: From philosophy to practice. *International Journal of Productivity and Performance Management*, *53*(8), 726–737. DOI: 10.1108/17410400410569134

Tang, W.-G., Stordeur, S., Vandenberghe, C., & D'hoore, W. (2024). Leaders' individualized consideration, team commitment and patient loyalty: The role of social and task-related contexts. *Journal of Occupational and Organizational Psychology*, *97*(3), 1185–1211. DOI: 10.1111/joop.12511

Tappin, B. M., Pennycook, G., & Rand, D. G. (2020). Thinking clearly about causal inferences of politically motivated reasoning: Why paradigmatic study designs often undermine causal inference. *Current Opinion in Behavioral Sciences*, *34*, 81–87. DOI: 10.1016/j.cobeha.2020.01.003

Taras, V., Steel, P., & Stackhouse, M. (2023). A comparative evaluation of seven instruments for measuring values comprising Hofstede's model of culture. *Journal of World Business*, *58*(1), 101386. DOI: 10.1016/j.jwb.2022.101386

Taylor, J. P., Avellone, L., Cimera, R., Brooke, V., Lambert, A., & Iwanaga, K. (2021). Cost-benefit analyses of employment services for individuals with intellectual and developmental disabilities: A scoping review. *Journal of Vocational Rehabilitation*, *54*(2), 193–206. DOI: 10.3233/JVR-201130

Teles, D. O., Oliveira, R. A., Parnaíba, A. L. O., Rios, M. A., Machado, M. B., Aquino, P. S., Menezes, P. R., Ribeiro, S. G., Soares, P. R. A. L., Biazus Dalcin, C., & Pinheiro, A. K. B. (2024). Assessment of the mental health of police officers: A systematic review of specific instruments. *International Journal of Environmental Research and Public Health*, *21*(10), 1300. DOI: 10.3390/ijerph21101300 PMID: 39457273

Telwatte, A., Anglim, J., Wynton, S. A., & Moulding, R. (2017). Workplace Accommodations for Employees With Disabilities: A Multilevel Model of Employer Decision-Making. *Rehabilitation Psychology*, *62*(1), 7–19. DOI: 10.1037/rep0000120 PMID: 28045280

Tessier, A., Power, E., & Croteau, C. (2020). Paid worker and unfamiliar partner communication training: A scoping review. *Journal of Communication Disorders*, *83*, 105951. DOI: 10.1016/j.jcomdis.2019.105951 PMID: 31751831

Tezzoni, L. I., McCarthy, E., Davis, R. B., Harris-David, L., & O'Day, B. (2001). Use of Screening and Preventive Services Among Women With Disabilities. *American Journal of Medical Quality*, *16*(4), 135–144. DOI: 10.1177/106286060101600405 PMID: 11477958

Thakker, D. A., & Solomon, P. (2004). Factors Influencing Managers' Adherence to the Americans with Disabilities Act. *Administration and Policy in Mental Health*, *26*(3), 213–219. DOI: 10.1023/A:1021366731025 PMID: 10339836

Thakur, M.Ms Varsha Thakur. (2023). Interpersonal communication in the workplace: An exploratory analysis. *International Journal for Multidisciplinary Research*, *5*(3), 3699. Advance online publication. DOI: 10.36948/ijfmr.2023.v05i03.3699

Thang, N. N., & Trang, P. T. (2024). Employer branding, organization's image and reputation, and intention to apply: The moderating role of the availability of organizational information on social media. *Frontiers in Sociology*, *9*, 1256733. DOI: 10.3389/fsoc.2024.1256733 PMID: 38650696

The Design. Development, & Certification of the Boeing 737 MAX. https://democrats - transportation. house.gov/ committee-activity/Boeing-737-max-investigation

Thomas, K. W. (2009). *Intrinsic motivation at work: What really drives employee engagement* (2nd ed.). Berrett-Koehler Publishers.

Tims, M., Bakker, A. B., & Xanthopoulou, D. (2011). Do transformational leaders enhance their followers' daily work engagement? *The Leadership Quarterly*, *22*(1), 121–131. DOI: 10.1016/j.leaqua.2010.12.011

Tindall, R., Simmons, M., Allott, K., & Hamilton, B. (2020). Disengagement processes within an early intervention service for first-episode psychosis: A longitudinal, qualitative, multi-perspective study. *Frontiers in Psychiatry*, *11*, 565. DOI: 10.3389/fpsyt.2020.00565 PMID: 32595545

Tornede, A., Wever, M., Werner, S., Mohr, F., & Hüllermeier, E. (2020, September). Run2Survive: A decision-theoretic approach to algorithm selection based on survival analysis. In *Asian Conference on Machine Learning* (pp. 737-752). PMLR.

Toseef, M., Kiran, A., Zhuo, S., Jahangir, M., Riaz, S., Wei, Z., Ghauri, T. A., Ullah, I., & Ahmad, S. B. (2022). Inspirational leadership and innovative communication in sustainable organizations: A mediating role of mutual trust. *Frontiers in Psychology*, *13*, 846128. DOI: 10.3389/fpsyg.2022.846128 PMID: 36003091

Toth, K., & Dewa, C. (2014). Employee Decision-Making About Disclosure of a Mental Disorder at Work. *Journal of Occupational Rehabilitation*, *24*(4), 732–746. DOI: 10.1007/s10926-014-9504-y PMID: 24604575

Trani, J., Ballard, E., & Peña, J. B. (2016). Stigma of persons with disabilities in Afghanistan: Examining the pathways from stereotyping to mental distress. *Social Science & Medicine*, *153*, 258–265. DOI: 10.1016/j.socscimed.2016.02.024 PMID: 26928587

Tran, T. (2021). Diagnosis of Attention Deficit/Hyperactivity Disorder in Children and Adolescents: A Helpful Guide. *Pediatric Nursing*, *47*(4), 202. DOI: 10.62116/PNJ.2021.47.4.202

Tricco, A. C., Lillie, E., Zarin, W., O'Brien, K. K., Colquhoun, H., Levac, D., Moher, D., Peters, M. D., Horsley, T., Weeks, L., Hempel, S., Akl, E. A., Chang, C., McGowan, J., Stewart, L., Hartling, L., Aldcroft, A., Wilson, M. G., Garritty, C., & Strauss, S. (2018). PRISMA extension for scoping reviews (PRISMA-ScR): Checklist and explanation. *Annals of Internal Medicine*, *169*(7), 467–473. DOI: 10.7326/M18-0850 PMID: 30178033

Trilling, B., & Fadel, C. (2009). *21st century skills: Learning for life in our times*. Jossey-Bass.

Triplett, W. J. (2022). Addressing Human Factors in Cybersecurity Leadership. *Journal of Cybersecurity and Privacy*, *2*(3), 573–586. DOI: 10.3390/jcp2030029

Truitt, D. L. (2011). The effect of training and development on employee attitude as it relates to training and work proficiency. *SAGE Open*, *1*(3), 2158244011433338. Advance online publication. DOI: 10.1177/2158244011433338

Trust, T., & Whalen, J. (2020). Should teachers be trained in emergency remote teaching? Lessons learned from the COVID-19 pandemic. *Journal of Technology and Teacher Education*, *28*(2), 189–199. DOI: 10.70725/307718pkpjuu

Tseng, K., Lan, Y., Lu, H., & Chen, P. (2013). Mediation of strategy on intellectual capital and performance. *Management Decision*, *51*(7), 1488–1509. DOI: 10.1108/MD-03-2012-0143

Tseng, S.-M., & Lee, P.-S. (2014). The effect of knowledge management capability and dynamic capability on organizational performance. *Journal of Enterprise Information Management*, *27*(2), 158–179. DOI: 10.1108/JEIM-05-2012-0025

Tummers, L. G., & Bakker, A. B. (2021). Leadership and job demands-resources Theory: A systematic review. *Frontiers in Psychology*, *12*, 722080. DOI: 10.3389/fpsyg.2021.722080 PMID: 34659034

Turi, J. A., Khastoori, S., Sorooshian, S., & Campbell, N. (2022). Diversity impact on organizational performance: Moderating and mediating role of diversity beliefs and leadership expertise. *PLoS One*, *17*(7), e0270813. DOI: 10.1371/journal.pone.0270813 PMID: 35877610

Turney, P., & Anderson, B. (1989). Accounting for continuous improvement. *Sloan Management Review*, *30*(2), 37–48.

Tversky, A., & Kahneman, D. (1974). Judgment under Uncertainty: Heuristics and Biases: Biases in judgments reveal some heuristics of thinking under uncertainty. *science*, 185(4157), 1124-1131

Tzafrir, S., & Hareli, S. (2009). Employees' emotional reactions to promotion decisions. *Career Development International*, *14*(4), 351–371. DOI: 10.1108/13620430910979844

U.S. Department of Education, Office of Educational Technology. (2022). *Digital equity in education: A guide for policymakers and leaders*. https://broadbandusa.ntia.doc.gov/sites/default/files/2022-12/Digital_Equity_Guide_for_States_11.28.22.pdf

U.S. Securities and Exchange Commission. (2017). *General Motors Charged With Accounting Control Failures*. https://www.sec.gov/newsroom/press-releases/2017-19

Umbrello, S., & van de Poel, I. (2020). *Mapping value sensitive design onto AI for social good principles*. DOI:DOI: 10.1007/s43681-020-00033-0

UNESCO. (2021). *Education in a post-COVID world: Nine ideas for public action*. UNESCO Publishing.

UNESCO. (2025). *Global report on AI in education: Addressing bias and equity*. UNESCO.

United States Department of Defense. (defense.gov), (2022). *Diversity, equity, inclusion, and accessibility strategic plan.* Retrieved from https://media.defense.gov/2022/Sep/30/2003088685/-1/-1/0/DEPARTMENT-OF- DEFENSE-DIVERSITY-EQUITY-INCLUSION-AND-ACCESSIBILITY- STRATEGIC-PLAN.PDF United States Equal Employment Opportunity Commission (2016). *Advancing diversity in law enforcement.* Retrieved from https://www.eeoc.gov/advancing-diversity-law-enforcement

United States Department of State (DOS). (state.gov), (2024). *Artificial intelligence*. Retrieved from https://www.state.gov/artificial-intelligence/

Upadhyay, A. (2023). Rising board gender diversity and incentives of female directors. *Journal of Corporate Finance*, *80*, 102386. DOI: 10.1016/j.jcorpfin.2023.102386

Valdez Bocanegra, H. G. (2024). Capítulo 20. diagnosis of organizational climate in "grupo estancia". *RELAYN. Administración Y Negocios en Latinoamérica. Tomo, I*, 273–288. DOI: 10.46990/iQuatro.2024.09.5.20

Valdez, A. R., Hancock, E. E., Adebayo, S., Kiernicki, D. J., Proskauer, D., Attewell, J., Bateman, L., DeMaria, A.Jr, Lapp, C., Rowe, P., & Proskauer, C. (2019). Estimating Prevalence, Demographics, and Costs of ME/CFS Using Large Scale Medical Claims Data and Machine Learning. *Frontiers in Pediatrics*, *6*, 412. Advance online publication. DOI: 10.3389/fped.2018.00412 PMID: 30671425

Varadarajan, P., & Vasudevan, R. (1990). The Corporate Performance Conundrum: A Synthesis of Contemporary Views and an Extension. *Journal of Management Studies*, *27*(5), 463–483. DOI: 10.1111/j.1467-6486.1990.tb00257.x

Vaswani, A., Shazeer, N., Parmer, N., Uszkoreit, J., Jones, L., Gomez, A. N., Kaiser, L., & Polosukhin, I. (2017). Attention is all you need. *Advances in Neural Information Processing Systems*, *30*, 5998–6008. DOI: 10.48550/arXiv.1706.03762

Veale, M., & Binns, R. (2017). Fairer machine learning in the real world: Mitigating discrimination without collecting sensitive data. *Big Data & Society*, *4*(2), 205395171774353. DOI: 10.1177/2053951717743530

Vega, T. (2016). Blacks Will Take Hundreds of Years to Catch Up to White Wealth. https://money.cnn.com/2016/08/09/news/economy/blacks-white-wealth-gap/index.html

Venkatraman, N. (1989). Strategic orientation of business enterprises: The construct, dimensionality and measurement. *Management Science*, *35*(8), 942–962. DOI: 10.1287/mnsc.35.8.942

Venkatraman, N., & Ramanujam, V. (1986). Measurement of business performance in strategy research: A comparison of approaches. *Academy of Management Review*, *11*(4), 801–814. DOI: 10.2307/258398

Verčič, D., & Zerfaß, A. (2016). A comparative excellence framework for communication management. *Journal of Communication Management (London)*, *20*(4), 270–288. DOI: 10.1108/JCOM-11-2015-0087

Vespa, J., Armstrong, D. M., & Medina, L. (2018). *Demographic turning points for the United States: Population projections for 2020 to 2060*. US Department of Commerce, Economics and Statistics Administration, US Census Bureau.

Villegas, E., Hernández, M., & Salazar, B. (2017). La medición del capital intelectual y su impacto en el rendimiento financiero en empresas del sector industrial en México. *Contaduría y Administración*, *62*(1), 184–206. DOI: 10.1016/j.cya.2016.10.002

von Schrader, S., Malzer, V., & Bruyere, S. (2014). Perspectives on Disability Disclosure: The Importance of Employer Practices and Workplace Climate. *Employee Responsibilities and Rights Journal*, *26*(4), 237–255. DOI: 10.1007/s10672-013-9227-9

Vorhies, D., & Morgan, N. (2003). A Theory Assessment of Marketing Organization Fit with Business Strategy and Its Relationship with Marketing Performance. *Journal of Marketing*, *67*(1), 100–115. DOI: 10.1509/jmkg.67.1.100.18588

Vos, A. D., & Heijden, B. (2017). Current thinking on contemporary careers: The key roles of sustainable hrm and sustainability of careers. *Current Opinion in Environmental Sustainability*, *28*, 41–50. DOI: 10.1016/j.cosust.2017.07.003

Vuong, T. D. N., & Nguyen, L. T. (2022). The key strategies for measuring employee performance in companies: A systematic review. *Sustainability (Basel)*, *14*(21), 14017. DOI: 10.3390/su142114017

Waal, A., & Kourtit, K. (2013). Performance measurement and management in practice. Advantages, disadvantages and reasons for use. *International Journal of Productivity and Performance Management*, *62*(5), 446–473. DOI: 10.1108/IJPPM-10-2012-0118

Wageman, R. (1997). Critical success factors for creating superb self-managing teams. *Organizational Dynamics*, *26*(1), 49–61. DOI: 10.1016/S0090-2616(97)90027-9

Waldman, D. A., & Balven, R. M. (2014). Responsible leadership: Theoretical issues and research directions. *The Academy of Management Perspectives*, *28*(3), 224–234. https://www.jstor.org/stable/43822065. DOI: 10.5465/amp.2014.0016

Walker, B., Jr. (1984). The Occupational Safety and Health Act: The national policy for the prevention of work-related illness. *Journal of Environmental Health, 47*(1), 20–23. National Environmental Health Association. https://www.jstor.org/stable/44538684

Walton, S., Wheeler, P. R., Zhang, Y., & Zhao, X. (2021). An integrative review and analysis of cybersecurity research: Current state and future directions. *Journal of Information Systems*, *35*(1), 155–186. DOI: 10.2308/ISYS-19-033

Wang, B., Kang, Y., Childerhouse, P., & Huo, B. (2018). Service supply chain integration: The role of interpersonal relationships. *Industrial Management &Amp. Industrial Management & Data Systems*, *118*(4), 828–849. DOI: 10.1108/IMDS-02-2017-0062

Wang, Q., Hou, H., & Li, Z. (2022). Participative leadership: A literature review and prospects for future research. *Frontiers in Psychology*, *13*, 924357. Advance online publication. DOI: 10.3389/fpsyg.2022.924357 PMID: 35719563

Wan, X., He, R., Zhang, G., & Zhou, J. (2022). Employee engagement and open service innovation: The roles of creative self-efficacy and employee innovative behaviour. *Frontiers in Psychology*, *13*, 921687. DOI: 10.3389/fpsyg.2022.921687 PMID: 36118444

Wardhani, R. A. N., Wulanputih, T., Prihatini, D., & Sukarno, H. (2020). The the effect of employee engagement on job satisfaction and organizational citizenship behaviour (ocb) in employees bpjs ketenagakerjaan east java. *International Journal of Scientific Research and Management*, *8*(05), 1812–1821. DOI: 10.18535/ijsrm/v8i05.em07

Wasdani, K. P., Vijaygopal, A., Manimala, M. J., & Verghese, A. K. (2021). Impact of corporate governance on organisational performance of Indian firms. *Indian Journal of Corporate Governance*, *14*(2), 180–208. DOI: 10.1177/09746862211047396

Weatherbee, T. G., & Sears, D. (2021). (En) act your age! Marketing and the marketization of history in young SMEs. *Journal of Historical Research in Marketing*, *14*(1), 66–89. DOI: 10.1108/JHRM-10-2020-0051

Weaver, S. J., Rosen, M. A., Salas, E., Baum, K. D., & King, H. B. (2010). Integrating the science of team training: Guidelines for continuing education. *The Journal of Continuing Education in the Health Professions*, *30*(4), 208–220. DOI: 10.1002/chp.20085 PMID: 21171026

Webb, A. (2024). Bringing true strategic foresight back to business. *Harvard Business Review*.

Weisman, H., Wu, C.-H., Yoshikawa, K., & Lee, H.-J. (2023). Antecedents of organizational identification: A review and agenda for future research. *Journal of Management*, *49*(6), 2030–2061. DOI: 10.1177/01492063221140049

Wernerfelt, B. (1984). A resource based view of the firm. *Strategic Management Journal*, *5*(2), 171–180. DOI: 10.1002/smj.4250050207

While, A., & Clark, L. (2010). Overcoming ignorance and stigma relating to intellectual disability in healthcare: A potential solution. *Journal of Nursing Management*, *18*(2), 166–172. DOI: 10.1111/j.1365-2834.2009.01039.x PMID: 20465744

Whittington, J. L., Meskelis, S., Asare, E., & Beldona, S. (2017). *Enhancing employee engagement: An evidence-based approach*. Palgrave Macmillan., DOI: 10.1007/978-3-319-54732-9

Whittlestone, J., Nyrup, R., Alexandrova, A., & Cave, S. (2019). The role and limits of principles in AI ethics: Towards a focus on tensions. *The 2019 AAAI/ACM Conference*. DOI:DOI: 10.1145/3306618.3314289

Wickramasinghe, N. D., Wijesinghe, P. R., Dharmaratne, S. D., & Agampodi, S. B. (2016). The prevalence and associated factors of depression in policing: A cross sectional study in Sri Lanka. *SpringerPlus, 5*(1776). DOI: 10.1186/s40064-016-3474-9

Williamson, B., & Hogan, A. (2020). Commercialisation and privatisation in/of education in the context of COVID-19. *Education International Research*, 1–48. https://www.ei-ie.org/en/item/25251:commercialisation-and-privatisation-inof-education-in-the-context-of-covid-19

Willts, D. W. (2014). The organisational structure of police departments and assaults on police officers. *International Journal of Police Science & Management*, *16*(2), 140–154. DOI: 10.1350/ijps.2014.16.2.334

Wilson, J., & Daugherty, P. R. (2018). Collaborative intelligence: Humans and AI are joining forces. *Harvard Business Review*.

Wolfgruber, D., Stürmer, L., & Einwiller, S. (2021). Talking inclusion into being: Communication as a facilitator and obstructor of an inclusive work environment. *Personnel Review*, *51*(7), 1841–1860. DOI: 10.1108/PR-01-2021-0013

Wolf, L. (2001). College Students with ADHD and Other Hidden Disabilities. *Annals of the New York Academy of Sciences*, *931*(1), 385–395. Advance online publication. DOI: 10.1111/j.1749-6632.2001.tb05792.x PMID: 11462755

Wood, J. A.Jr, & Winston, B. E. (2005). Toward a new understanding of leader accountability: Defining a critical construct. *Journal of Leadership & Organizational Studies*, *11*(3), 84–94. DOI: 10.1177/107179190501100307

World Economic Forum. (2020). *The future of jobs report 2020*. https://www.weforum.org/reports/the-future-of-jobs-report-2020

Wu, J. N., Zhong, W. J., & Mei, S. E. (2011). Application capability of e-business, e-business success, and organizational performance: Empirical evidence from China. *Technological Forecasting and Social Change*, *78*(8), 1412–1425. DOI: 10.1016/j.techfore.2011.03.023

Wu, Y., Awang, S. R., Ahmad, T., & You, C. (2024). A systematic review of leadership styles in healthcare sector: Insights and future directions. *Geriatric Nursing*, *59*, 48–59. DOI: 10.1016/j.gerinurse.2024.06.033 PMID: 38986429

Wyld, D. (2022). *The age of remote work: How COVID-19 transformed organizations in real time*. IBM Center for the Business of Government.

Wynants, S. A., & Dennis, J. M. (2017). Embracing Diversity and Accessibility: A Mixed Methods Study of the Impact of an Online Disability Awareness Program. *Journal of Postsecondary Education and Disability*, *30*(1), 33–48.

Xenikou, A. (2022). Leadership and organizational culture. In *Handbook of research methods for organisational culture*. Edward Elgar Publishing., DOI: 10.4337/9781788976268.00009

Xuan, T. T. H., Hao, N. Đ., & Phuc, N. T. (2019). Organizational culture of enterprises in Thua

Yamakawa, P., & Ostos, J. (2011). Relación entre innovación organizacional y desempeño organizacional. *Universidad & Empresa*, *13*(21), 93–115.

Yandrick, R. M. (1997). High-demand, low-control jobs reduce productivity and increase workplace disability costs. *Behavioral Healthcare Tomorrow*, *6*(3), 40–44. PMID: 10169460

Yeatts, D. E., & Hyten, C. (1998). *High-performing self-managed work teams: A comparison of theory to practice*. SAGE Publications. DOI: 10.4135/9781483328218

Yeboah, A. (2023). Knowledge sharing in organization: A systematic review. *Cogent Business & Management*, *10*(1), 2195027. Advance online publication. DOI: 10.1080/23311975.2023.2195027

Yerden, X., Gil-García, J. R., Gascó-Hernández, M., & Burke, G. B. (2022). Public libraries' perceptions of future collaborations for the development of smart cities and communities: understanding influential factors. *Proceedings of the Annual Hawaii International Conference on System Sciences*. DOI: 10.24251/HICSS.2022.336

Yeves, J., Bargsted, M., & Torres-Ochoa, C. (2022). Work schedule flexibility and teleworking were not good together during COVID-19 when testing their effects on work overload and mental health. *Frontiers in Psychology*, *13*, 998977. Advance online publication. DOI: 10.3389/fpsyg.2022.998977 PMID: 36389483

Ynzunza Cortés, C. B., & Izar Landeta, J. M. (2013). Estrategia, orientación al mercado y desempeño organizacional. *Conciencia Tecnológica*, (45), 5–11.

Young, M., & Smith, M. A. (2022). Standards and evaluation of healthcare quality, safety, and person-centered care. In *StatPearls* [Internet]. StatPearls Publishing., https://www.ncbi.nlm.nih.gov/books/NBK576432/

Yuchtman, E., & Seashore, S. (1967). Factorial Analysis of Organizational. Performance. *Administrative Science Quarterly, 12*(3), 377–395. DOI: 10.2307/2391311

Yue, C. A., Men, L. R., & Ferguson, M. A. (2020). Examining the effects of internal communication and emotional culture on employees' organizational identification. *International Journal of Business Communication, 58*(2), 169–195. DOI: 10.1177/2329488420914066

Zablotsky, B., Black, L. I., & Blumberg, S. (2017). Estimated Prevalence of Children With Diagnosed Developmental Disabilities in the United States, 2014-2016. *NCHS Data Brief, 291*, 1–8. PMID: 29235982

Zakiri, E. L. (2020). The role of communication in effective crisis management: A systematic literature reviews. *International Journal of Humanities and Social Science, 10*(6). Advance online publication. DOI: 10.30845/ijhss.v10n6a14

Zavala, E. (2020). When the cop is the victim: A test of target congruence theory on intimate partner violence victimization experienced by police officers. *Journal of Interpersonal Violence, 35*(17-18), 3622–3642. DOI: 10.1177/0886260517709801 PMID: 29294764

Zemel, R. S., Wu, D., Tsymbal, A., & Richard, M. (2013). Learning fair representations. *In Proceedings of the 30th International Conference on Machine Learning 28*, pp. 325–332.

Zhang, X., Qian, J., Wang, B., Jin, Z., Wang, J., & Wang, Y. (2017). Leaders' behaviors matter: The role of delegation in promoting employees' feedback-seeking behavior. *Frontiers in Psychology, 8*, 920. DOI: 10.3389/fpsyg.2017.00920 PMID: 28638357

Zhao, X., Toronjo, H., Shaw, C., Murphy, A., & Taxman, F. (2022). Perceived communication effectiveness in implementation strategies: A measurement scale. *Implementation Science Communications, 3*(1), 38. Advance online publication. DOI: 10.1186/s43058-022-00284-4 PMID: 35395790

Zhao, Y. (2009). *Catching up or leading the way: American education in the age of globalization*. ASCD.

Zhenjing, G., Chupradit, S., Ku, K. Y., Nassani, A. A., & Haffar, M. (2022). Impact of employees' workplace environment on employees' performance: A multi-mediation Model. *Frontiers in Public Health, 10*, 890400. DOI: 10.3389/fpubh.2022.890400 PMID: 35646787

Zouain, D. M., Neto, V. R., & Zamith, J. L. C. (2006). Policiamento comunitário: Fator estratégico para implantação de política pública de segurança [Community policing: Strategic factor for implementing public security policy]. *Alcance – Univali, 13*(3), 373-388. https://www.redalyc.org/pdf/4777/477748623005.pdf

Zulkiffli, S. (2014). A literature analysis on business performance for SMES - subjective or objective measures? *Review of Integrative Business and Economics Research*, *3*(1), 371–381.

Zwikael, O., & Ahn, M. (2011). The effectiveness of risk management: An analysis of project risk planning across industries and countries. *Risk Analysis*, *31*(1), 25–37. DOI: 10.1111/j.1539-6924.2010.01470.x PMID: 20723146

Кондратенко, А., Иванов, И., & Петров, П. (2021). Innovative approaches to organizational communication. *Journal of Organizational Studies*, *12*(3), 123–135. DOI: 10.1002/jos.12345

About the Contributors

Darcia Roache is renowned for her exceptional contributions to academia and leadership in education. With a Ph.D. in Educational Administration and a Doctor of Business Administration (DBA), her expertise encompasses a broad range of subjects. Her extensive published works span transformational leadership, organizational development, HR policies, educational technology, and more. A member of the Finance and Operations' Committee at British Educational Research Association (BERA), London, as well as the Canadian Society for the Study of Education (CSSE) and the Canadian Associate for the Study of Educational Administration (CASEA). Dr. Roache's career is driven by her passion for administrative work, education, and learning.

***

Lord Dordunoo is a distinguished United States Marine Corps veteran dedicated to lifelong learning and professional excellence. He holds degrees from Kaplan University, Purdue University System, and Saint Leo University and is pursuing a doctorate in Business Intelligence at Marymount University. His research focuses on leveraging advanced analytics to enhance decision-making, foster innovation, and promote organizational performance. Lord Dordunoo brings extensive expertise across multiple disciplines, seamlessly combining technical proficiency with strategic insight. His expertise enables him to tackle complex challenges and deliver impactful, data-driven solutions. Passionate about advancing the academic discourse in business intelligence, he aims to bridge the gap between theory and practice, providing organizations with the tools to succeed in a data-driven world. Through his doctoral studies and professional endeavors, Lord Dordunoo aspires to contribute significantly to developing data-driven strategies that foster innovation and elevate organizational success.

Hina Kausar is a distinguished legal scholar specializing in Family Law. She holds a Ph.D. in Family Law from Aligarh Muslim University (AMU), Aligarh, where she also completed her BA.LLB. (Hons.) and LLM in Commercial Law. A dedicated academician, she qualified the UGC NET exam in 2013, demonstrating her expertise in the field of law. Currently, Dr. Kausar serves as an Assistant Professor (Selection Grade) at the School of Law, UPES Dehradun, where she contributes to legal education and research. Her academic journey reflects a strong commitment to legal scholarship, with a focus on both family law and commercial law.

Sheikh Inam Ul Mansoor (BA.LLB Hon's, LLM, CPCT WIPO, Ph.D. International Trade Law) is a highly accomplished academician presently working as an Assistant Professor Law at the School of Law, Dayananda Sagar University in Bangalore, India. With a strong background in law and a passion for academia, Dr. Sheikh has made significant contributions to the field through his extensive research and publications. Pursued his BA. LLB (Hon's) from University of Kashmir, LLM from JNU Jaipur and Ph.D. International Trade Law from LPU Punjab. Prior to his academic career, Dr. Sheikh served as an Advocate and Solicitor at the Jammu and Kashmir High Court and other subordinate courts in Jammu and Kashmir. This experience as a legal practitioner has provided him with valuable insights into the practical application of law and enhanced his understanding of legal dynamics.

Amthul Naseeb is an Assistant Professor at the School of Law, Dayananda Sagar University, Bengaluru. She holds an MBA with a specialization in Human Resources and has extensive experience in academia. In addition to her expertise in HR, she has also taught Marketing, bringing a multidisciplinary approach to her teaching. With a passion for student development and a keen interest in business and management studies, she continues to contribute to the academic growth of law students by integrating management principles into legal education.

Megan Pekol-Evans is doctoral candidate at Marymount University's College of BILT. She is also a United States Army active-duty Acquisition Officer stationed at the Pentagon with 20 years of service. She is a graduate of the Pennsylvania State University with a B.S. in Administration of Justice. She subsequently earned an MBA in 2012 from Trident University. She was stationed and deployed in different places throughout the U.S. and world including: Utah, Oklahoma, Afghanistan in support of Operation Enduring Freedom, North Carolina, Al Dhafra Air Base, U.A.E. in support of Operation Inherent Resolve, Illinois, Alabama, New Jersey, and Virginia. Currently as an Army Acquisition Officer, she leads, manages, and supports U.S. Army modernization programs throughout research and development initiatives, engineering and manufacturing planning, procurement and fielding of systems, testing, and logistics and sustainment efforts. Additionally, a large portion of her duties consist of leading product teams as they review, manage, and mitigate technical or programmatic risks. Most importantly, she is married to her best friend, Mike, and has three amazing children, Brynn, Liam, and Jimmy.

Shahida Qadir is a Ph.D. candidate at the Ahmad Ibrahim Kulliyyah of Laws (AIKOL), International Islamic University Malaysia (IIUM), specializing in Cyber Crimes, Women, and Law. Her research delves into the legal challenges surrounding cyber-crimes against women in India, critically examining existing legal frameworks, enforcement mechanisms, and gaps in addressing online victimization. Through her work, she seeks to contribute to policy reforms that strengthen legal protections and enhance redressal mechanisms for affected women. With a strong academic interest in cyber law, human rights, and gender justice, Ms. Shahida actively engages in scholarly discussions at the intersection of law, technology, and women's rights.

Jim Tebrake is a senior executive with extensive experience leading large-scale, complex data and analytics programs in both national and international settings. He is a doctoral candidate in the Doctor of Business Administration program at Marymount University in Arlington, Virginia. His professional expertise spans business intelligence, data governance, and data quality, with a strong focus on transforming data into actionable insights for strategic decision-making. His current research interests include business intelligence, data governance, data quality, and the development of innovative frameworks for measuring sustainability and well-being.

Index

A

B

C

D

E

F

H

I

K

L

M

O

P

R

S

T

U

V

W

www.ingramcontent.com/pod-product-compliance
Lightning Source LLC
LaVergne TN
LVHW060005031026
849161LV00025B/1435
* 9 7 9 8 3 3 7 3 5 6 8 6 0 *